Ecological Modeling in Risk Assessment

Chemical Effects on Populations, Ecosystems, and Landscapes

Ecological Modeling in Risk Assessment

Chemical Effects on Populations, Ecosystems, and Landscapes

Edited by
Robert A. Pastorok
Steven M. Bartell
Scott Ferson
Lev R. Ginzburg

CRC Press
Taylor & Francis Group
Boca Raton London New York

CRC Press is an imprint of the
Taylor & Francis Group, an **informa** business

CRC Press
Taylor & Francis Group
6000 Broken Sound Parkway NW, Suite 300
Boca Raton, FL 33487-2742

First issued in hardback 2019

ISBN-13: 978-1-56670-574-5 (hbk)
ISBN-13: 978-0-367-39680-0 (pbk)
Library of Congress Card Number 2001038278

Library of Congress Cataloging-in-Publication Data

Ecological modeling in risk assessment : chemical effects on populations, ecosystems, and landscapes / Robert A. Pastorok ... [et al.], editors.
    p. cm.
    Includes bibliographical references.
    ISBN 1-56670-574-6 (alk. paper)
    1. Pollution—Environmental aspects—Simulation methods. 2. Ecological risk assessment. I. Pastorok, Robert A.

QH545.A1 E277 2001
577.27′01′1—dc21

2001038278

**Visit the Taylor & Francis Web site at
http://www.taylorandfrancis.com**

**and the CRC Press Web site at
http://www.crcpress.com**

Dedication

This book is dedicated to W.T. Edmondson (1916–2000), a pioneer in applying ecological data and practical mathematical models to solve environmental problems.

W.T. Edmondson in 1985 at the shore of Lake Washington, where for 45 years (1955–2000) he investigated the cause and consequences of eutrophication and provided information to the public that aided efforts to improve the condition of the lake. (Photo: Benjamin Benschneider, *The Seattle Times*. With permission.)

Preface

Ecotoxicological models have been applied increasingly to perform chemical risk assessments since the first models of this kind emerged about 25 years ago. The first ecotoxicological models were applied to very specific cases — for instance, cadmium contamination of Lake Erie or mercury contamination of Mex Bay, Alexandria. The models were inspired by the experience gained in ecological modeling and therefore contained good descriptions of ecological processes. Slightly later, the so-called fate models emerged, which were first developed by McKay and others. Such models described the distribution of a chemical in the atmosphere, the hydrosphere, the lithosphere, and the biosphere on the basis of the physical–chemical properties of the chemical. They were not able to give accurate and precise predictions about concentrations one would measure in nature, but they made it possible to compare the risks of two or more chemicals. They could therefore be applied to select which chemical among many to recommend for further environmental study.

The effect of a toxic chemical can in principle be exerted on all levels of the biological hierarchy, from cells to organs to organisms to populations to entire ecosystems. Ecotoxicological models have until now mainly been used to assess the risk to endpoints associated with individual organisms (e.g., survival, growth, and fecundity), but the need to apply models to evaluate risks at the population and ecosystem levels has been increasing (Kendall and Lacher 1994; Albers et al. 2001). Risks at higher levels of biological organization are not represented directly by effects on individual-level endpoints* because of the emergent properties of populations and ecosystems, including compensatory behavior (Ferson et al. 1996). Managing environmental risks and solving our current problems requires risk assessment at the population and ecosystem levels because reversing system-wide effects at a later stage is much more difficult (e.g., if a population is decimated or the structure of an ecosystem is completely changed). This volume acknowledges this need for a wider application of ecological models in environmental risk assessment and therefore reviews the available models, with an emphasis on models that could be applied to evaluate toxicological effects on populations, ecosystems, and landscapes.

We expect that, in the future, responsible ecological risk assessments of chemicals will rely on quantitative models of populations, ecosystems, and landscapes. For many chemicals, contaminated sites, and specific issues, ecological modeling in the context of a risk assessment could provide valuable information for environmental managers, policy-makers, and planners. Therefore, having a clear overview of the available models, which is the scope of this volume, is crucial.

In the *Introduction*, the authors give an overview of the current process of ecological risk assessment for toxic chemicals and of how modeling of populations, ecosystems, and landscapes could improve the status quo. The limitations of the hazard quotient approach based on individual-level endpoints are discussed. The role of ecological modeling is illustrated, especially in the context of evaluating the ecological significance of typical results from laboratory toxicity tests and the hazard quotient approach. Other introductory topics include deciding when to use ecological models, selecting models for application to specific assessments, various ways of expressing population-level risk, and steps in applying a population model to a chemical risk assessment.

Next, the *Methods* section contains a classification of ecological models and explains the differences between population, ecosystem, landscape, and toxicity-extrapolation models. The model evaluation process is described, and the evaluation criteria are defined.

The evaluation of models is organized by model type as follows: population models (scalar abundance, life-history, individual-based, and metapopulation), ecosystem models (food-web, aquatic, and terrestrial), landscape models, and toxicity-extrapolation models. Within each of the nine categories, individual models are described and evaluated. The descriptions include discussion of the mathematical approach used in the model, the conceptual structure of the model, endpoints,

---

* The specific meaning of *endpoint* depends on its context; there are model endpoints, toxicity test endpoints, or risk assessment endpoints. See *Glossary* entries for assessment endpoint, endpoint, and measurement endpoint.

treatment of uncertainty, and other information important for chemical risk assessments. The evaluation results and applications of the reviewed models are summarized in tabular form. Finally, an overview of the state of models within the category is applied, and selected models are recommended for further development and use in chemical risk assessment. More detailed profiles of the recommended models are provided.

The use of ecological models in environmental decision-making is constrained at present by the lack of understanding of such models by many managers and risk assessors. Therefore, the authors discuss ways to foster the use of ecological models to address toxic chemical problems, including recommendations for workshops and training.

Finally, results of the model evaluations and recommendations are summarized in the *Conclusions and Recommendations*. One of the primary views is that population and metapopulation models are well developed and applicable to many current ecological risk assessments. Recommendations for software development and training are also provided.

Lately, a new approach to modeling complex ecological systems has been developed called *structurally dynamic modeling* (Jørgensen 1997). These models can describe the changes in the properties of a system due to adaptation of organisms (genetic or physiological) or shifts in species composition when the prevailing environmental conditions are changed. Because the discharge of toxic substances sometimes implies very drastic changes in environmental conditions, structurally dynamic models are especially appropriate for ecological risk assessment. Nonetheless, this type of model has only been applied in 12 studies, and none involved ecotoxicological assessment. Therefore, including structurally dynamic models in the review of models that are applicable for chemical risk assessment is premature. However, such models should be evaluated further as more experience is gained in the use of this type of model for risk assessment. Ultimately, the challenge is not only to predict the responses of static assemblages of species to toxic chemicals but also to be able to consider adaptation and shifts in species composition — processes that we know ecosystems experience.

Sven E. Jørgensen
Robert A. Pastorok

References

Albers, P.H., G.H. Heinz, and H.M. Ohlendorf (Eds.). 2001. Environmental contaminants and terrestrial vertebrates: effects on populations, communities, and ecosystems. SETAC Special Publication Series. Society of Environmental Toxicology and Chemistry, Pensacola, FL.

Ferson, S., L.R. Ginzburg, and R.A. Goldstein. 1996. Inferring ecological risk from toxicity bioassays. *Water Air Soil Pollut.* 90:71–82.

Jørgensen, S.E. 1997. *Integration of Ecosystem Theories: A Pattern.* Kluwer Academic Publishers, Dordrecht.

Kendall, R.J. and T.E. Lacher, Jr. 1994. Wildlife toxicology and population modeling: integrated studies of agroecosystems. Proceedings of the Ninth Pellston Workshop, July 22–27, 1990. SETAC Special Publication Series. Society of Environmental Toxicology and Chemistry. Lewis Publishers, Boca Raton.

Acknowledgments

This book was based on a draft report completed under a project funded by the American Chemistry Council (ACC). Authors of individual chapters are listed under chapter titles. All authors contributed to the *Profiles of Selected Models*, the *Initial Screening of Ecological Models*, and the *Summary*. We thank Janos Hajagos of Applied Biomathematics* and Steave Su and Craig Wilson of Exponent for assistance in searching for and compiling information on ecological models. In addition to the authors, several other individuals contributed to the draft report. Erin Miller of The Cadmus Group contributed to the chapter on *Aquatic Ecosystem Models*. Dreas Nielsen of Exponent provided insightful review comments throughout the project and facilitated a workshop on ecological modeling. Ellen Kurek of Exponent was technical editor and production assistant. Betty Dowd and Mary Bilsborough of Exponent prepared graphics. Marie Cummings, Eileen McAuliffe, and Lillian Park of Exponent were responsible for word processing of the manuscript. Coreen Johnson was production supervisor.

A workshop was held in Fairmont, Montana, on May 17–18, 2000, to review preliminary results of the evaluation of ecological models and to develop recommendations for further methodological development. The results of the workshop were summarized in a series of recommendations from the expert review panel (Jørgensen et al. 2000). We would like to especially thank the members of the expert review panel for their participation in the workshop and for reviewing drafts of the manuscript. These members are Lawrence Barnthouse of LWB Environmental, Donald DeAngelis of the National Biological Service, John Emlen of the U.S. Geological Survey, Sven Jørgensen of the Royal Danish School of Pharmacy (panel chairperson), John Stark of Washington State University, and Kees van Leeuwen of RIVM/CSR, the Netherlands.

Members of the project monitoring team for ACC were James Clark of Exxon Mobil Biomedical Sciences, Donna Morrall of Procter & Gamble, Susan Norton of the U.S. Environmental Protection Agency, and Ralph Stahl of the Corporate Remediation Group, DuPont Engineering (project manager for ACC). Robert Keefer of Keefer Associates was the project administrator for ACC. Their assistance throughout the project is much appreciated. Other participants in the model evaluation workshop included John Fletcher of the University of Oklahoma, Tim Kedwards of ZENECA Agrochemicals, and Steve Brown of Rohm and Haas.

We are especially grateful to the many developers of ecological models, who have undoubtedly spent long hours in front of the computer screen to explore the best ways of representing ecological systems. Several individuals provided helpful comments or draft text for specific models reviewed herein, including:

Daniel Botkin (University of California) — JABOWA (co-author of draft text)
Marcus Lindner (University of Alberta) — FORSKA
Joao Gomes Ferreira (IMAR — Institute of Marine Research, Portugal) — EcoWin2000
Don Vandendriesche (USDA Forest Service) — FVS (author of draft text)
Aaron Ellison (Mount Holyoke College) — Disturbance to wetland plants model
Glen Johnson (New York State Department of Health) — Multi-scale landscape model
Ferdinando Villa (University of Maryland) — Island disturbance biogeographic model
Richard Park (Eco Modeling) — AQUATOX
Alexy Voinov (University of Maryland) — Patuxent watershed model
Chuck Hopkinson (Marine Biological Laboratory, Woods Hole) — Barataria Bay model

Finally, Rob Pastorok would like to thank Thomas C. Ginn of Exponent, Clyde E. Goulden of the Philadelphia Academy of Natural Sciences, John M. Emlen of the U.S. Geological Survey, and Robert T. Paine of the University of Washington for inspiration throughout the journey leading to this work. Their scientific insights and unrelenting spirit in seeking understanding of the natural world have guided many ecologists and modelers.

* Applied Biomathematics is a registered service mark.

About the Editors

Robert A. Pastorok, Ph.D., is a managing scientist at Exponent, a consulting firm specializing in risk assessment and failure analysis. He has 30 years of experience as an ecologist with expertise in analyzing the risks of toxic chemicals in the environment. Dr. Pastorok obtained his Ph.D. in zoology from the University of Washington in 1978. After teaching population modeling and ecology courses at the university level, he entered the environmental consulting field. For more than 20 years he has applied ecological concepts in assessing and solving complex environmental problems. He has supported the U.S. Environmental Protection Agency, state agencies, and private industry in developing risk analysis models, toxicity testing methods, and chemical guidelines for soil, sediment, and surface water. His current interests are in applying population dynamics and landscape ecology theory to risk assessment models for wildlife. He is senior editor for ecological risk assessment for the journal *Human and Ecological Risk Assessment* and associate editor for ecosystems and communities for the online publishing entity *The Scientific World*.

Steven M. Bartell, Ph.D., earned his Ph.D. in limnology and oceanography from the University of Wisconsin, Madison. Dr. Bartell's primary research and technical interests include ecosystem science, ecological modeling, and ecological risk assessment. Dr. Bartell has conducted extensive basic and applied research concerning the effects of nutrients, herbicides, organic contaminants, toxic metals, radionuclides, sediment resuspension, and habitat alteration on the ecological integrity of aquatic plants, invertebrates, and fish. He has directed, designed, and performed ecological risk assessments for a variety of physical, chemical, and biological stressors in aquatic and terrestrial ecosystems. Dr. Bartell has authored more than 100 technical publications concerning ecology, environmental sciences, and risk assessment. He is a principal author of the books *Ecological Risk Estimation* and the *Risk Assessment and Management Handbook*. Dr. Bartell currently serves on the editorial boards of *Risk Analysis*, *Human and Ecological Risk Assessment*, and *Chemosphere*. He is a two-term member of the U.S. Environmental Protection Agency Science Advisory Board (SAB) Ecological Processes and Effects Committee. Dr. Bartell also participates as a member of the U.S. EPA/SAB Executive Committee's Subcommittee that addresses the use of ecological models in support of environmental regulations.

Scott Ferson, Ph.D., is a senior scientist at Applied Biomathematics, a research firm specializing in methods for ecological and environmental risk analysis. His research focuses on developing reliable mathematical and statistical tools for ecological and human health risk assessments and on methods for uncertainty analysis when empirical information is very sparse. Dr. Ferson holds a Ph.D. in ecology and evolution from the State University of New York at Stony Brook. He is an author of *Risk Assessment for Conservation Biology* and editor of the collected volume *Quantitative Methods for Conservation Biology*. He is author of the forthcoming book *Risk Calc: Risk Assessment with Uncertain Numbers*. He has written more than 60 other scholarly publications, including several software packages, in environmental risk analysis and uncertainty propagation. His research has addressed quality assurance for Monte Carlo assessments, exact methods for detecting clusters in small data sets, backcalculation methods for use in remediation planning, and distribution-free methods of risk analysis appropriate for use in information-poor situations.

Lev R. Ginzburg, Ph.D., has been professor of ecology and evolution at State University of New York at Stony Brook since 1977. He founded Applied Biomathematics in 1982. Dr. Ginzburg's scholarly research in trophic interactions in food chains has sparked a controversial revision of the fundamental equations used for modeling food chain dynamics. He has published widely on theoretical and applied ecology, genetics, and risk analysis and has produced six books and more than 100 scientific papers. In 1982, Dr. Ginzburg was primary author of one of the seminal papers inaugurating the field of ecological risk analysis.

Contributing Authors

H. Resit Akçakaya
Applied Biomathematics
Setauket, New York
e-mail: resit@ramas.com

Steven M. Bartell
The Cadmus Group
Oak Ridge, Tennessee
e-mail: sbartell@cadmusgroup.com

Steve Carroll
Applied Biomathematics
Setauket, New York
e-mail: admin@ramas.com

Jenée A. Colton
Exponent Environmental Group
Bellevue, Washington
e-mail: coltonj@exponent.com

Scott Ferson
Applied Biomathematics
Setauket, New York
e-mail: scott@ramas.com

Lev R. Ginzburg
Applied Biomathematics
Setauket, New York
e-mail: lev@ramas.com

Sven E. Jørgensen
Royal Danish School of Pharmacy
Department of Analytical and Pharmaceutical
    Chemistry
Environmental Chemistry
Copenhagen, Denmark
e-mail: sej@mail.dfh.dk

Christopher E. Mackay
Exponent Environmental Group
Bellevue, Washington
e-mail: mackayc@exponent.com

Robert A. Pastorok
Exponent Environmental Group
Bellevue, Washington
e-mail: pastorokr@exponent.com

Stan Pauwels
Abt Associates, Inc.
Cambridge, Massachusetts
e-mail: stan.pauwels@gte.net

Helen M. Regan
National Center for Ecological Analysis and
    Synthesis
University of California Santa Barbara
Santa Barbara, California
e-mail: regan@nceas.ucsb.edu

Karen V. Root
Applied Biomathematics
Setauket, New York
e-mail: kroot@ramas.com

Acronyms and Abbreviations

ACR	acute-to-chronic ratio
AEE	analysis of extrapolation errors
AF	application factor
ALEX	analysis of the likelihood of extinction
ATLSS	across-trophic-level system simulation
CASM	comprehensive aquatic system model
CATS-4	contaminants in aquatic and terrestrial ecosystems-4
CCC	criteria continuous concentration
CDF	cumulative distribution function
CEL HYBRID	coupled Eulerian–Lagrangian hybrid model
CIFSS	California individual-based fish simulation system
CITES	Convention on International Trade in Endangered Species
CO_2	carbon dioxide
CV	coefficient of variation
DEB	Dynamic Energy Budget
EC50	median effect concentration
EPA	U.S. Environmental Protection Agency
EPRI	Electric Power Research Institute
ESA	Endangered Species Act
ERSEM	European regional seas ecosystem model
FORET	Forests of Eastern Tennessee
FCV	final chronic value
FORCLIM	forest climate model
FORMIX	forest mixed model
FORMOSAIC	forest mosaic model
FVS	forest vegetation simulator
GAPPS	generalized animal population projection system
GEM	general ecosystem model
GIS	geographic information system
GMCV	genus mean chronic value
GBMBS	Green Bay mass balance study
HC_p	hazardous concentration for a population
HCS	hazardous concentration for sensitive species
HOCB	hydrophilic organic compound bioaccumulation model
IBP	International Biological Programme
IFEM	integrated fates and effects model
INTASS	interaction assessment model
IUCN	The World Conservation Union
LANDIS	landscape disturbance and succession
LC50	median lethal concentration
LC01	the 0.1% response in a toxicity test
LD50	median lethal dose
LEEM	Lake Erie ecosystem model
LERAM	littoral ecosystem risk assessment model
LOEL	lowest-observed-effects level
MATC	maximum acceptable toxicant concentration
NA or n/a	not applicable
NOEC	no-observed–effect concentration

NOEL	no-observed-effect level
NOYELP	Northern Yellowstone Park model
OFFIS	Oldenburger Forschungs- und Entwicklungsinstitut für Information-Werkzeuge und Systeme
ORGANON	Oregon growth analysis and projection
PAH	polycyclic aromatic hydrocarbon
PATCH	program to assist in tracking critical habitat
PC	personal computer
PCB	polychlorinated biphenyl
QSAR	quantitative structure–activity relationship
QWASI	quantitative water, air, and sediment interaction model
RAMAS	risk analysis and management alternatives software
SAGE	system analysis of grassland ecosystems model
SALMO	simulation by means of an analytical lake model
SD	standard deviation
SF	scaling factor
SIMPDEL	spatially explicit individual-based simulation model of Florida panthers and white-tailed deer in the Everglades and Big Cypress landscapes
SIMPLE	sustainability of intensively managed populations in lake ecosystems
SIMSPAR	spatially explicit individual-based object-oriented simulation model for the Cape Sable seaside sparrow in the Everglades and Big Cypress landscapes
SLOSS	single reserve of equal total area
SPUR	simulation of production and utilization of rangeland
SWACOM	standard water column model
TEEM	terrestrial ecosystem energy model
TNC	The Nature Conservancy
UF	uncertainty factor
UFZ	Umweltforschungszentrum Leipzig–Halle Sektion Ökosystemanalyse
ULM	unified life model
USDA	U.S. Department of Agriculture
USFWS	U.S. Fish and Wildlife Service
WESP	workbench for modeling and simulation of the extinction of small populations

Contents

CHAPTER 1

Introduction

Robert A. Pastorok

Ecological risk assessment for toxic chemicals has become an important part of the decision-making process for managing environmental problems (Suter 1993; U.S. EPA 1998). Risk assessments are used to evaluate environmental problems associated with past, ongoing, and potential future practices. For example, risks to plants, invertebrates, amphibians, reptiles, birds, and mammals are considered in the evaluation of chemical contamination at hazardous waste sites under the Superfund program administered by the U.S. Environmental Protection Agency (EPA) and under similar programs in most U.S. states, in Canada, in Europe, and in other countries throughout the world. In pesticide regulatory programs, ecological risk assessments are used to evaluate new chemicals as part of the registration process or new uses for already registered pesticides. Risk assessments also support environmental decisions about siting new facilities, about waste discharges, and about remedial actions to clean up or treat contaminated areas.

Despite the important role that ecological risk assessments play in supporting decisions about toxic chemical issues, many assessments done in support of environmental regulatory programs rely on simplistic approaches and fail to incorporate basic ecological information and modeling capabilities. Typically, an ecological risk assessment for toxicants relies on comparison of some exposure estimate for each chemical of interest with a corresponding toxicity threshold for individual-organism endpoints such as survival, growth, or reproductive potential (e.g., fecundity). This comparison is often accomplished by calculating a hazard quotient, which is simply the exposure estimate divided by the toxicity threshold. In many cases, the toxicity threshold selected for a given chemical is a no-observed-effect level (NOEL) or a lowest-observed–effects level (LOEL), and the complete dose–response curve is unknown. Arbitrary uncertainty factors or other simple toxicity-extrapolation methods are often applied to translate an available toxicity threshold into the endpoint of interest (e.g., extrapolation from acute to chronic exposures, or from one species to another) (Chapman et al. 1998).

Ecologists and statisticians have pointed out the limitations of current ecological risk assessment approaches like the hazard quotient, especially when uncertainties in the exposure and toxicity estimates are unquantified (Barnthouse et al. 1986; Landis and Yu 1995; Warren-Hicks and Moore 1998; Kammenga et al. 2001). Yet, ecological risk assessors continue to rely primarily on point estimates of hazard quotients, often with conservative assumptions about exposure of organisms to toxic chemicals. This approach was originally intended as a screening method (Barnthouse et

0--56670-574-6/01/$0.00+$1.50

al. 1986) and may produce misleading results because of compounding conservatism (Burmaster and von Stackelberg 1989; Cullen 1994).

Many ecologists recognize the value of population and ecosystem modeling as applied to ecological risk assessment for toxic chemicals (e.g., Barnthouse et al. 1986; Emlen 1989; Bartell et al. 1992; Ferson et al. 1996; Barnthouse 1998; Forbes and Calow 1999; Landis 2000; Snell and Serra 2000; Suter and Barnthouse 2001; Sample et al. 2001). Such ecological models are used to translate the results of fecundity and mortality measures in toxicity tests on organisms to estimate effects on population, ecosystem, and landscape endpoints. Examples of ecological endpoints to be considered in risk modeling include species richness, population abundance or biomass, popu-lation growth rate or reproductive output, population age structure, and productivity. Ecological models can be used to address two critical questions in ecotoxicology (Kareiva et al. 1996): (1) how does population growth rate change as a function of toxic chemical concentration, and (2) how rapidly can a population recover from an impact due to transient exposure to a toxic chemical? Nevertheless, estimation of effects beyond individual-level endpoints is rare in current chemical risk assessments (Landis 2000).

Further development and use of ecological models with population, ecosystem, and landscape endpoints are clearly needed to increase the value of chemical risk assessments to environmental managers. For example, Landis (2000) noted that loss of habitat and invasion of exotic species are typically identified as the major issues for natural resource management. He argued further that toxic chemical contamination alters the use of habitats by species and is thereby a major contributor to current ecological problems. For example, in aquatic systems, this places chemical contamination on a par with dams, siltation, destruction of riparian areas, and the introduction of non-native species that compete strongly with indigenous species. Contamination may act as a barrier to species migration, lower the rate of population growth, cause behavioral modifications, or reduce important food resources for species of concern (Landis 2000). All of these factors may have effects on the population level that cannot be directly predicted from hazard quotients based on individual-level traits. Forbes and Calow (1999) evaluated laboratory toxicity data for a wide range of aquatic species in the context of population dynamics theory and considered the relative sensitivity of population growth rate and individual-level traits to toxic chemicals. These authors found that the population growth endpoint was usually less sensitive but sometimes more sensitive than individual-level endpoints. They also found no consistent pattern with respect to which individual-level traits were most or least sensitive to toxicant exposure. Kammenga et al. (2001) evaluated the effects of cadmium and pentachlorophenol on laboratory populations of soil invertebrates and found that hazard quotients for individual-level endpoints could not be used directly to predict population-level effects.

OBJECTIVES

We report here the results of a critical evaluation of ecological-effects models that are potentially useful for chemical risk assessment and recommend further development of selected models. The selected models were identified on the basis of their relatively high ratings with respect to eight evaluation criteria. The criteria included model realism and complexity, prediction of relevant ecological endpoints, treatment of uncertainty, ease of estimating parameters, degree of model development, regulatory acceptance, credibility, and resource efficiency. A workshop was held in Fairmont, Montana, on May 17–18, 2000, at which a panel of experts in ecological modeling (Jørgensen et al. 2000) reviewed preliminary results of the model evaluations and helped refine recommendations for further methodological development. This book extends the excellent work of Jørgensen et al. (1996) by including more ecological models, by classifying models, and by explicitly evaluating models with respect to specific performance criteria.

The objectives of this book are to:

- Conduct a critical evaluation of ecological-effects models that are potentially useful for chemical risk assessment
- Rank the various candidate models on the basis of evaluation criteria such as scientific support, regulatory acceptance, state of development, and ability to predict relevant assessment endpoints
- Recommend selected models for further evaluation and testing

The most promising ecological models may be evaluated further by implementing them with available data or by comparing model predictions with field data collected specifically for testing the models.

For our purposes, an *ecological model* is a mathematical expression that can be used to describe or predict ecological processes or endpoints such as population abundance (or density), community species richness, productivity, or distributions of organisms. Ecological models typically deal with endpoints at the population, ecosystem, or landscape level, which are directly relevant to natural resource managers. Models that address only toxic chemical transport, fate, and exposure (e.g., the predictive bioaccumulation models of Gobas 1993 and Traas et al. 1996) are not considered ecological models in our review — although such models may be combined with relationships describing toxic chemical effects to produce an ecosystem model, which can be used in the context of a risk assessment. Many ecological models that predict ecosystem and landscape endpoints also include submodels that describe environmental transport, fate, and exposure.

As defined here, *ecological-effects models* include ecological models (i.e., those with population, ecosystem, or landscape endpoints as state variables) as well as toxicity-extrapolation models. Toxicity-extrapolation models do not address population demographics or other aspects of a species' ecological role, but they are used within the context of ecological modeling to translate toxicity thresholds between species, endpoints, or exposure durations (i.e., acute vs. chronic) or to derive toxicity thresholds protective of communities (OECD 1992; Aldenberg and Slob 1993).

We discuss the selection and use of ecological models in the context of ecological risk assessment in the next section.

THE PROCESS OF ECOLOGICAL MODELING FOR CHEMICAL RISK ASSESSMENT

The U.S. EPA (1992) defined ecological risk assessment as "a process that evaluates the likelihood that adverse ecological effects may occur or are occurring as a result of exposure to one or more stressors." This definition allows for risk assessment to be conducted at various levels within a hierarchy of biological endpoints, from individual organisms to populations, communities, ecosystems, and landscapes (Figure 1.1). Most toxicity data are developed for endpoints expressed as effects on individual organisms, such as mortality, fecundity, age at reproduction, growth, behavior, or physiological responses. Typical risk assessments focus on individual-level effects and either ignore higher-level effects or only qualitatively discuss the potential for adverse effects on populations. Such risk assessments consist of an exposure assessment for individuals, an effects assessment for one or more individual-level endpoints based on available toxicity data, and a risk characterization (Figure 1.2). The information addressed in each step of the assessment is summarized below:

1. Problem Formulation — The physical features, general distribution of chemicals, and ecological receptors (plants and animals) in the study area are described using existing data. In a preliminary analysis, chemicals of potential concern, physical stressors, ecological receptors, and endpoints to be considered in the assessment are identified. A conceptual model of the chemical exposure pathways is developed, and risk assessment questions and objectives are defined.

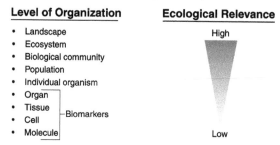

Figure 1.1   Hierarchy of biological endpoints.

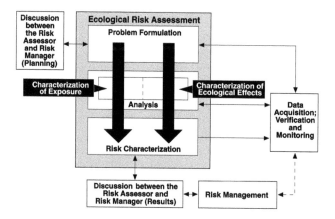

Figure 1.2   U.S. Environmental Protection Agency framework for ecological risk assessment. From U.S. EPA, 1998.

2. Exposure Analysis — The ecological receptors likely to contact chemicals of potential concern, the means of contact (e.g., ingestion, dermal contact), and the magnitude and frequency of exposure are identified and described.
3. Effects Analysis — Potential effects of the chemicals of potential concern on organisms and toxicity thresholds or exposure–response relationships are developed.
4. Risk Characterization — Results of the exposure and effects analyses are combined to evaluate the likelihood of adverse effects on ecological receptors. The degree of confidence in the risk estimates and the most important sources of uncertainty are also described. The ecological significance of any identified risks is described.

Barnthouse et al. (1986), Bartell et al. (1992), Calabrese and Baldwin (1993), Suter (1993), U.S. EPA (1998), and many others provide detailed guidance on the process of ecological risk assessment.

Limitations of the Hazard Quotient Approach

The risk characterization step often involves calculating a "risk estimate" as a hazard quotient, which is simply the estimated exposure divided by a toxicity threshold. A measured NOEL or LOEL for the individual-level endpoint of interest is typically used as the toxicity threshold. For example, Menzie et al. (1992) estimated hazard quotients for songbirds exposed to DDT through their diet at a contaminated site in Massachusetts. These authors applied several dietary scenarios and divided estimated exposure values by toxicity thresholds for survival and reproductive endpoints from the literature. Because the hazard quotient approach fails to incorporate exposure–response relationships, it can only indicate whether effects on individuals are expected, not the magnitude of effects. That is, the magnitude of a hazard quotient greater than one cannot be used to reliably estimate the severity of the toxicity response. Moreover, the results are difficult to interpret when the hazard quotient for one endpoint (e.g., mortality) conflicts with that for another endpoint (e.g., reproduction), as was the case in the risk assessment by Menzie et al. (1992). Use of a hazard quotient approach can lead to ambiguous results, and inferences about population-level effects are unwarranted regardless of results based on individual-level endpoints (as is typically the case). Examples from the songbird risk scenarios presented by Menzie et al. (1992) are discussed here to illustrate these limitations of the simple hazard quotient approach.

For their hypothesized songbird with a diet consisting of litter invertebrates (Diet A), Menzie et al. (1992) measured a DDT concentration of 132.4 mg/kg dry weight in invertebrates at the site and divided this exposure estimate by the lowest observed NOEL of 10 mg/kg for the songbird survival endpoint to obtain a hazard quotient of approximately 13. The same exposure estimate (132.4 mg/kg dry weight in invertebrates) was also divided by an estimate of the LOEL (350 mg/kg) for bird survival to obtain a hazard quotient of 0.37. The range of hazard quotient values (0.37 to 13) therefore encompassed a value of 1.0, which is typically the benchmark for consideration of whether the risk is ecologically significant and requires further investigation. Given the range of hazard quotients estimated by Menzie et al. (1992), the results are ambiguous, and it is impossible to determine the extent of the risk to individual songbirds. As is typical of most risk assessments that rely on hazard quotient approaches, the risk to populations was not even considered. Given the uncertainty, an environmental manager would be likely to expend additional resources to further investigate risks at this site. (For the purposes of our argument, we will ignore the inconsequential fact that additional data had already been collected as part of a well-balanced study design at this particular site.)

Using an approach similar to that used in the example just discussed for the survival endpoint, Menzie et al. (1992) also calculated hazard quotients ranging from approximately 1.3 to approximately 66 for reproductive effects in songbirds that eat exclusively litter invertebrates. In this case, where some hazard quotient values are very high and all are greater than 1.0, the potential risk to individuals needs to be addressed further. But is the risk to individuals high enough to be cause for concern about adverse effects at the population level? Population-level effects are, after all, the ultimate concern for responsible management actions for most songbirds (those that are not species of special status such as threatened and endangered species). From the hazard quotient approach, one usually does not have enough information to make a management decision. The hazard quotient always should be interpreted in light of the assumptions used in estimating exposure and the confidence in toxicity thresholds, but these considerations make it impossible to extrapolate to population-level endpoints based simply on a qualitative evaluation of the risk assessment results. Probabilistic assessments (e.g., MacIntosh et al. 1994; Sample and Suter 1999) improve the situation somewhat but still cannot provide meaningful information for a population- or higher-level assessment. Such insight can be obtained only from the application of ecological models to the risk assessment problem.

We focused in this section on the inability of the hazard quotient approach to provide useful information for determining risks to populations in an ecological risk assessment. Cullen (1994) expressed additional concerns about the use of deterministic hazard quotients, even for assessing risks to individuals, in analyses that use conservative values for each of the exposure variables and the toxicity threshold. Using an analytical approach, Cullen (1994) showed that even when using a reasonable-maximum exposure approach, the point estimate of the hazard quotient was hyper-conservative and that the conservatism was variable relative to a fixed upper percentile (e.g., the

95th percentile) of the risk distribution derived in a probabilistic analysis (Monte Carlo). Other authors (Burmaster and von Stackelberg 1989; Chrostowski et al. 1991; Copeland et al. 1993; Sample et al. 1999) have also demonstrated compounding conservatism in deterministic hazard quotients for reasonable maximum exposures. Moreover, Cullen (1994) showed that the conservatism of the hazard quotient estimate increases as the variance in the distributions for exposure and toxicity reference values increases, as the level of conservatism in point estimates of input variables increases, and as the number of input variables increases. Thus, in a conservative analysis, deterministic hazard quotients cannot be used to reliably rank chemicals, identify priority receptors, or identify problem areas of the site because the degree of conservatism is inconsistent among the elements ranked. At best, deterministic hazard quotients based on conservative assumptions about exposure and toxicity can only be used to screen out chemicals, receptors, or site areas that are clearly not a problem (when the hazard quotient is considerably less than 1).

Role of Ecological Modeling in Chemical Risk Assessment

The primary purpose of ecological models in a specific risk assessment is to evaluate the *ecological significance* of observed or estimated effects on individual organisms. Essentially, ecological models predict the responses of population, ecosystem, and landscape endpoints to perturbations in ecological components, which may include individual-level endpoints such as mortality or fecundity. For example, using a life-history model such as the Leslie matrix (Leslie 1945; Caswell 2001), one can estimate the population growth rate or the temporal dynamics of population abundance from estimates of survivorship and fecundity for individual age classes of organisms. Chemical effects can be modeled by perturbing the age-specific mortality and fecundity values on the basis of knowledge about changes in these parameters obtained from toxicity test results.

Assessment and Measurement Endpoints

In the language of ecological risk assessors, an ecological model can be used to extrapolate a measurement endpoint to an assessment endpoint. *Assessment endpoints* are defined as environmental characteristics or values that are to be protected (e.g., wildlife population abundance, species diversity, or ecosystem productivity) (U.S. EPA 1998). *Measurement endpoints* are quantitative expressions of an observed or measured biological response, such as the effects of a toxic chemical on survivorship or fecundity, related to the valued environmental characteristic chosen as the assessment endpoint. In some cases, the measurement endpoint is the same as the assessment endpoint (e.g., when benthic macroinvertebrate communities are surveyed directly in a stream to assess species richness). When these endpoints differ, a model must be used to express their relationship quantitatively. Essentially, the mathematical model is used to precisely define the relationship as well as assumptions and uncertainties in the extrapolation between measurement and assessment endpoints.* The primary measurement endpoints for a chemical risk assessment are related to the survival, growth, and reproduction of exposed organisms (U.S. EPA 1998). These endpoints are used in most standardized toxicity tests and in the development of EPA ambient water quality criteria, wildlife criteria, and sediment quality criteria. Moreover, such endpoints can be quantitatively related to changes in population numbers and structure. Although other endpoints (such as enzymatic responses and histological lesions in individual organisms) may indicate chemical exposure, they do not necessarily indicate adverse effects on populations, communities, or ecosystems, and therefore cannot be used easily in ecological modeling.

---

* The output of an ecological model typically corresponds to one or more assessment endpoints. The model may also provide probabilistic risk estimates derived from simulation of multiple scenarios (e.g., Monte Carlo). Ways of expressing risk from the output of a population model are discussed later in this chapter (see *Steps in Ecological Modeling for a Chemical Risk Assessment, Define Ecological Modeling Objectives*).

Risk Questions and Applications of Ecological Models

Understanding the role of ecological models in risk assessment and the basis for selecting specific models for a particular assessment is also important. Three kinds of general questions are addressed in ecological risk assessments for toxic chemicals (Jørgensen et al. 2000):

1. What is the ecological risk associated with new chemicals or products and their uses?
2. What are the ecological impacts and risks associated with past uses of chemicals and products?
3. What are appropriate clean up criteria for soil, water, sediment, and air; which clean up or restoration options will be most effective at reducing risk; and what is the residual risk after clean up?

The first question addresses risk assessment to support notification and registration activities for chemicals (e.g., pesticides). In this type of assessment, a generic ecological model would be used. The model has to have a structure flexible enough to accept alternative parameterizations to represent the characteristics of the different kinds of habitats, receptors, and chemical release scenarios expected. The second and third questions deal with assessing the ecological effects of previous releases of chemicals, remediation measures, and restoration of habitats. These two questions drive most of the risk analyses done under the EPA Superfund program as well as similar assessments by individual state agencies and countries other than the U.S. To address the second and third questions at a specific contaminated site requires models precisely specified for the conditions at the site (e.g., chemicals, habitats, and receptors of interest). For example, ecological models might be used to relate predicted initial impacts of a clean up technique (e.g., damage to intertidal zones in Alaska from hosing of beaches after the Exxon Valdez spill) and residual contamination (e.g., oil left on Alaskan beaches after the clean up) to population-level effects (e.g., changes in population abundances of selected intertidal invertebrate species). Alternatively, eco-logical-effects models could be used to develop generic environmental criteria. In the past, toxicity-extrapolation models and exposure models have been used to develop environmental criteria (Stephan et al. 1985; van Leeuwen 1990; OECD 1992), but population, ecosystem, and landscape models have not been applied in this way.

So far, we have discussed the use of ecological modeling in evaluating the ecological significance of risks to individual organisms. In addition to its use in risk characterization, ecological modeling may also be very useful in earlier phases of an assessment. For example, population models may be useful for evaluating the relative effects of different kinds of stressors (e.g., physical vs. chemical stressors) and thereby inform the design of field studies developed early in an assessment. Population modeling can also help interpret observed fluctuations in population abundance to determine the extent of natural variability in populations and the possible sources of the fluctuations (e.g., effects on birth rate vs. effects on death rate). Predator-prey and food-web modeling may provide information about the keystone species in a community, which could be critical for selecting receptors and endpoints for the risk assessment. Ecological models can also be used to inform management decisions about remedial actions. For example, results of modeling may help define hypotheses about system behavior before and after remediation of a contaminated site (e.g., in defining the age or size structure of a "recovered" population or the structure of a "recovered" community that was previously affected by contamination at a site).

Finally, the use of ecological modeling in determining the ecological significance of estimated risks is not limited to the population level. If a true population-level risk estimate is derived as part of the risk characterization, then use of higher-level models (ecosystem and landscape models) may be appropriate to evaluate the population risk estimates. Whether each ecological model is a part of the fundamental risk estimation or a part of the evaluation of ecological significance depends on the objectives of the assessment. Generally, models used as a fundamental part of risk estimation would be applied to a wider array of exposure scenarios or over a greater spatial extent than those used simply to evaluate ecological significance. The latter use can focus on a limited number of scenarios from which the results can be extrapolated to interpret other risk estimates.

Tiered Assessments and Model Structure

The level of detail required for a given risk assessment depends on environmental management objectives, the complexity of the site or the behavior of the chemical contaminants, and the difficulty of adequately describing exposure, toxicity, and other properties of the chemicals. An ecological risk assessment can be conducted in tiers with the most basic analyses conducted first (U.S. EPA 1998). For example, an initial *screening-level* risk assessment is conducted that uses available data and conservative assumptions about exposure and toxicity. From the results of this screening-level assessment, chemicals, habitats, and species of potential concern are identified, and decisions are made about additional data collection. In the next tier, more realistic models are used, and additional data may be collected that will better define the relationship between chemical concentrations and adverse effects. As the analysis becomes more complex (proceeding from screening-level to higher-tier assessments), the role of ecological models in the assessment is likely to become more important, and the complexity of the models applied will increase.

Many currently available ecological models do not incorporate functional relationships to address toxicity. Such models may need some structural modification (addition of complexity) to incorporate the effects of chemical toxicity, which we term *explicit modeling of toxicity*. Eventually, validating the modified models by using case study data may be necessary. Alternatively, modifying the structure of ecological models to incorporate functional relationships to account for toxicity explicitly may not be required, especially in screening-level assessments. If some information is available, but complete exposure–response curves are not available, implicitly modeling the effects of chemicals may be preferable. This modeling can be accomplished by varying the parameters of a model within a series of runs representing different impact scenarios, which we term *implicit modeling of toxicity*. For example, one could decrease the population birth rate by 20% in a scalar population model to account for a chemical's toxic effect on the basis of toxicity test results. Different levels of potential toxicity would be simulated by varying the percentage decrease in birth rate in a series of model runs. Whether the effects of toxic chemicals are modeled explicitly or implicitly, running the ecological effects model for the baseline case (with no effect of the chemical) is always important.

Rationale for Ecological Modeling

Ecological modeling does require additional costs and special expertise that many risk assessors currently do not possess. However, population modeling is a cost-effective approach for addressing many risk assessment issues (Ferson et al. 1996; Barnthouse 1998), and ecosystem models have been shown to be predictive of at least the general behavior of systems affected by toxic chemicals (Bartell et al. 1992). In a review of laboratory toxicity tests that compared the results for individual-level endpoints with those for population-level endpoints, Forbes and Calow (1999) concluded that the basic population growth parameter, r, integrates potentially complex interactions among life-history traits and thereby provides a more relevant measure of ecological impact than individual-level endpoints. Currently, many ecological risk assessments are limited by a failure to consider population, ecosystem, or landscape endpoints. Forbes and Calow (1999) concluded that failure to consider endpoints above the individual-organism level often leads to an overestimation of risk but in some cases may lead to an underestimation of risk. Moreover, Kammenga et al. (2001) used population matrix models to evaluate the effects of cadmium or pentachlorophenol on soil invertebrates in laboratory exposures and found that exposure to toxicants increased the sensitivity of organisms to other stressors that affect vital rates other than the ones affected by the toxicants.

Thus, ignoring population-level effects and focusing only on individual-level endpoints can lead to inaccurate risk estimates and errors in environmental management decisions.

Misjudgment of risks can result in overregulation that wastes resources in addressing apparent problems that are not really important or to underregulation that may cause the environment to suffer adverse effects because the risk was underestimated. Either way, an error made in risk estimation may lead to inefficiency. Thus, ecological modeling may be viewed as a form of insurance against poor decisions. Of course, this view does not imply that ecological models are fault-free or that poor decisions cannot be based on the results of ecological modeling. As these kinds of models are used more in risk assessment, evaluating the kinds of errors that can be made and developing modeling approaches for avoiding such errors will become more important. Adaptive management (Holling 1978; Walters 1986) is necessary with or without the use of ecological models, but we believe that such models can enhance decision-making and reduce errors made in estimating and interpreting risks.

Deciding When to Use an Ecological Model

The decision whether to do some sort of ecological modeling as part of a risk assessment depends partly on the objectives of the assessment, which are related to the complexity and ecological importance of the habitat, the magnitude and extent of contamination, the mode of action of the contaminants of concern, the species present or of special concern, the interests of the stakeholders, future uses contemplated for the site, and costs of current or future actions. Basing decisions on ecological endpoints rather than toxicity to individual organisms requires agreement among stakeholders that population-level or ecosystem-level effects are of greatest importance to overall ecological risks. In certain conditions, such agreement may not be achieved or may not even be appropriate (e.g., the presence of threatened or endangered species). In most cases, risk assessments that currently rely on simplistic hazard quotient approaches would benefit from the application of an ecological model, even if only a screening assessment were done.

Many ecological risk assessors believe that population-level effects can be inferred from an analysis of hazard quotient results for individual-level endpoints. A sample line of reasoning is that if a preliminary risk assessment using realistic exposure assumptions and individual-level endpoints shows an extremely high level of risk of reproductive effects or mortality, then risks at the population level are likely. Thus, they see no need to do ecological modeling to characterize the risks as ecologically significant. For example, if a 75% chance of complete reproductive failure is predicted for >50% of the adults in the population based on realistic exposure and toxicity assessments, then aren't population-level effects likely? Or if a hazard quotient value is greater than 1000 for individual-level endpoints (e.g., survival or reproduction) in a plausible exposure case, then shouldn't one expect population-level effects? However, these kinds of inferences have little or no scientific basis because population-level processes may compensate for adverse effects on individuals (EPRI 1982, 1996; Ferson et al. 1996; Tyler et al. 1997). The life history and ecology of a species can strongly influence the effects of toxic chemicals at the population level. For some species, application of a stressor can actually decrease the risk of a population decline, which is counterintuitive. For example, using RAMAS®*/age (Ferson and Akçakaya 1988) and the Ricker (1975) density-dependence function to model a brook trout (*Salvelinus fontinalis*) population at a site in Michigan, Ferson et al. (1996) showed that simulating a 20% decrease in mean fecundities actually resulted in a *lowering* of the risk of population decline. Even a 50% decrease in mean fecundities did not substantially affect the risk of decline. However, decreasing mean fecundities by 75% resulted in a crash of the simulated population. From analysis of empirical results of toxicity testing, Forbes and Calow (1999) showed that percentage response of population endpoints might be greater or less than that of individual-level endpoints. Although inferences about

* RAMAS is a registered trademark of Applied Biomathematics. Applied Biomathematics is a registered service mark.

population effects from individual effects may appear reasonable, predicting the qualitative effects of toxic chemicals or other stressors at the population level is difficult, and predicting quantitative effects without some ecological modeling is impossible.

In some circumstances, ecological modeling would not benefit a risk assessment. For example, when a worst-case analysis using individual-level endpoints shows that the risk is negligible, application of population and higher-level models would not be warranted. Conversely, suppose quantitative field evidence indicates severe effects on the abundance of target species from chemical contaminants. In this case, application of ecological models may not be necessary during the baseline risk assessment. However, such models may still prove useful for evaluating the mechanisms of population-level effects (i.e., through a decrease in birth rate or an increase in mortality as a result of toxicity). After the baseline risk assessment, ecological models may aid in deriving remedial action goals, in assessing natural recovery (Glaser and Connelly 2000), in planning restoration strategies, or in developing monitoring programs (e.g., Urban 2000).

Selecting Ecological Models for Application to Specific Risk Assessments

Many of the ecological models discussed in this report could be applied to any of the three questions posed in ecological risk assessments (see *Role of Ecological Modeling in Chemical Risk Assessment*). Nevertheless, the selection of the best model(s) to apply to a specific problem depends on the risk question and management objectives derived in relation to that question (Figure 1.3). To select models, the ecosystem, chemicals, and endpoints of interest must be defined. After these have been selected, a screening-level assessment would be conducted, preferably with a population-level perspective and quantitative uncertainty analysis. The screening-level assessment generally uses exposure estimates and available toxicity thresholds for the selected endpoints to assess whether a detailed risk assessment is warranted. In many cases, using a population model in the screening-level assessment before using more detailed models would provide valuable information.

If further ecological-effects modeling is warranted, population, ecosystem, and landscape endpoints and modeling objectives would be defined (Figure 1.3). The final selection of ecological models then depends on the desired level of detail in the analysis and the ease with which such models can interface with the chemical fate and transport model used. Chemical fate and transport models (e.g., Barber et al. 1991; Gobas 1993; Traas et al. 1996) often provide the basic chemical distribution and concentration data needed for estimating exposure. Linking the fate and transport model with the ecological model provides a comprehensive risk assessment approach. Toxicity-extrapolation methods support parameterization of ecological models by providing interendpoint and interspecies extrapolations, specifically when toxicity thresholds or exposure–response relationships are not available for species of interest. Whatever models are ultimately selected to address an ecological risk assessment issue, the strengths and weaknesses of the models should be summarized (Figure 1.3). In most cases, a quantitative uncertainty analysis would also prove beneficial in model implementation as well as management decision-making (Barnthouse et al. 1986; Warren-Hicks and Moore 1998).

The properties of a chemical may influence the selection of ecological models as well as fate and transport models (Jørgensen et al. 2000). For example, a model with a detailed description of water-column processes is appropriate for evaluating the effect of a water-soluble compound on an aquatic ecosystem, whereas a model with a more detailed description of sediment processes is better suited for evaluating the effects of a lipophilic compound. In the second case, a more detailed description of the food-web would be required because of the high bioaccumulation potential of the lipophilic compound and probable transfer though food-web pathways.

The model selection process described in Figure 1.3 is necessarily general because model selection depends on the specific assessment endpoints and receptors of interest, which in turn may initially depend on the quality and quantity of available data. Thus, model selection is usually

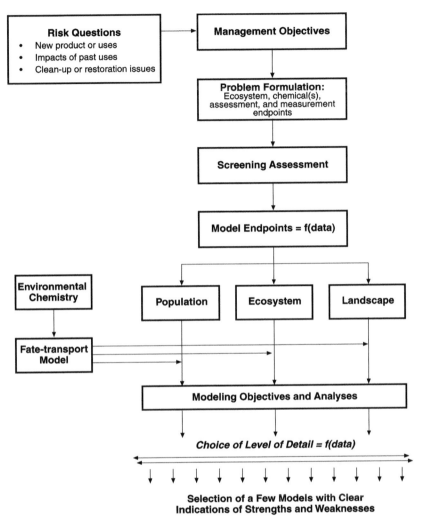

Figure 1.3   Process for selecting ecological models to be used for a specific risk assessment. Note: Toxicity-extrapolation models may be used during any phase of the risk assessment with population, ecosystem, or landscape models. (Adapted from Jørgensen et al. (2000). Improvements in the Application of Models in Ecological Risk Assessment: Conclusions of the Expert Review Panel. Prepared for the American Chemistry Council, Washington, D.C. With permission.)

site- or issue-specific (Jørgensen et al. 2000). Moreover, the complexity of the model selected depends on the level of realism and precision desired as well as the quality and quantity of data. When the desired endpoints require using a model for which not all data are available, missing data must be collected, which entails additional costs. Alternatively, the endpoints may be reformulated to make them less ambitious (Jørgensen et al. 2000).

Steps in Ecological Modeling for a Chemical Risk Assessment

Examples of applications of population models and other ecological models to assess risks of toxic chemicals are common in the literature (Barnthouse et al. 1986; Bartell et al. 1992; also see tabular summaries of model applications in later sections on the results of the model evaluations). However, outside of user manuals for these models, little guidance exists on the step-by-step process for applying ecological models in a risk assessment context. We summarize below the steps in

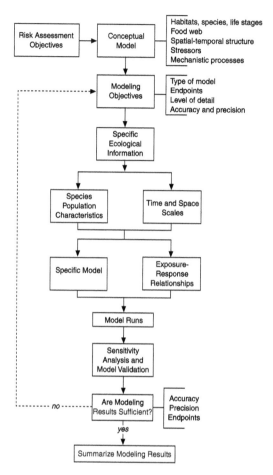

Figure 1.4   Steps in implementing a population model in the context of a chemical risk assessment. Note: Consideration of risk assessment objectives in developing a conceptual model and modeling objectives may include review of risk assessment results for individual organism endpoints. See Figures 1.3, 1.6A, and 1.6B for additional information about selecting specific ecological models.

implementing a population model to assess toxic chemical risk (Figure 1.4). Hilborn and Mangel (1997) and Jackson et al. (2000) provide guidance on how to develop and apply ecological models to support ecological research programs.

Develop Conceptual Model

A conceptual model should be developed to guide the application of any ecological model used for risk assessment. Often, this conceptual model is very similar to conceptual models developed during the problem formulation phase or the exposure assessment phase of the risk assessment. Generally, the conceptual model is a diagram with boxes for state variables (environmental components such as sediment, surface water, macrophytes, macroinvertebrates, phytoplankton, zooplankton, planktivorous fish, and piscivorous fish for a lake ecosystem model) and arrows between the boxes to illustrate the flows of energy, carbon, or contaminants (Figure 1.5). For the Comprehensive Aquatic System Model (CASM*) (DeAngelis et al. 1989) that is used for chemical risk assessment, the conceptual model illustrates food-web relationships among representative biotic groups within a lake, as well as potential exposure pathways (Figure 1.5). Some conceptual

---

* CASM is a proprietary product of The Cadmus Group. Trademark registration is in process.

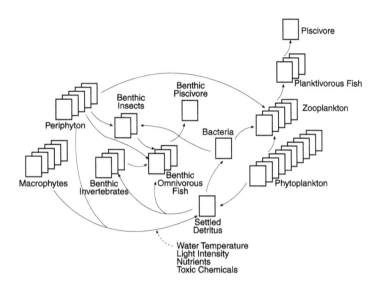

Figure 1.5   Structure of CASM. (From DeAngelis et al. 1989. Effects of nutrient recycling and food-chain length on resilience. *Am. Nat.* 134(5):778–805. Reprinted with permission of University of Chicago Press.)

models may also show ecological relationships such as competition or parasitism. For example, in Figure 1.5, competition between benthic insects and benthic omnivorous fish may be inferred because both groups feed on periphyton. However, in this conceptual model, most of the competitive relationships are among species within the broader ecological groups (as illustrated in Figure 1.5 by the multiple boxes indicating several species of benthic omnivorous fish, zooplankton, planktivorous fish, and other groups). Conceptual models also help to identify interactions among top predators (e.g., the piscivores in Figure 1.5), intermediate predators (e.g., the planktivorous fish in Figure 1.5), and other species and thereby facilitate interpretation of possible trophic cascades (Hairston et al. 1960; Carpenter et al. 1985; Carpenter and Kitchell 1988). For representing a stage-structured population, the conceptual model may be simply an illustration of the various life stages of an individual of the species of concern and the points at which environmental factors, including contaminants, may affect an individual. For a spatially explicit population model, the conceptual model would show the geographic distribution of local populations within the metapopulation.

In developing a conceptual model for a mathematical representation of a population, the risk assessor needs to answer several questions:

1. What are the key environmental components (habitats, species, life stages) to be modeled?
2. What are the relationships among those components?
3. Do characteristics of the population vary in space and time?
4. Are other species or physical stressors in addition to chemical stressors likely to influence the population?
5. What mechanistic processes are important enough that they should be included in the population model?
6. What currency should be used to describe component interactions (e.g., number of individuals, biomass, energy, carbon)?

Information for answering some of these questions may already be available from other activities conducted as part of the risk assessment. For example, evaluating extrinsic factors that may confound population model results (e.g., variation in predation pressure or risk of extreme weather events that affect populations drastically) and determining if any of these factors need to be added to the model are important. Information on these extrinsic factors may have already been considered at least qualitatively as part of the analyses of exposure and effects. Shoemaker (1977) provides further guidance on how to develop a conceptual model as the basis for an ecological model.

Define Ecological Modeling Objectives

The objectives of ecological modeling in the context of a chemical risk assessment are determined partly by the objectives of the assessment. The first decision concerns whether only a screening-level model is needed or the modeling effort will be more extensive. Second, the habitat context of the modeling effort needs to be decided. Is it a freshwater lake, a river, or a pond? Or is it a terrestrial system, and if so what kind (desert, grassland, forest)? Habitat types should have been defined in the problem formulation stage of the risk assessment, and each habitat has unique characteristics that may influence the choice of a model. Next, the type of model needs to be determined. For example, if risk to a particular population is of concern, then the modeling focuses on population-level models. If empirical data on population effects are already available, then modeling of the food-web or the ecosystem may be desired. After the general level of modeling is determined, the specific type of model should be selected. Is a spatially explicit model needed, or can the input parameters and results be averaged across space? If a population model is to be used, does it have to be age- or stage-structured? Would a scalar model suffice? Or would an individual-based population model be potentially useful (e.g., in the case of assessing risks to very small populations of threatened and endangered species). Should density-dependent effects be included in the model? Should the model be deterministic or stochastic?* Specific categories of ecological models and their characteristics are summarized in Figures 1.6A and 1.6B and discussed in detail later.

After the type of model is selected, several decisions need to be made regarding the endpoints of the analysis. Some aspects of the model endpoints will correspond to assessment endpoints in the risk characterization. However, the expression of the output of the simulation varies depending on the objectives and the audience for the final results. There are two basic kinds of model endpoints: state variables and risk estimates. *State variables* are expressed as population, ecosystem, or landscape indicators, such as organism abundance, species richness, or landscape fragmentation index, respectively. Output from a single run of a population model, for example, might be the population abundance for each time-step of the simulation, from which a population trajectory can be derived. *Risk estimates* can be derived from the model output for state variables in several ways, but the most common is to run the simulation multiple times in a Monte Carlo analysis to account for variability and uncertainty in input variables as well as initial conditions (Figure 1.7). Risk estimates derived from multiple runs (Monte Carlo) of a population model may be expressed in many ways (Spencer and Ferson 1997a,b,c) (Figure 1.8):

- Interval decline risk — the probability of a population declining by as much as a given percentage of its initial value *at any time during the period of prediction*
- Interval extinction risk — the probability of a population falling as low as a given abundance *at any time during the period of prediction*
- Terminal decline risk — the probability of a population being as much as a given percentage lower than its initial value *at the end of a simulation*
- Terminal extinction risk — the probability of a population being as low as a given abundance *at the end of a simulation*

---

* A deterministic model has no random components, whereas a stochastic model includes at least one random factor. Stochastic models attempt to reproduce realistic variation in model parameters or state variables. Thus, unlike a deterministic model, they yield different results each time they are run, even when the initial conditions and the values of model parameters (other than the random terms) are fixed. We use the term *stochastic model* to refer to any model with random error terms built into the model. Another approach to incorporating stochasticity in the results is to select values for model parameters and initial conditions of state variables from probability distributions specified by the modeler, such as in a Monte Carlo analysis (Nisbet and Gurney 1982).

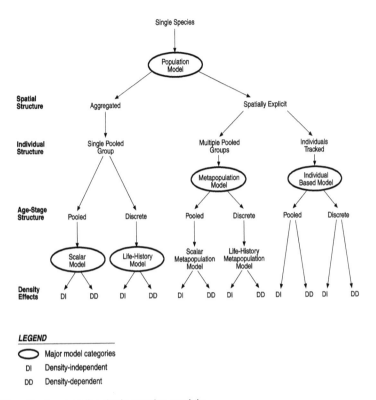

Figure 1.6A   Classification tree for single-species models.

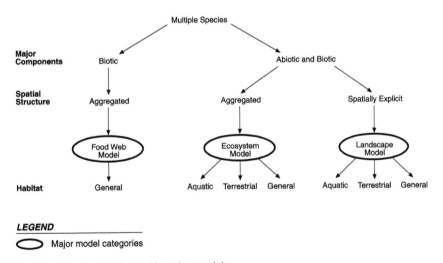

Figure 1.6B   Classification tree for multispecies models.

- Interval explosion risk — the probability of a population equaling or exceeding a given abundance *at any time during the period of prediction*
- Terminal explosion risk — the probability of a population being as great as or greater than a given abundance *at the end of a simulation*
- Time to extinction — the time required by a population to decrease to less than a given threshold abundance
- Time to explosion — the time required by a population to exceed a given threshold abundance

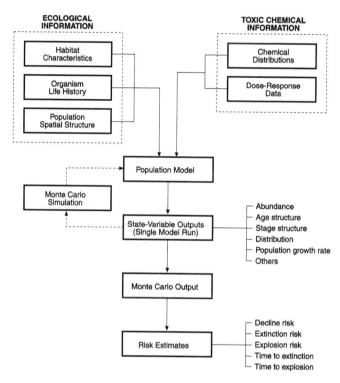

Figure 1.7 Use of a population model to develop chemical risk estimates. Note: See also Figure 1.8 for examples of how population-level risk is expressed.

The selection of a specific expression for the risk estimate depends partly on the objectives of the assessment and partly on available information for the species being modeled.

For each model endpoint, a desired level of prediction accuracy and precision should be specified. By doing so, the modeler may judge the performance of the model when predictions are compared with available data during any validation. The way in which the desired level of precision and accuracy is to be expressed depends on the nature of the prediction and of the validation data that are feasible to collect. The most useful expression of accuracy is likely to be in terms of the probability that predicted values of the endpoints and validation data are drawn from the same statistical population.* The type of statistical test needed to assess this probability depends on the nature of the model and the validation data. The coefficient of variation of the mean prediction (the standard deviation divided by the mean expressed as a percentage) is a useful expression of precision. Another one is the statistical power to distinguish a predicted result from some critical value of the assessment endpoint. Note that for most species, any model validation would most likely be done on predictions of state variables rather than the risk expressions listed. Long-term replicate observations of population behavior under similar conditions are difficult if not impossible to obtain from laboratory experiments or field surveys, precluding validation of the risk estimates per se.

Each of these modeling choices needs to be considered in light of the management objectives and the specific questions of the risk assessment. Finally, the modeling exercise and the other parts of the risk assessment may be iterative because one of the primary uses of ecological models is to help formulate predictions and to identify data needs or knowledge gaps (Jackson et al. 2000). Just as the modeler needs to be aware of the objectives of the overall risk assessment, the results of any simulations may initiate a reconsideration of those objectives.

* Note that the term *population* is used here in the statistical sense to refer to a collection of items, in this case endpoint values — not to the biological population.

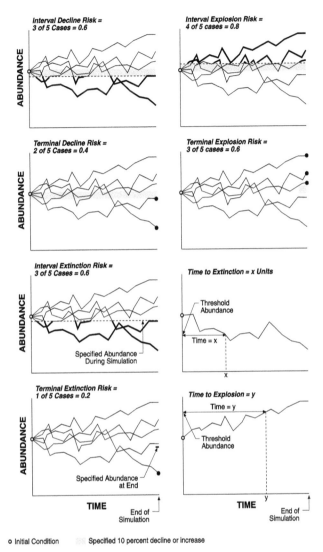

Figure 1.8   Examples of risk expressions for population-level risk assessment. Note: A small number of simulation runs are shown for clarity. In practice, many more runs would be used to derive risk estimates.

Compile Ecological Information for Species and Life Stages of Concern

Before implementing an ecological model in a risk assessment, one must gather all relevant ecological information on the receptor species and life stages selected for the assessment. If the model is used to select receptor species and life stages, then this step may be an iterative process. Information to be compiled from the literature typically includes the geographic and habitat distribution of the receptor, the life history, foraging behavior, migratory behavior, diet composition and preferences for particular foods, breeding and nursery behavior, and the general lifestyle (aerial, surface dwelling, or subterranean; freshwater, marine, or terrestrial). The level of detail needed for a particular topic depends on the modeling objectives.

The population characterization described in the next section should begin before any new data collection because it may result in the identification of data needs that might otherwise be overlooked. Identification of a feasible modeling approach may reasonably be one of the very first steps in the process of defining objectives, to be followed by organization of the available data and then collection of additional data (if necessary). Often all three of these may need to be pursued more or less simultaneously if several different modeling approaches are feasible (or desirable).

Develop Population Characterization

After the modeling objectives have been defined and available information on the populations of interest has been compiled, the information should be organized in a form that will help guide the development of the model and facilitate use of the data. For example, for an age-structured population model, an appropriate life-history table should be developed for species of concern. The life-history table consists of a schedule of births and deaths according to age classes or stages within the population. Scenarios would also be developed for the distribution of the species among habitat types and possibly their entire geographic distribution, depending on the spatial scale of the assessment. If the population model were spatially explicit, the analyst would need to specify metapopulation structure. The analyst would also define initial conditions for population size, density, age structure, and so forth based on an estimate for the exposed population at a contaminated site. Alternative scenarios might also be developed for a hypothetical population at the site in the absence of contamination by chemicals by considering the available information on the species in reference areas.

Define Time and Space Scales of Interest for the Population Model

For most chemical risk issues, the population is defined as the group of individuals potentially exposed to the chemicals of possible concern. After the species and life stages representative of receptors for the assessment are selected, the geographic areas and habitats of interest are initially identified by considering the extent of contamination associated with a specific industrial site, a chemical spill, or use of a chemical product (e.g., spraying of an insecticide on an orchard or agricultural field). The population to be considered in the risk assessment would then include those individuals potentially in contact with the areas and habitats identified in the previous step. The migratory behavior and foraging distribution of individual receptors are the primary factors to consider in defining the potentially exposed population. For larger-bodied wildlife species at contaminated sites, the number of individuals in the potentially exposed population may be quite small. For example, Sample and Suter (1999) evaluated the potential for several individual osprey to be exposed to contaminants in specific subreaches of the Clinch River/Poplar Creek system at the Oak Ridge Reservation by considering the locations of known nests in the region and the foraging radius of an osprey. (In this case, however, a population model was not needed for the assessment because effects on individuals of this federally listed Threatened and Endangered Species were the endpoint for the assessment.)

Considering the potential distribution of the receptor based on available habitats within and surrounding the contaminated area — not only the current distribution of observed individuals — is also appropriate. For example, Clifford et al. (1995) estimated exposures of burrowing owls at Rocky Mountain Arsenal by using multiple foraging (exposure) ranges overlaid on a geographic information system (GIS) representation of the site. The foraging ranges represented all possible individual ranges within the appropriate habitat, which was essentially homogeneous grassland. For each scenario, the foraging ranges were laid side-by-side on a GIS map of the site, and a series of scenarios was developed by a *lagging procedure*. Lagging is the process by which exposures of receptors potentially inhabiting an area (at present or in the future) are estimated by assuming equal probability of an individual centering its foraging range on any point in the site. Thus, a habitat analysis to determine potential distributions of receptors becomes an essential part of exposure assessment. This analysis also forms the basis for defining the potentially exposed population to be considered in ecological modeling.

The temporal scale of any population modeling effort is determined based on the modeling objectives, the life history of the receptor, the natural dynamics of the population, and the data available to run the model. Essentially, two factors need to be decided to address temporal scale — the time-step of the model and the duration of the simulation. The time-step depends on the expected temporal frequency and duration of toxic chemical exposures as well as the temporal scale of events in the life history of the receptor. For example, when do toxic chemical exposures occur, and how long do they last in relation to breeding periods, nursery periods, migrations, and other life-history events? The duration of the simulation depends on the chemical of interest, the risk assessment issues, and confidence in the model structure and parameterization. As confidence in the model increases, longer durations are appropriate. However, in most cases, running a population model for more than 30 to 50 years may be more of an exercise in wild speculation than a realistic prediction.

Select a Specific Model and Software to Implement the Model

Various ecological models are already available for use in chemical risk assessment, and the general purpose of this book is to evaluate representative models relative to their potential usefulness. Each one has its own advantages and limitations. The mathematics used often relates to the conceptual model of the system. Differential equations may be used to describe continuous processes (e.g., in a spatially aggregated model of a population with overlapping generations). Difference equations treat time in discrete intervals and are typically used for populations with discrete life stages (e.g., insects with egg, larval, and adult stages) or non-overlapping generations. Matrix algebra is most convenient for representing life-history characteristics in age- or stage-structured populations. GISs may be appropriate for spatially explicit models.

To select a mathematical representation appropriate for the objectives of the analysis, the analyst must address several questions. For example, do the equations address all the environmental components in the conceptual model? Has the model been previously applied to a problem similar to the one under consideration in this risk assessment? Has the model been validated? Is the level of detail sufficient for the problem at hand without being unnecessarily complicated? Can the model be implemented with the available data? Or can sufficient data be collected in a cost-effective way to run simulations for this risk assessment? If the analyst has trouble answering these questions, the objectives of the simulation exercise may not be focused enough. The structure of a model should be consistent with both the questions asked and the measurement endpoints for the risk assessment. The model selected should be as simple as possible while still satisfying the objectives. An overly complex model is unnecessarily difficult to parameterize and calibrate.

Many population models are available for use in evaluating chemical risk, but each assessment has its own unique characteristics that may preclude directly applying it. The user may have to modify the structure of the model or, in the case of an unusual species or site, build a new one. Jackson et al. (2000) provide guidance on creating ecological models. Given the availability of many types of population models, building one from scratch should not be necessary. However, many representations of populations do not include explicit equations to link dose–response relationships to critical parameters, such as fecundity and survivorship, that affect population growth rate (see detailed evaluations in later sections). Thus, the user may have to modify an available model to suit the objectives of a specific risk assessment. Even if the model does not need structural modification, parameters need to be set to appropriate values for the population and site of concern. Reviewing other applications is a good way to find appropriate equations for certain processes or to choose parameter values. For certain processes (e.g., feeding relationships, growth processes), the same equations are common to many different applications.

For more information on creating and modifying ecological models, the reader should consult some of the extensive works on this topic such as *Models in Ecology* (Maynard Smith 1974); *Mathematical Ecology: An Introduction* (Hallam and Levin 1986); *Handbook of Ecological Param-*

eters and Ecotoxicology (Jørgensen et al. 1991); *Fundamentals of Ecological Modeling* (Jørgensen 1994); *Wildlife Toxicology and Population Modeling: Integrated Studies of Agroecosystems* (Kendall and Lacher 1994); *Handbook of Environmental and Ecological Modeling* (Jørgensen et al. 1996); *The Ecological Detective: Confronting Models with Data* (Hilborn and Mangel 1997); and *Matrix Population Models: Construction, Analysis and Interpretation* (Caswell 2001).

The selection of software for ecological modeling depends partly on the experience of the modeler and how much user friendliness is desired. Excellent software packages are available for many types of ecological models, especially population models (e.g., RAMAS [Ferson and Akçakaya 1988], Populus [Alstad et al. 1994a,b; Alstad 2001], RAMAS Ecotoxicology [Spencer and Ferson 1997b], and the Ecotox simulator [Bledsoe and Megrey 1989; Bledsoe and Yamamoto 1996]). These and other software packages are discussed in later sections. Many of these programs offer options for customizing model structure.

An alternative to software with fixed equations for simulating ecological systems is to use a model-building program such as STELLA (High Performance Systems 2001), ModelMaker (ModelKinetix 2001), or MATLAB (MathWorks 2001). STELLA has a graphical interface that simplifies model building through a three-step method in which model equations are automatically generated. In the first stage, called *mapping*, icons are selected to represent environmental components, and relationships among them are defined in a graphical representation of input parameters. In the second phase, called the *model phase*, the software automatically creates equations that are needed to run the model. The third step, the *actual simulation*, produces results that can be viewed as graphs, tables, or an animation. STELLA emphasizes interrelationships among compartments rather than modeling a random collection of complex variables.

Like STELLA, ModelMaker has a graphical interface that allows the user to add compartments, material flows, influences, and other model components. Mass balance equations are automatically generated for model compartments, and users can edit or enter the governing equations for compartments and other components. Distributions can be specified for input parameters (i.e., to create stochastic models), and a great deal of control can be exerted over model time-steps and numerical integration techniques. Submodels can be used within a larger model framework. ModelMaker can be used to optimize model variables (for example, to find the best fit to calibration data) and can be used to perform sensitivity analyses. Simulation results can be viewed in a variety of graphical and tabular formats, and statistical summaries can be prepared.

MATLAB is essentially a mathematical computing and visualization program that allows more control than STELLA of the details of the modeling. MATLAB is used for data acquisition, data analysis and exploration, visualization and image processing, algorithm prototyping and development, modeling and simulation, and programming and application development. MATLAB has built-in interfaces that allow the user to quickly access and import data from instruments, files, and external databases and programs. According to the software developers (MathWorks 2001), integrating external routines written in C, C++, FORTRAN, and Java with MATLAB applications is also easy.

When building a model, substantial effort is needed for testing and debugging even apparently simple models to ensure calculations are done correctly. For the novice modeler, using available models may be best. Commercial software packages with user-friendly interfaces and predeveloped modeling options (e.g., RAMAS or VORTEX) have undergone extensive testing and debugging, allowing the user to avoid the effort of starting from scratch.

Link Exposure–Response Relationships to Input Variables

The next step in implementing an ecological model is the derivation of exposure–response relationships for chemicals of potential concern from available literature or site specific observations of toxic effects. Because this is a major step in the ecological risk assessment and considerable guidance is available on how to analyze for toxicity and exposure–response relationships (e.g., Barnthouse et al. 1986; Calabrese and Baldwin 1993; Suter 1993 and references therein), this topic is not addressed further here (but see the related topic of toxicity extrapolation models in *Results of Model Evaluations*). If the person implementing the population simulations is not the primary person responsible for the entire risk assessment, then the team should discuss how exposure–response relationships are used throughout the risk assessment, including in the population model.

After exposure–response relationships have been developed, they need to be linked to changes in life-history parameters and the input variables for the population model. Some models already have generic exposure–response relationships that only require parameterization for the specific issues of interest. If the model does not already include exposure–response functions, the analyst may choose to modify the equations to include such relationships or simply manually perform the operations that transfer the output of the exposure–response analysis to the input of the population model. As noted above, we have termed the structural approach *explicit modeling* of toxicity and the manual approach *implicit modeling* of toxicity. Whichever method is used, the modeler must be aware of the temporal and spatial distribution of chemicals of potential concern and how data on chemical concentrations will be summarized for use in exposure–response relationships. This step is already a major part of the exposure analysis phase of the risk assessment, so discussions between the risk assessor and the modeler and a consistent approach to data reduction are essential. If a spatially explicit population model is used, then the spatial structure of the exposure data needs to correspond to the input requirements of the model.

Perform Simulations using Estimates of Species, Exposure, and Toxicity Parameters

The analyst must next define the values for all parameters in the model. This is one of the critical steps of modeling, and care must be taken to justify the choice of parameter values as well as to quantify their uncertainty to the extent possible. In addition to parameter estimation, the analyst needs to set initial conditions for all input variables, such as chemical concentrations in space and time, initial density of individuals in the population, age- or stage-structure, spatial distribution of individuals, fecundity or birth rates, mortality rates, and so forth. Which variables need to be assigned values depends on which ones are included in the specific simulation. Model parameterization is an iterative process and may include information obtained from comparing the results of initial simulations to available data for output variables. Parameters may then be adjusted in a model calibration exercise. Consideration of the details of parameter estimation and calibration is beyond the scope of our discussion, and the reader should consult the ecological modeling textbooks cited earlier for more information. Skalski and Smith (1994) provide useful insights about integrating field sampling techniques, parameter estimation, measurement of population responses to toxic chemicals and other environmental factors, and population modeling.

Perform Sensitivity Analysis and Validate the Model

Sensitivity analysis and uncertainty analysis should be done on the model to evaluate the level of confidence to be placed in model results. In a sensitivity analysis, model parameters and the initial values of state variables are systematically varied to determine how the model results respond to changes in these values. Sensitivity analysis can guide further data collection — for example, when the analyst finds that the results are most sensitive to changes in a variable whose value is poorly defined (Jackson et al. 2000). Sensitivity analysis may be done by changing one variable at a time, by systematic sampling, or by random sampling (Swartzman and Kaluzny 1987; Hanski 1994). Changing one variable at a time is the simplest approach, but it does not account for the effects of interactions among variables, which can be assessed through the approaches using systematic or random sampling. Uncertainty analysis may include an evaluation of model uncertainty, in which the equations are changed to other plausible functions to see how different the output is from that produced by the basic model.

If possible, only ecological models that have been validated against data independent of the current risk assessment problem should be used for risk assessment. However, this type of validation is a tall order that is rarely filled. Various degrees of validation exist. In some cases, selecting an available model that has already been validated may be possible, but too often this is not the case. Model validation is usually done by comparing qualitative and quantitative predictions of the model with experimental data or field observations. The desired level of agreement between predictions and data should be specified before the model validation exercise on the basis of the modeling objectives and the risk assessment issues. In most cases, predicting within approximately 50% of the measured value may be sufficient for current practice in chemical risk assessment because of the large uncertainties in other parts of the assessment (e.g., exposure estimates and toxicity thresholds).

Sensitivity analysis and model validation results may suggest the need to refine the model. Most simulation exercises are iterative, and models can almost always be improved with more data.

Summarize Simulation Results

The final step in ecological modeling for a risk assessment is summarizing the simulation results in the risk characterization. Results from population or other models should be described as part of risk estimation. The relationship between risk endpoints at various levels of biological organization (Figure 1.1) should be discussed. As with other results of the risk assessment, the precision of any ecological modeling exercise should be clearly presented as part of the risk characterization.

A detailed example of the application of a life-history model to assess risk from a toxic chemical to a fish population is provided in Appendix A. The example shows the kind of data needed to parameterize fish population models and explores the population-level consequences of exposure of brook trout to the pesticide toxaphene. Although the example is relatively simple, it illustrates the kinds of data and decisions needed to implement an ecological model for chemical risk assessment. It also illustrates some of the limitations of fish population modeling for chemical risk assessment. These limitations include the perception of substantial effort to parameterize even a relatively simple model, the site specificity of the model, the difficulty of quantifying density-dependence using field data, the paucity of published life-cycle toxicity data for fishes, and the difficulty of dealing with contaminant mixtures. However, many of these limitations are faced in any risk assessment regardless of whether a population model is applied. When data exist or reasonable estimates of model variables can be made, population modeling represents a highly flexible and cost-effective approach to forecasting ecologically relevant responses to toxic chemicals.

CHAPTER **2**

Methods

Robert A. Pastorok and H. Resit Akçakaya

The general approach is to initially identify a variety of models for potential use in ecological risk assessment, to narrow this list on the basis of criteria such as usability and relevance, and then to conduct a more detailed evaluation of the most promising models (Figure 2.1). Thus, a two-step evaluation procedure is used. The first step is a simple analysis to identify models that meet minimum requirements. The second step is a detailed evaluation of models that pass the first step.

COMPILATION AND REVIEW OF MODELS

Models that have been applied to ecological risk assessments for aquatic and terrestrial habitats are considered in our review. Basic ecological models used to describe populations, ecosystems, or landscapes are also included, even if others do not currently consider them for use in ecological risk assessment. Both mathematical formulations that comprise the models themselves and selected software packages used to implement them (e.g., VORTEX by Lacy 1993; ULM by Legendre and Clobert 1995) are considered. Although our review emphasizes user-friendly software environments that are designed to support population and ecosystem modeling, generalized modeling software such as STELLA (High Performance Systems 2001), ModelMaker (ModelKinetix 2001), and MATLAB (MathWorks 2001) can be used to implement all model types discussed herein.

Compilation and Classification of Models

Ecological-effects models were identified through literature searches, personal contacts with scientists involved with ecological modeling, and a review of model compilations (e.g., Morrison et al. 1992; Jørgensen et al. 1996; SAIC 1996; Hilborne and Mangel 1997; ANL 1998; ORNL 1998). A search over the Internet was conducted, including a search of the database on ecological models maintained by the University of Kassel in Germany (http://dino.wiz.uni-kassel.de/model_db/server.html) and other web sites with general information on ecological modeling (Table 2.1). A call for models was posted on relevant list servers (e.g., ecotoxicology, enveng, riskanal) and published in the newsletter of the Society for Environmental Toxicology and Chemistry.

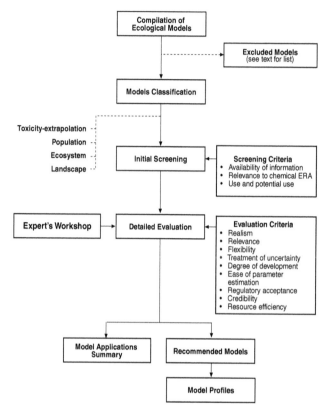

Figure 2.1   Process for evaluating ecological models for use in chemical risk assessment. Note: Full classification scheme for ecological models is shown in text.

Examples of categories of models for which information was compiled include:

Population Models
    Scalar abundance models
        Malthusian growth
        Logistic growth
        Stock-recruitment
        Stochastic differential equation models
        Stochastic discrete-time models
        Equilibrium exposure models
        Bioaccumulation and population growth models
    Life-history models
        Deterministic age- or stage-based matrix models
        Stochastic age- or stage-based matrix models
    Individual-based models
    Metapopulation models
        Occupancy — incidence function models
        Occupancy — state transition models
        Other models

Ecosystem Models
    Food-web models (biotic only)
        Population-dynamic food-chain models
        Predator–prey models

Complex ecosystem models (abiotic and biotic)
    Aquatic models
        Marine and estuarine models
        Lake models
        River models
        Wetland models
    Terrestrial models
        Desert models
        Forest models
        Grassland models
        Rangeland models
        Island models
    General models

Landscape Models
    Aquatic models
        Marine and estuarine models
        Freshwater and riparian models
        Wetland models
    Terrestrial models
        Forest models
        Grassland models
        Island models
        Multi-scale models
    General models

Toxicity-Extrapolation Models
    Species-sensitivity distribution models
    Interendpoint extrapolation models
    Interspecies extrapolation models

Our use of three general categories of ecological models (population, ecosystem, and landscape models) generally corresponds to the model classification scheme of Jackson et al. (2000; their Table 2.1), although these authors separate food-web models into their own categories of "multi-species models." We have included food-web (community) models under ecosystem models because food-web models are often the basis for ecosystem models, and the only difference is the inclusion of abiotic components in the ecosystem model. Jackson et al. (2000) also provide a matrix classification of ecological models in which the basic features of aggregated models, structured models, and individual-based models within each major category of models are defined.

Definition of General Model Categories

Population, ecosystem, and *landscape models* are true ecological models in that they predict endpoints beyond the individual-organism level. *Population models* typically deal with the dynamics of the abundance or distribution of single species, sometimes with explicit descriptions of endpoints in time and space. For simplicity, single-species models are identified as population models, and multispecies models are defined as ecosystem or landscape models, regardless of whether they include abiotic factors. Thus, *ecosystem models* are mathematical expressions that are intended to describe ecological systems composed of interacting species. These models incorporate species dynamics (individual or population) as well as specific biological interactions (predator–prey, competition, dependence, etc.) to predict ecosystem endpoints such as species richness or the productivity of a multispecies assemblage. They include food-web models, defined here as simple multispecies models, generally without abiotic environmental variables, and complex ecosystem

Table 2.1 General Internet Web Site Resources for Ecological Modeling[a]

Home Page	Address	Host	Resources
WWW—server for ecological modeling	http://dino.wiz.uni-kassel.de/ecobas.html	University of Kassel, Germany	Covers many ecological models for terrestrial and aquatic systems. Includes the simulation software, contacts for model acquisition/use, and literature citations
Bibliography of ecological modeling textbooks	http://homepage.ruhr-uni-bochum.de/Michael.Knorrens child/embooks.html	Michael Knorrenschild, Fachhochschule Bochum, Germany	Bibliography of literature on ecological modeling
System dynamics and simulation bookmarks	http://home.earthlink.net/~tomfid/sdbookmarks. html#Spatial	Tom Fiddaman, Ventana Systems	Links to web sites with information on publications, software, and contacts for simulation modeling of all kinds
Ecological modeling	http://www.env.duke.edu/lel/env303/em_ contacts.html	Landscape Ecology Laboratory, Duke University	Links to web sites on government agencies, academic programs, and individual models
Ecosystem simulation	http://www.isima.fr/~ecosim/	ECOSIM	Home page for ECOSIM, a program on various aspects of simulating complex systems.
Individual-based models	http://www.red3d.com/cwr/ibm.html	Craig Reynolds, Sony Computer Entertainment America	Many links to information about individual-based modeling approaches and simulation software
Integrative ecological models	http://www.anserc.org/research/projects/ coastes/ integrative_ecological_models.htm	COASTES: COmplexity And STressors in Estuarine Systems (National Oceanic and Atmospheric Administration)	Food web and risk analysis models for coastal systems
CAMAS Home Page	http://www.agralin.nl/camase/	Agralin	Agro-ecosystem models
Spatial Ecology Modeling and Analysis Software	http://userzweb.lightspeed.net/~jpthomas/ssoft. html	JT Connection	Links to resources for spatial modeling of ecological systems
International Society for Ecological Modeling Home Page	http://ecomod.tamu.edu/~ecomod/isem.html	International Society for Ecological Modeling	Information on the International Society for Ecological Modeling and their activities
Society for Mathematical Ecology Home Page	http://www.smb.org/	Society for Mathematical Ecology	Information on the Society for Mathematical Ecology and their activities

[a] The list of web sites is this table is not intended to be comprehensive but should provide access to the rich resources on ecological modeling that are available on the worldwide web. The links shown in this table are current at the time of publication, but web site addresses change frequently.

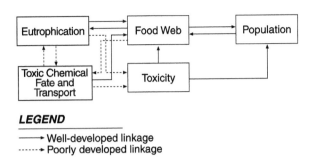

Figure 2.2   Major components of an integrated chemical fate and effects modeling framework. (Modified from Koelmans et al. 2001. Integrated modeling of eutrophication and contaminant fate and effects in aquatic ecosystems: A review. *Water Res.*)

models, which include abiotic variables as well as biological resources. Because of the inherent complexity of such models, they tend to be specific to either a geographic area or a type of ecosystem. Spatially explicit, multispecies models, which generally include abiotic factors, are defined as *landscape models*, whereas spatially explicit models of single-species populations are defined as *metapopulation models*.

Toxicity-extrapolation models are simple empirical, sometimes statistical, methods for extrapolating toxicity thresholds or for ordering species sensitivity to toxic chemicals. Although extrapolation models may not be considered to be ecological models per se, they are commonly used to develop data inputs such as toxicity thresholds or dose-response relationships applied in such models.

In addition to the model types listed, we also identify several methods for classifying species or specific populations according to their status or risk of extinction. Because these rule-based systems are not mathematical models with ecological endpoints such as species richness or abundance, they are somewhat different from the models included in the main evaluation. Nevertheless, classification systems are commonly used by organizations worldwide in managing biological resources and are potentially useful in ecological risk assessments for toxic chemicals. A separate review of classification systems is included in Appendix B.

Koelmans et al. (2001) summarized the major components of integrated chemical fate and effects models as applied to aquatic ecosystems. We have adapted their approach to illustrate the main components of an integrated modeling framework: a eutrophication model, a toxic chemical fate and transport model, a toxicity model, and a food-web model (Figure 2.2). Some of the linkages between these components, as shown by the solid lines in Figure 2.2, have been well developed. Others, as shown by the dashed lines in Figure 2.2, are not as well developed or have not been included in many models.

The major categories of ecological models defined above can be clarified by interpreting them relative to Figure 2.2. Population models are linked to form food-web models. An ecosystem model results from the combination of a food-web model with a eutrophication model when the output of the latter eutrophication model sets nutrient conditions that may affect food-web structure, biological community characteristics, or component populations. The combination of a food-web model with a toxic chemical fate and transport model without consideration of toxicity is a bioaccumulation model, which is not considered an ecological model here. A true integrated chemical fate and effects model includes all the components shown in Figure 2.2. An ecosystem model or a fully integrated fate and effects model implemented in a spatially explicit fashion would be considered a landscape model.

Other Model Types

Accounting for changes in species composition by using multispecies models is possible, but they may fail to predict adaptive or dynamic structural changes in ecosystems (Jørgensen 1992, 2000;

Jørgensen and de Bernardi 1997). Two ways of describing these ecological changes exist: (1) by including expert knowledge as part of artificial intelligence algorithms in the model or (2) by applying goal-oriented functions that describe relationships among model variables or parameter changes corresponding to shifts in community composition or species relative abundances. Ecosystem models that incorporate artificial intelligence algorithms to discern dynamic structural changes are still under development and were not considered practical for application to many chemical risk assessments at present. Although the application of explicit goal-oriented functions is also rare in current models, most ecosystem models evaluated incorporate some implicit goal of Malthusian or logistic population growth for individual species.

Most of the ecosystem models reviewed incorporate multiple interactions among various factors to predict a system's future state from past or present conditions. Goal-oriented hypotheses that involve maximizing resource utilization by individuals often underlie functions that describe these interactions. However, maximization of resource utilization is a complex process within the context of a multispecies (ecosystem) model. For example, there are almost always requirements for more than one type of resource (i.e., use of multiple prey species by a predator, use of nitrogen and phosphorus by plants, need for food and nesting sites to support avian population productivity, etc.). Furthermore, many of the resources are utilized by more than one species, leading to interspecific competition. Ultimately, there are physical limitations on both resource use efficiency and ultimate structure (e.g., photosynthetic efficiency and maximum tree size). Interactions between these functions essentially describe dynamic structural changes in ecosystems expressed as shifts in dominant species or changes in species composition. Limitations in the predictive power of the models to describe such structural changes are based on the scope of the model and the availability of reasonable data for parameterization.

Other model types and formulations that might be defined by others as ecological models were excluded from the initial compilation because they did not predict relevant ecological endpoints or they addressed spatial or temporal scales beyond the scope of interest (Figure 2.1). Excluded models or other formulations and the reason for their exclusion are included in the following list:

- Scale beyond the scope of interest
  - Global models (e.g., climate change) and large-region oceanic models
- Ecological endpoints limited
  - Biogeochemical, chemical fate (including bioaccumulation), and hydrodynamic models
  - Agricultural single-species models (some agricultural food-chain models may be included because of value in pesticide risk assessment)
  - Water quality and eutrophication models with only chlorophyll *a* or phytoplankton biomass as the main biological endpoint
  - Genetic models
  - Watershed erosion, nonpoint source pollution, and land use models
- Indices or statistical endpoints only
  - Landscape indices that only describe landscape pattern without model components to predict endpoints such as species distributions, abundances, or vulnerability to extinction
  - Community indices (e.g., index of biological integrity, Ephemeroptera/Plecoptera/Tricoptera Index)
  - Habitat suitability (or selection) models
  - Simple statistical models of population, community, and ecosystem data without predictive use (e.g., lognormal distributions of species abundances)

To exclude the models and indices listed from the current project in no way indicates that such models are not useful in ecological risk assessment. Indeed, some of the excluded approaches are essential to the risk assessment process (e.g., biogeochemical, chemical fate, and bioaccumulation models; water quality and eutrophication models). Such models typically form the basis for the exposure analysis; thus, the population and higher-level models reviewed here must be linked to them. Nevertheless, the focus of the current project is on models that can be used to predict population, ecosystem, or landscape effects from endpoints at lower levels of biological organi-zation, especially from effects on individual organisms. The toxicity-extrapolation models are also included here because these approaches are currently in widespread use in ecological risk assess-ment to estimate toxicity thresholds. In the evaluation of extrapolation models, quantitative structure–activity relationships (QSARs) were excluded because a review of QSARs is a large undertaking beyond our scope.

Initial Selection of Models

An initial review of models helped to identify those models that should be evaluated in detail (Figure 2.1). Several criteria were used during the initial review to select models for further evaluation. First, sufficient information about the model had to be available from publications, unpublished reports, or web sites to be able to describe the model inputs and endpoints, the basic modeling approach, model equations, and past uses of the model. Second, the model had to address ecologically relevant endpoints (generally at the population, ecosystem, and landscape level) and have some potential relevance to chemical risk assessment. Third, the past uses of the model or potential uses needed to match the general objectives of ecological risk assessment (e.g., in terms of emphasis on ecological components and endpoints). After the initial review was completed, a list of models was developed for inclusion in the detailed evaluation.

Selection of relevant endpoints is critical to the model evaluation process because endpoints are related to model uses. General endpoints are identified for each model included in the screening. Examples of ecological endpoints to be considered in evaluating models include:

- Population abundance — the total number or density (number per unit area or unit volume) of individuals of a given species in a given location
- Biomass of a population or community — the total mass (or mass per unit area or volume) of living organisms in the population or community
- Reproductive output of a population — measures of the yield of newborns; (e.g., fecundity, number of individual young produced by a female or breeding pair, or number of recruits to the adult stage per time period)
- Population growth rate — the rate at which numbers of individuals are added to the population over time
- Age structure — the relative proportions of different age- or stage-classes in the population
- Species richness — the total number of species in a location or the number per unit area or volume
- Trophic structure — the relative proportions of different feeding types in the community (e.g., primary producers, herbivores, primary carnivores, secondary carnivores, detritivores, etc.)
- Population or community productivity — the rate of production of living biomass in a population or community
- Spatial distribution of organisms — the locations or spatial arrangement of organisms in a popu-lation or community

Note that the endpoints just listed are state variables in the model or may be derived directly from model output for related state variables (e.g., the model may yield abundances of various species, and the user would group species and estimate trophic structure). Risk estimates are then derived from information about state variables, either by comparing contaminated and reference (or control) conditions or by implementing multiple runs of the model (e.g., in a Monte Carlo analysis) (Figures 1.7 and 1.8).

Detailed Evaluation of Models

The approach to evaluating models recognizes that choosing the right model for ecological risk assessment entails a tradeoff between flexibility (e.g., the power to address endpoints of concern) and practicality (e.g., data requirements). This tradeoff depends on the availability of data and the questions addressed. The questions addressed greatly influence the type of models and level of analysis that are appropriate. For the detailed evaluation, model characteristics and predictive power are considered relative to the model endpoints, which serve as surrogates for questions to be addressed and model uses. Model endpoints are used as surrogates for model uses because identifying all current and future uses of models is impossible. Because the uses of a model depend directly on the endpoints, the value of a model in terms of its ability to predict the identified endpoints is evaluated on the basis of a series of criteria for the detailed evaluation (Figure 2.1).

The criteria for detailed evaluation of models include:

- Scientific criteria
  - Realism and complexity (i.e., whether key processes are included and how they are represented)
  - Prediction of relevant assessment endpoints and utility relative to regulatory compliance
  - Flexibility
  - Treatment of uncertainty
  - Degree of development, consistency, and validation
  - Ease of parameter estimation
- Political and economic criteria
  - Regulatory acceptance
  - Credibility (e.g., prevalence of users, availability of published reviews)
  - Resource efficiency

Next, we describe the criteria used in evaluating ecological-effects models. Each criterion description includes a definition of low, medium, and high scores for the criterion. Criteria for evaluating extrapolation models are somewhat distinct from those used to evaluate the other models. Therefore, the interpretation of the criteria for use in evaluating extrapolation models is also described when it differs from those for other model types. Note that the ratings applied to ecological models in later sections apply only in the context of ecological risk assessment. Models rated low on some or many criteria may be perfectly good models for other environmental management issues or for ecological research. The ratings are also generally applied to the models as they are reported in the literature or as they have been used. The reader should keep in mind that the ratings for a model could change in some cases if relatively minor structural modifications were made to a model, or if the model were used with other complementary models. Any modeler should always carefully evaluate the information on models contained herein and make his or her own judgment about the value of a given model as applied to a specific ecological risk assessment problem.

Realism

Does the model incorporate key processes that are known to be important in the system it simulates? Does the model incorporate key factors that affect each of these processes? Is the model structure detailed enough to use all the data that is available and relevant in typical cases? Are the *assumptions* realistic with respect to the ecology of the system?

For extrapolation models, opportunities to increase realism include the following design choices: using data from tests with more realistic conditions (e.g., dietary exposure, a flow-through system), incorporating relationships between toxicity and physical characteristics, using chronic vs. acute exposure data, selecting uncertainty factors, and selecting protection level (e.g., the severity of effects or percentage of a community affected).

Low: the model incorporates only one or a few key factors and processes. The structure of the model is not detailed enough to use all the data that are available and relevant in typical cases. Assumptions are not realistic with respect to the ecology of the system. For extrapolation models, a rating of *low* is assigned when most or all of the following design selections made for this model are unrealistic:

application of acute instead of chronic exposure test results; no incorporation of relationships between toxicity and physical habitat characteristics; nonselective inclusion of toxicity data regardless of test conditions; and selection of mortality endpoints instead of sublethal effect endpoints.

Medium: the model combines several key factors and processes with functions that have been observed in these systems. The structure of the model is detailed enough to use most of the relevant data. Most assumptions are realistic with respect to the ecology of the system. For extrapolation models, a rating of *medium* is assigned when several options of the design selections listed are unrealistic.

High: the model combines many key factors and processes with functions that have been observed in these systems. The structure of the model is detailed enough to use the typical data that would be relevant for an assessment without being overly complex. All assumptions are realistic with respect to the ecology of the system. For extrapolation models, a rating of *high* is assigned when most of the design selections listed are realistic.

Relevance

How related are the model results to the endpoints that are used in ecological risk assessments (i.e., primarily population-, community-, or ecosystem-level endpoints)? If the model results are not relevant endpoints, can endpoints of interest be calculated easily from model results? Does the model have parameters and functions that can be used to analyze effects of toxic chemicals, physical habitat disturbance, or both?

Low: model results are expressed in terms of metrics (measures) that are irrelevant in population-, community-, or ecosystem-level risk analyses. The model does not have parameters that can be modified to analyze the effect of factors such as toxicity.

Medium: model results are expressed in terms of metrics (measures) that are sometimes used in population-, community-, or ecosystem-level risk analyses. The model has at least one parameter that can be modified to analyze the effect of factors such as toxicity.

High: model results are expressed in terms of metrics (measures) that are commonly used in population-, community-, or ecosystem-level risk analyses. The model has several parameters that can be modified to analyze the effect of factors such as toxicity or includes explicit functions that account for toxic effects.

Flexibility

Can the model accept alternative formulations? How easy is it to model different life histories, different spatial or temporal distributions of organisms, or different systems (e.g., various habitat types)? How easy is it to represent differences between species and populations with different structures, functions, selections, or parameter values that are allowed within the model? How easily can the model be applied to species and systems other than those for which it was originally developed?

Low: the model has few parameters, functions, or alternative structures that can be modified to represent different situations. Model structure applies only to a specific species, a specific type of organism (e.g., river fish population), or a specific type of ecosystem (e.g., forest). For extrapolation models, the method only applies to a single species or community type and cannot be applied elsewhere.

Medium: the model can represent a range of ecological components (e.g., alternative life histories) with user-specified structures, functions, and parameter values. The model can be applied to a group of species (e.g., vertebrates, invertebrates, or plants) in freshwater, marine, or terrestrial systems. For extrapolation models, the method applies to more than one large taxonomic group. Species or individual differences in sensitivity cannot be accounted for within the parameters of the model (i.e., interspecies toxicity information is pooled).

High: the model can represent a wide range of ecological components (e.g., alternative life histories) with user-specified structures, functions, and parameter values. The model can be applied to vertebrate, invertebrate, and plant species in freshwater, marine, and terrestrial systems. For a toxicity extrapolation, the model has the ability to incorporate species or individual differences in sensitivity.

Treatment of Uncertainty

Does the model incorporate uncertainties? How easy is it to use the model to separate effects of natural variability and measurement uncertainties? Does it use application factors or other uncertainty factors?

Low: the model ignores variability and uncertainties or uses arbitrary uncertainty factors.
Medium: the model allows running scenarios, bounding analysis, or rudimentary Monte Carlo simulations.
High: the model includes comprehensive Monte Carlo simulations (or another detailed probabilistic approach) and allows sensitivity analysis.

Degree of Development and Consistency

How easy is it to understand what the model does? How easy is it to get help for a new application? Are there any errors in the model equations? Is software available to run the model? How many errors/bugs are in the model software? How easy is it to identify wrong or meaningless input values? Has the model been tested or validated?

Low: the model is not implemented as software, or the software is not easy to customize or does not have sufficient documentation or technical support. The model fails checks of consistency (e.g., unit and dimensional), does not do feasibility checks, and has not been tested or validated.
Medium: the model is implemented as software that is easy to apply to a new system and has sufficient documentation. It is consistent or checks the validity and feasibility of user-input parameters. For extrapolation models, the model has been tested or validated once or twice.
High: the model is implemented as software that is easy to apply to a new system and has good documentation and technical support. It is consistent and checks the validity and feasibility of user-input parameters. For extrapolation models, the model has been tested or validated multiple times.

Ease of Estimating Parameters

How easy is it to estimate the required model parameters (i.e., state variables and coefficients) given the type of data typically available? Are accepted sampling or statistical methods for estimating model parameters included? How much data is required for an unbiased estimation of parameters? Do key model parameters have intuitive biological meanings? For extrapolation models, are appropriate toxicity data readily available for all the relevant taxonomic groups as the basis of extrapolation?

Low: the model requires parameters that are hard or impossible to estimate given typical data sets. Unbiased estimation of model parameters requires extensive data sets. Some key parameters are not easily interpreted. For extrapolation models, toxicity data are difficult to acquire.
Medium: some model parameters are hard to estimate given typical data sets or require data collection with new protocols. Unbiased estimation of model parameters requires large data sets. Model parameters are intuitive and can be interpreted biologically. Toxicity data are readily available for extrapolation models.
High: model parameters are easily available from the literature or are easy to estimate with established protocols. Model parameters are intuitive and can be interpreted biologically.

Regulatory Acceptance

How likely are regulatory agencies to accept results of an analysis with the model? Have any regulatory agencies adopted this extrapolation method?

Low: the model is not used by any regulatory agency.

Medium: the model is used by a U.S. or foreign regulatory agency but has not been formally adopted. For extrapolation models, the method has at least been reviewed and found acceptable for use by a regulatory agency.

High: the model was adopted by a U.S. or foreign regulatory agency; or it is used by several regulatory agencies in different countries and by international organizations; or the model has been used in a regulatory setting (to assess mitigation, or for endangered species listing, recovery planning, etc.).

Credibility

Does the model have much scientific and technical credibility?

Low: the model is not widely known in academia; few publications or users have applied the model to different cases.

Medium: the model is known in academia; several publications or users have applied the model to different cases.

High: the model is widely known in academia and has been positively reviewed in the literature; many publications or users have applied the model to different cases.

Resource Efficiency

How much time and effort would be needed to apply the model in a particular case? Because none of the extrapolation techniques requires considerable resources or software for its application, this criterion was not applied to the review of extrapolation models.

Low: applying the model to a particular case requires programming, testing, and debugging; or the data requirements are such that in almost all cases, available data would not be sufficient, and additional data collection is necessary.

Medium: applying the model to a particular case requires minor changes to the program and recompilation in addition to data analysis and parameter estimation. Many cases may require additional data collection.

High: applying the model to a particular case does not require any programming. Some cases will require additional data collection, but in many cases, available data are sufficient.

SELECTION OF MODELS FOR FURTHER DEVELOPMENT AND USE

As described, a separate rating was developed for individual models within each model subcategory with respect to scientific criteria and political–economic criteria. Ratings of models for each evaluation criterion were compared across models within a model subcategory, and recommendations about which models should be further developed for use in the near term were developed (Figure 2.1). For example, metapopulation models were compared, and the best ones were selected for further development. Similarly, aquatic ecosystem models were compared, and recommendations were made for further development. However, when developing ratings, metapopulation models were not compared with other types of population models or with ecosystem and landscape models.

Profiles of the recommended models were prepared by describing model attributes according to the scheme in Reinert et al. (1998) as follows: stressor, type of ecosystem, temporal resolution, spatial resolution, geographic scale, biological scale (e.g., individual, population, community), time scale, model type (on the basis of the classification provided), level of detail (i.e., tier of the risk assessment), and endpoints.

CHAPTER 3

Results of the Evaluation
of Ecological Models: Introduction

Robert A. Pastorok

The results of the screening evaluation are shown in Appendix C. A total of 29 population models, 34 ecosystem models, and 19 landscape models are identified for detailed evaluation; 13 toxicity-extrapolation models are included in the detailed evaluation.

The selected models are reviewed and rated on a scale of low, medium, or high for a set of nine defined criteria (listed in the earlier *Methods* section). On the basis of the results of this evaluation, a subset of models is recommended for further development and use in ecological risk assessment. Classification systems are addressed in Appendix B.

The results of the detailed evaluation are discussed according to model categories that follow.

- Population models — scalar abundance
- Population models — life history
- Population models — individual-based
- Population models — metapopulation
- Ecosystem models — food webs
- Ecosystem models — aquatic
- Ecosystem models — terrestrial
- Landscape models — aquatic and terrestrial
- Toxicity-extrapolation models

Recall that detailed evaluations are made within each model subcategory (scalar abundance, life history, metapopulation, aquatic ecosystem, terrestrial landscape). The ratings should be interpreted with caution and are only comparable within each evaluation table and within a model subcategory. Individual extrapolation models may have scored high relative to other extrapolation models, but the tabulated results cannot be compared with the scores for population or ecosystem models. Similarly, the tabulated results for population models cannot be compared with the scores for ecosystem models or landscape models. Because many current models are still under development, their ratings with respect to the evaluation criteria may change. Persons interested in a particular model should always check with those developing and applying the model to obtain the most current information on model structure, uses, and availability of software.

Before the presentation of evaluation results for each category, examples of state-variable endpoints are listed for the given type of model. As discussed earlier, various kinds of risk estimates can be derived from any one of these endpoints (Figure 1.7).

CHAPTER 4

Population Models — Scalar Abundance

Scott Ferson

This chapter reviews models of single-species population dynamics that represent abundance with a single, scalar dimension without consideration of population age or stage structure. These models are the simplest possible representations of the biological processes that govern population growth, and they ignore much of the complexity of the ecological world. However, more complicated models are not necessarily better. In the case of population growth, these simple models may be able to capture the essential features with low expenditures for empirical data collection and minimal model calibration.

A primary use for scalar population abundance models is in performing screening assessments of the ecological risks of chemicals. Because screening assessments usually must be done in data-poor situations, scalar models might be especially appropriate. Whenever the assessments are generic (not parameterized for a particular site), the simple scalar models may provide sufficient detail for modeling effects of toxic chemicals on biological populations. Example endpoints for scalar population models are:

- Abundance of individuals
- Population growth rate

We review the following scalar population abundance models (Table 4.1):

- Malthusian population growth models (May 1974; Poole 1974; Alstad 2001)
- Logistic population growth models (May 1974; Poole 1974; Alstad 2001)
- Stock-recruitment population models (Beverton and Holt 1959; Ricker 1975, 1954; Alstad 2001)
- Stochastic differential equation models (Poole 1974; Goel and Richter-Dyn 1974; May 1974; Ginzburg et al. 1982; Iwasa 1998; Tanaka 1998; Hakoyama 1998)
- Stochastic discrete-time models (Poole 1974; Capocelli and Ricciardi 1974; Goel and Richter-Dyn 1974; May 1974; Ginzburg et al. 1982; Ferson 1999)
- Equilibrium exposure models (Hallam et al. 1983a,b)
- Bioaccumulation and population growth models (Bledsoe and Megrey 1989; Bledsoe and Yamamoto 1996; Spencer and Ferson 1997b).

1-56670-574-6/01/$0.00+$1.50

Table 4.1 Internet Web Site Resources for Scalar Population Models

Model Name	Description	Reference	Internet Web Site
Malthusian population growth models	Models of exponential population growth	Alstad (2001); Poole (1974); May (1974)	http://www.arcytech.org/java/population/facts_math.html
Logistic population growth models	Basic population models incorporating density dependence	Alstad (2001); Poole (1974); May (1974)	http://www.arcytech.org/java/population/facts_math.html
Stock-recruitment population models	Population models incorporating density dependence used for natural resource management	Ricker (1954, 1975), Beverton and Holt (1959)	http://www.fw.umn.edu/FW5601/ALAB/stockrec/stockrec.htm
Stochastic differential equation models	Basic population models incorporating random variability — continuous form	Poole (1974); Goel and Richter-Dyn (1974); May (1974); Ginzburg et al. (1982); Iwasa (1998); Tanaka (1998); Hakoyama and Iwasa (1998)	N/A*
Stochastic discrete-time models	Basic population models incorporating random variability — discrete form	Poole (1974); Capocelli and Ricciardi (1974); Goel and Richter-Dyn (1974); May (1974); Ginzburg et al. (1982); Ferson (1999)	N/A*
Equilibrium exposure models	Scalar model that combines population dynamics with toxicant chemistry	Hallam et al. (1983a,b)	http://www.baltzer.nl/emass/contents/2-1,2.html#ema208m
Bioaccumulation and population growth models	Scalar models that consider population dynamics as well as the kinetics and bioaccumulation of toxic chemicals	Spencer and Ferson (1997b); Bledsoe and Yamamoto (1996); Bledsoe and Megrey (1989)	http://www.ramas.com/ http://www.wiz.uni-kassel.de/model_db/mdb/ecotox.html

* N/A = not available.

The categories are obviously not disjoint. For instance, logistic models are a special case of stock-recruitment models. Moreover, the deterministic Malthusian, logistic, and stock-recruitment models have stochastic analogs that are covered in the fourth and fifth categories.

Many scalar population models were not explicitly considered in this review (Appendix C, Table C.4). Many of those are simply applications of reviewed models to particular species and are therefore merely special cases that do not merit additional review. Agricultural crop models and harvest models (Getz and Haight 1989) were excluded by the design of the project. In addition, within the review process, we concluded that neither branching process models, nor chaos models, nor time-delay models were relevant to ecological assessment of the effects of chemicals on populations.

Although chaos theory seems to experience regular revivals in popularity within the ecological literature (May 1974), there is no evidence for the existence of chaos in real ecological systems. In fact, Ellner and Turchin's (1995) survey of the Lyapunov exponents in a wide variety of population abundance data sets showed that virtually all were convergent rather than divergent (the latter would be expected according to chaos theory). Whether the nonexistence of chaos in ecological systems is the result of selection pressures or some other kind of constraint, use of dynamic chaos models in chemical risk assessment is not warranted at present.

In contrast, time-delay models have their champions (e.g., Maynard Smith 1974; Royama 1992; Turchin and Taylor 1992; Turchin 1995). However, these models are too immature for serious applications. They have four limitations. First, time-delay models commonly have several more parameters than other comparable models without time delays. Second, the data needed to parameterize the models are time-series data and are therefore especially difficult to gather. Third, because they are essentially regression coefficients, the parameters of a time-delay model are generally not biologically interpretable. This drawback leads to the fourth and most serious limitation — biologists will not know how to perturb the parameters to model the effects of chemicals. For these reasons, we expect that, at least in the next 5 to 10 years, time-delay models will not be useful in assessments of the effects of chemicals on biological populations.

MALTHUSIAN POPULATION GROWTH MODELS

Malthusian population growth is named for Thomas Malthus, an English minister whose famous essay (Malthus 1798) expressed concern that geometric increase in the human population would be restrained by finite arithmetic increase in food and space resources and would result in starvation and epidemic disease. Ecologists use the word Malthusian to describe the natural geometric or exponential increase in populations in habitats in which resources are relatively plentiful. Thorough treatments of the theoretical underpinnings and derivation of the Malthusian model of population growth are given by many authors (e.g., Slobodkin 1961; Wilson and Bossert 1971; Poole 1974; May 1974; Maynard Smith 1974; McNaughton and Wolf 1979; Ginzburg and Golenberg 1985; Gotelli 1998).

Malthusian population growth can be formulated in either continuous time or discrete time. In continuous time, it is usually written as:

$$dN/dt = rN$$

where N denotes the population abundance and r is the instantaneous rate of increase. The constant r is sometimes called the Malthusian parameter (Leslie and Ranson 1940; cf. Hairston et al. 1970; Fenchel 1974). The solution to this differential equation is the exponential trajectory:

$$N(t) = N(0) \exp(rt)$$

where $N(t)$ is the population abundance at time t and $N(0)$ is the starting abundance. If the value of r is less than zero, the population declines over time asymptotically to zero. If the value of r is larger than zero, the population grows exponentially to infinity. If the value of r is precisely zero, the population abundance is precariously balanced at a constant value. A recursive expression along this exponential trajectory through time is:

$$N(t+1) = N(t) \exp(r).$$

The Malthusian population growth model can also be formulated in a discrete-time version. In this case, it is normally derived from the recursive expression:

$$N_{t+1} = R\,N_t$$

where N_t is the population abundance at time t whose values range across the positive integers. The constant R is the reproductive rate, a dimensionless number representing the average number of daughters per female during that female's entire lifetime. This formulation is often used for species with nonoverlapping generations such as annuals or univoltine insects. Mathematically, the constant R corresponds to $\exp(r)$ in the continuous-time formulation. The solution is the geometric growth equation:

$$N_t = R^t\,N_0\,,$$

and the associated difference equation is:

$$\Delta N / \Delta t = (R - 1)\,N.$$

Realism — LOW — The Malthusian model captures the typical demographic behavior of small populations. However, it ignores most of the complexity that biologists consider important, including density dependence, demographic and environmental stochasticity, internal structure in age or stage classes, dispersal and migration, and developmental delays between birth and reproductive maturity. The asymptotic decline predicted by $r < 0$ is probably unrealistic for harsh environments where population declines may be more abrupt.

Relevance — MEDIUM — The model produces predictions about population abundance at specified future times. Although this endpoint is directly relevant to population-level ecotoxicological analyses, the model cannot produce population-level risks, such as the probability of a decline. The model has only a single parameter, which can be perturbed to address the effects of chemicals on the species (e.g., Snell and Serra 2000).

Flexibility — LOW — There are only two formulations (discrete and continuous). Although it is commonly applied to species that have overlapping generations, this application is somewhat subtler. Because only one parameter is used, representing differences among species is difficult unless they are manifested in terms of the growth parameter.

Treatment of Uncertainty — LOW — The classical Malthusian model of population growth is deterministic. Uncertainty and variability in the model parameters are not considered. Stochastic versions of Malthusian growth are considered later in this chapter.

Degree of Development and Consistency — HIGH — The model has been implemented in several convenient software packages including both Populus (Alstad 2001) and RAMAS (Ferson and Akçakaya 1988). It is simple to implement in a spreadsheet or even a handheld calculator and is well understood by all ecologists.

Ease of Estimating Parameters — HIGH — Only one parameter is used in a Malthusian growth model. The parameter can be estimated straightforwardly from census data or from knowledge about reproductive capacity and juvenile survivorship. Ecologists could model impacts of chemicals on the value of this parameter.

Regulatory Acceptance — HIGH — Government agencies at both the state and federal levels are very likely to accept reasonable applications of the model so long as they make only short-term projections that avoid asymptotic behavior.

Credibility — HIGH — The model is prominently described in virtually every textbook of ecology and environmental science. It has been used in innumerable biological studies. It is one of the very few biological models that is sometimes characterized as a law.

Resource Efficiency — HIGH — It is very easy and straightforward to apply the model to a particular case, whether by using existing software, specially written software, or *ad hoc* spreadsheets. Existing census data about a species may often suffice to parameterize the model.

LOGISTIC POPULATION GROWTH MODEL

The logistic population growth model is sometimes called the Verhulst–Pearl model after Verhulst (1838), who first described it in a biological context, and Pearl, who was largely responsible for its wide popularity in mathematical biology (Pearl and Reed 1920; Pearl 1925). Almost every ecology textbook covers the logistic model in considerable detail (e.g., Lotka 1924; Allee et al. 1949; Andrewartha and Birch 1954; Slobodkin 1961; Wilson and Bossert 1971; Maynard Smith 1974; Poole 1974; Pielou 1977; Ginzburg and Golenberg 1985; Gotelli 1998). The differential equation that leads to logistic growth is:

$$dN/dt = r\,N\,(1 - N/K)$$

where K is the carrying capacity (the asymptote that the population size eventually reaches). In Figure 4.1, a hypothetical logistic growth curve is plotted. It has a carrying capacity of 1000 (individuals or individuals per unit area). The rate at which the population approaches this size is determined in part by the value of r, which is sometimes called the low-density rate of increase.

There are different ways to discretize the logistic equation. The proper discretization (which cannot produce negative abundances) is the Ricker formulation (Ricker 1975). The simplistic discretization (in which dN/dt is just replaced by $\Delta N/\Delta t$) is pathological and exhibits chaotic behavior for a wide range of parameter values (May 1974; Gotelli 1998).

The most important feature of the logistic population growth model is that it does not project to infinite population sizes. In this respect, it is the answer to the concern originally voiced by Thomas Malthus about what happens when a population can no longer grow at an exponential rate. Although this gross qualitative feature of the logistic model of population growth is widely accepted by ecologists, the specific model formulation is unrealistic in several respects. The most important is that it neglects the stochasticity (noisiness) exhibited by real populations. One of the most important early applications of the logistic population growth model was to human populations in countries around the world (Pearl and Reed 1920; Lotka 1924).

Realism — LOW — The model is much more realistic than the Malthusian population growth model because it recognizes that population growth cannot continue forever. However, its realism is still very limited. Although its qualitative features — one stable equilibrium abundance (the carrying capacity) and one unstable equilibrium at zero — are perhaps reasonable for many populations, the symmetry of the particular functional form is almost certainly unrealistic. The model cannot represent Allee effects or multiple stable states in general. Because it is deterministic, the model also ignores the stochasticity that is a very salient feature of the natural world.

Relevance — MEDIUM — The model produces predictions about population abundance at arbitrary future times. Although this endpoint is directly relevant to population-level ecotoxicological analyses, the model cannot produce population-level risk assessments such as the probability of a decline. Although no explicit functions to account for toxic chemical effects are included in the model, the model parameters can be perturbed to address the effects of chemicals on the species.

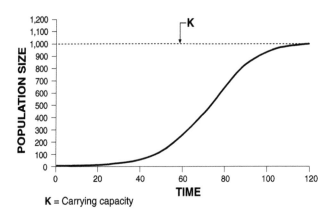

Figure 4.1   Example of logistic population growth curve.

Flexibility — LOW — The model essentially has only two formulations (discrete and continuous), although the model can be rescaled in various ways so that its two parameters have different interpretations. Nevertheless, because only two degrees of freedom are available, representing differences among species is difficult unless they manifest themselves strictly and obviously in terms of the two parameters.

Treatment of Uncertainty — LOW — The classical logistic population growth model is deterministic. Uncertainty and variability in the model parameters are not considered. Stochastic versions of the logistic growth model are considered later in this section.

Degree of Development and Consistency — HIGH — The model has been implemented in many convenient software packages, including Populus (Alstad 2001) and RAMAS (Ferson and Akçakaya 1988), and it is very widely used around the world for numerous purposes, including assessment of the effects of toxic chemicals.

Ease of Estimating Parameters — HIGH — There are only two parameters. Statistical techniques for estimating their values from empirical data are well established. Both parameters are biologically meaningful; therefore, ecologists could model the effects of chemicals on their values.

Regulatory Acceptance — HIGH — The model has been used by many regulatory agencies. Government agencies at both the state and federal levels are very likely to accept applications of the model.

Credibility — HIGH — The model is prominently described in virtually every textbook of ecology. It has been used in innumerable biological studies, including several focused on the effects of chemicals.

Resource Efficiency — HIGH — Applying the model to a particular case is very easy. Appropriate software already exists and, in any case, developing special-purpose implementation is straightforward. Census data or even semiquantitative life-history information about a species could often suffice to parameterize the model.

STOCK-RECRUITMENT POPULATION MODELS

In the Malthusian model, the per capita population growth rate is a constant. In stock-recruitment models of population growth, the per capita growth rate is some (often nonlinear) function of the population abundance or density. This dependence of the growth rate on the population size is called *density dependence* by academic ecologists. Applied ecologists commonly refer to the dependence as *compensation* or *depensation*, depending on whether there are more or fewer offspring than a Malthusian model would suggest. Stock-recruitment models generalize the logistic function to represent the different forms of density dependence of different species.

Several models of density dependence have been suggested to represent density dependence in different species. One example is the Beverton–Holt function (Beverton and Holt 1959; Ricker 1975):

$$N_{t+1} = 1 / (\rho + k / N_t),$$

where ρ is the reciprocal of the largest possible recruitment, and k is a nonnegative parameter governing the strength of the density dependence. This equation is often used to model territorial species that partition resources inequitably or when intensive reproduction leads to diminishing returns in terms of the number of offspring produced. The Beverton–Holt model has an asymptote at $1/\rho$. Another example is the Ricker model (Ricker 1975, 1954):

$$N_{t+1} = \alpha \, N_t \, \exp(-\beta \, N_t),$$

where α is a positive number representing the low-density population growth rate, and β is a positive number determining the strength of density dependence. This model is often used when the returns on parental investment are not just diminishing but actually worsening so that large populations yield fewer offspring than a smaller population would. This function is useful for modeling density dependence in species whose individuals share critical resources equally. If the density dependence is strong enough, the Ricker model can produce dynamic chaos (May 1974). Both the Beverton–Holt and Ricker functions, as well as other models, such as the logistic and the Shepherd function, are widely used in fisheries modeling. Many other functions are commonly used in other areas of ecology, such as the Hassell model (Begon and Mortimer 1981):

$$N_{t+1} = N_t \, \exp(r) / [\, 1 + ((\exp(r/b) - 1)/K) \, N_t]^b,$$

where the value of b determines the intensity of density dependence. This function is often used to model the density dependence of plant and insect species. Recently, the "ceiling" model of density dependence (e.g., Akçakaya et al. 1999) in the following form has become very popular:

$$N_{t+1} = N_t \, \exp(r) \text{ for } N_t < K, \text{ and}$$

$$N_{t+1} = K \text{ for } N_t \geq K$$

This model is a combination of Malthusian growth (which is simply $N_{t+1} = N_t \, \exp(r)$) and zero reproduction once the carrying capacity has been reached.

In many species, mutualisms exist among the organisms within a biological population. For instance, isolated trees at timberline always succumb to the harsh environment that a grove of trees can withstand because the individuals in the grove partially shelter each other. When such mutualisms are an important aspect of the natural history of a species, the population growth rate can decline precipitously when the population density becomes too low, resulting in local extinction. For example, when the density of fish becomes so low in a lake that individuals cannot easily locate mates, the population density is likely to decrease even further. Andrewartha and Birch (1954) called these phenomena *underpopulation*, but they are more commonly known today as depensation (Butterworth 2000) or *Allee effects* (Gotelli 1998). Allee described the phenomena in several species (Allee and Bowen 1932; Allee 1938), including goldfish exposed to colloidal silver. He found that a single animal invariably died when it experienced a certain concentration that seemed innocuous to a group of fish that were exposed to the same concentration. To represent Allee effects, the stock-recruitment functions should have smaller and smaller values near zero. Haefner (1996) suggested one of several possible formulations of a stock-recruitment relation that could be used to model Allee effects.

> **Realism — MEDIUM —** In general, stock-recruitment models can be more realistic than the logistic model because they can incorporate the nonlinear relationships that may characterize density dependence. The models can also represent Allee effects (e.g., Haefner 1996) and multiple stable states if appropriate. However, because these models are deterministic, they still ignore the stochasticity of the real world.

Relevance — MEDIUM — The models produce predictions about population abundance at arbitrary future times. Although this endpoint is directly relevant to population-level ecotoxicological analyses, the deterministic models cannot produce population-level risk assessments such as the probability of a decline. Although no explicit functions to account for toxic chemical effects are included in the model, the model parameters can be perturbed to address the effects of chemicals on the species.

Flexibility — MEDIUM — Stock-recruitment models have been derived for a variety of life histories. They have been applied to a variety of species, including fish, mammals, insects, and birds. In general, stock-recruitment models can be arbitrarily complex scalar functions and may exhibit multiple stable states (e.g., Steele and Henderson 1984). However, the formulations are deterministic and are generally limited to two, three, or, at most, four parameters.

Treatment of Uncertainty — LOW — Stock recruitment models are usually deterministic. Uncertainty and variability in the model parameters are not considered. Stochastic versions of these models are considered later in this chapter.

Degree of Development and Consistency — MEDIUM — Although stock-recruitment models have been widely used in the fisheries literature, their uses have been more limited in other branches of ecology. Some software packages implement popular formulations (e.g., Ferson and Akçakaya 1988). Populus (Alstad 2001) even allows users to specify arbitrary formulations (with its Interaction Engine). However, the application of these models to new systems can be complicated, and the models can be tricky to use because no checks on feasibility of parameter values are provided.

Ease of Estimating Parameters — HIGH — Density dependence has been considered difficult to parameterize. The fundamental problem is that natural populations can rarely be experimentally manipulated to explore the full spectrum of their reproductive responses over a wide range of population densities. Nevertheless, estimation techniques are relatively straightforward, and the literature is replete with examples of their application.

Regulatory Acceptance — HIGH — The results from stock-recruitment models are likely to be accepted by regulatory agencies if the model used provides a reasonable depiction of density dependence for the species. The models have been widely used, especially by agencies dealing with fisheries or other aquatic or marine resources.

Credibility — HIGH — The models are widely known in academia and have been used in probably hundreds of applications.

Resource Efficiency — HIGH — Very modest investments are required to use stock-recruitment models in a new situation. *Ad hoc* programming effort is unnecessary or at worst minor. Parameterizations for many common species have been published.

STOCHASTIC DIFFERENTIAL EQUATION MODELS

Differential equation models of stochastic population growth were described by Poole (1974), Goel and Richter-Dyn (1974), May (1974), and Pielou (1977). Nisbet and Gurney (1982) applied a diffusion approximation to the problem. Ginzburg and colleagues (1982) were the first to treat the problem from a risk-analysis perspective by using a Malthusian model of population growth of the form:

$$dN/dt = (r + \sigma\xi(t))N$$

where σ is the magnitude of stochasticity, and ξ denotes white noise. They provided analytical solutions for the risk of population decline as a function of σ, time, N_O (the starting abundance), and the threshold N_C. Lande and Orzack (1988) extended their result to age-structured populations by applying them to the eigenvalue of the transition matrix. Lande (1993) derived the mean extinction time as a stationary solution of a diffusion equation under density independence except near the carrying capacity. Iwasa (1998; Iwasa and Hakoyama 1998; Iwasa and Nakamaru 1999; Hakoyama and Iwasa 2000; Hakoyama et al. 2000; Iwasa et al. 2000) used a different formulation in which two sources of randomness were introduced additively to the logistic equation:

$$dN/dt = rN(1 - N/K) + \sigma\xi_e(t)N + \xi_d(t)\sqrt{N}$$

where σ is the magnitude of environmental stochasticity, and the ξ symbols denote separate white noises for environmental and demographic stochasticity. They provided expressions for mean time to extinction that can be computed by means of numerical integration techniques. Hakoyama and Iwasa (1998) applied the method to study the effect of unspecified toxic chemicals on freshwater fish. Tanaka (1998) used it to study the effects of several heavy metals on a variety of invertebrate species.

Despite these achievements, the study of stochastic differential equations of population growth is generally very difficult. The introduction of any nonlinearity to model density dependence usually renders the problem analytically intractable. Practical numerical results generally require *ad hoc* programming effort involving advanced numerical techniques.

Realism — HIGH — These models range in complexity from generalized Malthusian population growth Models with no density dependence to arbitrary nonlinear density dependence relationships. Therefore, these models can be realistic in terms of qualitative equilibrial features. Even more importantly, however, these stochastic models incorporate the variability commonly seen in the natural world, either as random noise or as noise with autocorrelation structures (e.g., positive exponentially declining autocorrelation to simulate realistic environmental noise). For scalar models, this is the most important component of realism.

Relevance — HIGH — The models produce predictions about population abundance at arbitrary future times. Because the models are stochastic, they can also produce risk-analysis summaries such as mean time to decline to a specified size. These endpoints are directly relevant to population-level ecotoxicological analyses. Although no explicit functions to account for toxic chemical effects are included in the model, the model parameters can be perturbed to address the effects of chemicals on the species.

Flexibility — MEDIUM — Because the scalar versions of stochastic differential equation models have a relatively simple structure, they can be widely applied to disparate biological species and to multiple populations (Iwasa and Nakamaru 1999). However, the models have relatively few parameters or alternate formulations that could distinguish subtle differences among different species.

Treatment of Uncertainty — HIGH — These models are stochastic and can therefore incorporate natural variability or scientific uncertainty.

Degree of Development and Consistency — LOW — Although understanding how the models work is fairly straightforward, the mathematical expertise needed to apply them to new systems is a barrier to their wide application. The models are not implemented as software.

Ease of Estimating Parameters — MEDIUM — Hakoyama and Iwasa (1999) describe a bootstrap technique for estimating parameters. However, because time-series data in biology are usually only available for very short periods, parameters in general may be difficult to estimate. Model parameters are generally easily interpretable by ecologists.

Regulatory Acceptance — HIGH — Stochastic differential equation models have been used several times by regulatory agencies. Interest among agencies in the risk-analysis perspective they afford is very strong. The likelihood of regulatory acceptance is fairly high.

Credibility — MEDIUM — These models are well known in academic circles. Because they have been the focus of concentrated attention only fairly recently, few publications about them have appeared so far.

Resource Efficiency — MEDIUM — Applying these models to a new case might require *ad hoc* programming. The computational technology for using these models is not widely distributed among researchers, and a local expert is usually needed.

STOCHASTIC DISCRETE-TIME MODELS

Discrete-time models of stochastic population growth are described by Poole (1974), Capocelli and Ricciardi (1974), Goel and Richter-Dyn (1974), May (1974), and Tuljapurkar (1990). Unlike the stochastic differential equation models discussed in the previous section, discrete-time models admit direct numerical solutions with straightforward Monte Carlo techniques for computer simulation. Consequently, they are much more convenient for practical ecological risk assessments (Ferson 1999). Computer simulation also permits the modeler to estimate many more kinds of summary statistics than the number that was easy to obtain from the continuous models. In particular, it is fairly easy to estimate:

- Population size as a function of time (means or distributions at each time-step)
- Time to population decline (or recovery) to a given level (mean or full distribution)
- Probability of population decline (or recovery) to a given level within a specified time horizon

The last of these summaries is especially useful because the probability of population decline is precisely the risk of interest in ecological risk assessments. Fagan et al. (1999) demonstrated the use of several different models of density dependence within discrete-time stochastic models of population growth and explained how they could be parameterized from the limited data usually available to ecologists.

Figure 4.2 illustrates the different perspective offered to modelers by a stochastic model in contrast with that offered by a deterministic one. (Similar graphs could have been obtained from using the continuous differential equation models.) The graphs depict the recovery of a hypothetical population according to a deterministic Malthusian model (Figure 4.2A) and two different versions of a stochastic population model (Figures 4.2B and 4.2C). Figure 4.2B illustrates the use of a distribution around the growth rate as a model of scientific uncertainty about the rate's true value. Figure 4.2C illustrates the use of a distribution for the growth rate at each time-step of the simulation to model the environmental stochasticity experienced by the population. In principle, both scientific uncertainty and environmental stochasticity can be incorporated into a single, comprehensive model (which might even include other sources of noise, such as demographic stochasticity).

Realism — HIGH — These models range in complexity from generalized Malthusian population growth with no density dependence to arbitrary nonlinear density dependence relationships. Therefore, they can be realistic in terms of qualitative equilibrial features. Because the models are stochastic, they can incorporate the variability of the natural world.

Relevance — HIGH — The models produce predictions about population abundance at arbitrary future times. Because the models are stochastic, they can also produce risk-analysis summaries such as risk of population decline, chance of recovery, and mean time to decline or recovery. These endpoints are directly relevant to population-level ecotoxicological analyses. Although no explicit functions to account for toxic chemical effects are included in the model, the model parameters can be perturbed to address the effects of chemicals on the species.

Flexibility — MEDIUM — Because the scalar versions of discrete-time stochastic models of population growth have a relatively simple structure, they can be widely applied to disparate biological species. These models have relatively few parameters or alternate formulations that could distinguish subtle differences among different species. However, because they are usually implemented via computer simulation, they can usually be generalized with minimal effort.

Treatment of Uncertainty — HIGH — These models are stochastic and can therefore incorporate natural variability or scientific uncertainty. Because discrete scalar models are comparatively simple, they accommodate sophisticated uncertainty partitioning techniques.

Degree of Development and Consistency — MEDIUM — Understanding how these models work is easy. Their implementation in software is relatively straightforward. Because the models can be studied in computer simulations, detecting nonsense results (such as negative abundances) would be easy. Software that implements stochastic discrete-time models of Malthusian and density-dependent population growth is currently being developed with funding from the Electric Power Research Institute (EPRI).

Ease of Estimating Parameters — HIGH — Fagan et al. (1999) describes techniques for estimating parameters for Malthusian and density-dependent discrete-time models with stochasticity from

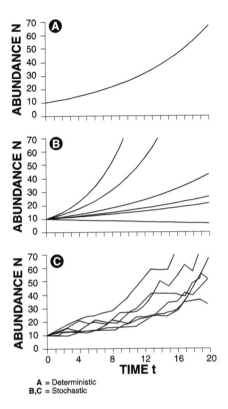

Figure 4.2 Comparison of deterministic and stochastic versions of the Malthusian population growth model. © 1994 by Applied Biomathematics. With permission.

census data. Hakoyama and Iwasa's (1999) approach is also applicable to such models. Depending on precisely how the stochasticity is modeled, straightforward least-squares approaches may be equivalent to maximum-likelihood approaches. Model parameters are often easily available from the literature or from well-known allometries such as that of Fenchel (1974). Parameterization is especially easy if modelers are satisfied with using a bounding approach that produces conservative estimates. Model parameters have straightforward interpretations.

Regulatory Acceptance — HIGH — Stochastic models have been used several times by regulatory agencies, and the interest among agencies in the risk-analysis perspective they afford is currently very strong. The likelihood of regulatory acceptance of discrete-time stochastic models of population growth is fairly high.

Credibility — MEDIUM — Stochastic models are well known in academia, although so far relatively few publications about their use have appeared.

Resource Efficiency — HIGH — Applying these models to a new case might require *ad hoc* programming. The computational technology for using these models is not widely distributed among researchers, and a local expert is usually needed.

EQUILIBRIUM EXPOSURE MODEL

Hallam et al. (1983a,b) describe an interesting scalar model that combines population dynamics with toxicant chemistry. The model consists of a system of deterministic differential equations whose underlying biology is logistic population growth. The model assumes that uptake of the toxicant is primarily by direct, transdermal absorption and that uptake indirectly via food is negligible. The formulation models the birth and death functions as explicit functions of the toxicant concentration. The toxicant concentration is equilibrial and does not itself fluctuate over time except through a forcing function. The model has fairly modest data requirements. The only things that are needed are the dose–response functions for birth and death and the toxicant concentration as it varies throughout time.

Figure 4.3 illustrates a possible transition between a thriving biological population with a large equilibrium abundance (panel A) and a population that is locally extinct because of the effects of toxicity (panel D). The isocline for the environmental concentration of the toxicant as a function of population density appears as a solid curve, and the isocline for the tissue concentration function appears as a thin curve. Analysis of the model reveals that it can admit multiple stable equilibria (shown as highlighted dots). The number and nature of the equilibria change qualitatively as the thin isocline smoothly falls below the solid one in the sequence from Figure 4.3A to Figure 4.3D. As a consequence, the model can portray important behaviors that are sometimes seen in nature, such as the Allee effects of goldfish in colloidal silver solutions (Allee and Bowen 1932; Allee 1938) mentioned previously. This sort of model could in principle be useful for screening assessments that forecast qualitative population-level effects and for planning bioremediation strategies. The model would be especially useful for these purposes if it were generalized to include stochasticity of various kinds.

This model has not been refined in the last 15 years. Hallam himself has since abandoned the approach in favor of individual-based models because of the difficulty of aggregating effects data to the population level as needed for the dose–response functions (Hallam 2000, pers. comm.).

Realism — HIGH — This model includes the crucial connection between population dynamics and chemistry. Although the model is deterministic, it could probably be generalized to account for environmental stochasticity. Assuming that exposures are constant because environmental concentrations of the chemical have equilibrated is probably not always realistic, especially in the case of temporally fluctuating emissions. However, this assumption is probably reasonable in many cases because the time scale of chemical phenomena is much shorter than that of population dynamic processes. The model can incorporate arbitrarily complex density dependence and generally offers a rather high degree of realism.

Relevance — HIGH — The model is directly relevant to the endpoints used in population-level ecotoxicological risk assessment. Because it addresses the concentration of toxicant in biological tissue, it is considerably more relevant than the population models considered heretofore in this section. The model takes as input two dose–response functions that directly model the impacts of chemicals on the population.

Flexibility — HIGH — The model can accept alternate formulations of the dose–response functions. In principle, the logistic formulation of population growth used in the model could be replaced by an arbitrary stock-recruitment relation if such generality were important. The model should be applicable to a wide variety of organisms in different environments.

Treatment of Uncertainty — LOW — The model is a set of deterministic differential equations. It does not provide for variability or parametric uncertainties. However, the model is simple and compact enough to be embedded in a Monte Carlo shell for this purpose.

Degree of Development and Consistency — LOW — Understanding how the model works is fairly easy. The model has not yet been implemented as software. No feasibility checks on input parameters or consistency checks on output have yet been specified. Implementation as software would require numerical methods for solving differential equations.

Ease of Estimating Parameters — MEDIUM — The model's parameters and functions are intuitive and can be interpreted biologically. Nevertheless, Hallam (2000, pers. comm.) suggests that the needed dose–response functions are hard to estimate given typical data sets.

Regulatory Acceptance — MEDIUM — Although the model itself is not used by a regulatory agency, all of the conceptual pieces from which it is built are widely used by such agencies.

Credibility — LOW — The model is not widely known in academia, although it would not be alien to an academic audience in any way. Very few descriptions of the particular approach have been published. The primary limiting assumption — That toxicant concentration in the environment has already equilibrated — is not widely accepted.

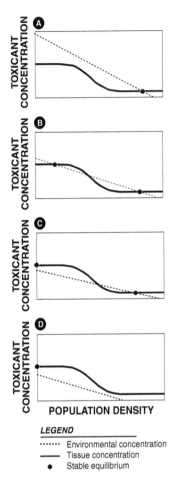

LEGEND

········ Environmental concentration
——— Tissue concentration
● Stable equilibrium

Figure 4.3 Equilibrium exposure model. Note: The lines in the figure represent isoclines to which the toxicant concentration tends to move. For example, if the environmental concentration of the toxicant is above the dotted straight line, the concentration tends to decrease, whereas if it is below the line, it tends to increase (similarly for the tissue concentration of toxicant in relation to the bold line). Stable equilibria are designated as dots where the two isoclines cross. In panels B and C, note that there is a third, unstable equilibrium. Redrawn from Hallam et al. (1983a, Figure 2). Effects of toxicants on populations: a qualitative approach, I, equilibrium environmental exposure, *Ecol. Modeling*, 18:291–304; and 1983b, II, first order kinetics, *J. Math. Biol.*, 18:25–37.

Resource Efficiency — HIGH — Modest time and effort would be required to apply the model to a particular case. Although a formal analytical solution could be cumbersome, graphical analysis of conservatively estimated functions in a screening assessment could be very efficient.

BIOACCUMULATION AND POPULATION GROWTH MODELS

In addition to the population dynamics of biological species, the models in this section consider bioaccumulation and chemical kinetics (rather than assuming them to be equilibrial). Examples of such models are the risk assessment software RAMAS Ecotoxicology (Spencer and Ferson 1997b) and the Ecotox simulator (Bledsoe and Yamamoto 1996; Bledsoe and Megrey 1989). These programs were developed on the basis of differential equations and, although they are intended to explicitly model the effects of toxic chemicals within a food web, both can be used as unstructured models of a single species.

RAMAS Ecotoxicology imports data from standard laboratory bioassays, incorporates these data into the parameters of a population model, and performs a risk assessment by analyzing population-level differences between treatment and control (or reference) samples. Bioassays for assessing the impact of toxic chemicals on natural systems are usually expressed in terms of individual-level endpoints such as growth, survivorship, and fecundity. RAMAS Ecotoxicology translates such results into a forecast of their likely consequences at the population level. The software is essentially a framework in which users construct models by using a variety of components. For instance, users can choose among several models of density dependence, including Malthusian (none), logistic, Ricker, and Beverton–Holt. Expressions for dose– (or concentration–) response relationships include Weibull, probit, and logit functions. Three models of chemical uptake kinetics are supported, including (1) first-order kinetics with mass balance, (2) unbalanced first-order kinetics (which assumes a huge environmental pool of the chemical), and (3) equilibrium concentration (a simple, multiplicative function of the environmental concentration).

Ecotox is a very similar DOS-based program that implements a bioenergetic population food-web model described by Bledsoe and Megrey (1989). The implementation does not model density dependence directly but instead allows the phenomenon to emerge from a mechanistic accounting of the processes that produce it. In a scalar application, this implies Malthusian population growth. Although it does not now explicitly account for environmental variability or scientific uncertainty, it is currently being embedded within a Monte Carlo shell for this purpose (Bledsoe 2000, pers. comm.). Users can set constants to represent additional mortality factors, an external resource, and the timing and decay rate for the environmental toxicant. The model accounts for population growth, bioenergetics, and toxicology and yields estimates of the body burden of toxicant and the resulting weight and number of individuals.

Realism — HIGH — The models incorporate key population-dynamic and chemical processes, including density dependence, chemical uptake, bioaccumulation, and toxicant kinetics. These models incorporate relatively few assumptions because all the complexity is driven by inputs supplied by the user.

Relevance — HIGH — The models yield results that are directly relevant in population-level ecotoxicological risk assessment, including population abundance, risk of population decline, chance of recovery, and concentration of toxicant in the environment and within the tissues of organisms. Several parameters can be modified to describe the impacts of toxicity.

Flexibility — HIGH — Considered together, the models permit alternate formulations of the dose–response functions. They also support several different models of population growth. The models should be applicable to a wide variety of organisms in different environments.

Treatment of Uncertainty — HIGH — RAMAS Ecotoxicology solves its equations within a Monte Carlo shell that explicitly handles both environmental variability and parametric uncertainty. Ecotox is now being modified to address uncertainty in a Monte Carlo framework.

Degree of Development and Consistency — HIGH — Although the models are relatively straightforward applications of numerical methods for solving differential equations, they would probably be considered fairly complex by an average ecologist or toxicologist. However, because they have been implemented in user-friendly software, users are insulated from this complexity. Little or no programming effort is required to use the models. The models use feasibility and consistency tests, such as checks on the positivity of abundance and agreement of units given for input parameters.

Ease of Estimating Parameters — MEDIUM — The limitation on parameter estimation is generally the paucity of appropriate data rather than statistical or interpretational difficulties. If abundant data are available, the necessary parameters are relatively straightforward to estimate. However, given the data that are typically available, these estimations may be quite difficult in some situations. Model parameters are fairly intuitive and can be interpreted biologically.

Regulatory Acceptance — MEDIUM — Ecotox was developed in collaboration with scientists at California EPA, and RAMAS Ecotoxicology has been used by several state and federal regulatory agencies in the U.S. and abroad. However, neither model has yet been formally adopted by any regulatory agency.

Credibility — LOW — Although the methods used in the models are themselves well accepted, neither model is widely known in academia. The models have had few applications, and very few publications cover them.

Resource Efficiency — MEDIUM — In principle, applying the model to a particular case should not require any programming. However, both models were developed relatively recently, and applications to new life histories or management situations may suggest further software development. Typically, applications to new cases will require additional data collection or at least marshalling relevant data from literature sources.

DISCUSSION AND RECOMMENDATIONS

In general, scalar population models are relevant, efficient, and well accepted for ecological risk assessment (Table 4.2). The two kinds of models that are recommended for further development among the candidates reviewed for scalar population models are stochastic discrete-time and stochastic differential equation models (Table 4.2). Stochastic scalar population models, preferably in discrete-time formulations, should be further developed for use in screening-level ecological risk assessment. Allometries and other macroecology databases should be developed to enrich the parametric realism of the population growth models.

Scalar population models have been applied many times in ecological research (Table 4.3). We do not attempt to provide a comprehensive review of these applications but present a representative sample to convey the breadth of the applications. Although our review of stochastic models distinguishes between continuous and discrete versions, applications of stochastic models are pooled in this table. We know of no applications of the approaches of Hallam et al. (1983a,b). Several well-known studies are excluded from this table because they involve age- or stage-structure (e.g., Speed 1993), spatial distribution (e.g., Bailey 1968), predator–prey or more general trophic interactions (e.g., Spencer et al. 1999), ecosystem dynamics (e.g., Minns et al. 1986), physiological or individual-based models (e.g., Hallam et al. 1996), or combinations of such structuring elements (e.g., Kooijman and Metz 1984). These more complicated models are covered in later chapters.

The primary practical use of scalar population models in ecological risk assessments would almost certainly be for screening-level assessments. Scalar abundance models in general — and the recommended subcategories of discrete and continuous stochastic models especially — are well suited for this use. They can be used to estimate population-level risks, yet they are simple to understand and practical because they are easy to parameterize. They are also widely accepted, have reasonably good credibility, and are amenable to comprehensive treatments of uncertainty. These advantages mean that an assessment can be made in a conservative way that is practical and appropriate for the screening level. The simplicity of the models also implies that they can be embedded in a fully probabilistic framework that is most appropriate for risk assessments.

One of the oldest and most important debates in ecology has been whether natural populations are regulated by density dependence. The dominant school of thought holds that feedback mechanisms embodied in density dependence keep populations in check and regulate their relative densities (Slobodkin 1961). The competing school of thought held that random events (e.g., extreme weather, landslides) actually govern the abundances of the species in the natural world (Andrewartha and Birch 1954). Turchin (1995) has argued convincingly that this debate becomes less fundamental and more a matter of semantics once the role of natural stochasticity is recognized and incorporated into our models. The recommended stochastic models may include density dependence or omit it, according to the purpose and time scale of the risk assessment, the magnitude of expected impacts, and the available empirical information. If density dependence and stochasticity are not incorporated (e.g., as in simple Malthusian growth models), then the use of the population model is very limited (in comparative risk assessment to evaluate relative risks of different chemicals or relative risks at contaminated sites and reference sites).

Table 4.2   Evaluation of Population Models — Scalar Abundance Models

Model	Reference	Realism	Relevance	Flexibility	Treatment of Uncertainty	Degree of Development	Ease of Estimating Parameters	Regulatory Acceptance	Credibility	Resource Efficiency
								Evaluation Criteria		
Malthusian growth	Poole (1974); May (1974); Alstad et al. (1994a,b); Alstad (2001)	low	medium	low	low	high	high	high	high	high
Logistic growth	Poole (1974); May (1974); Alstad et al. (1994a,b); Alstad (2001)	low	medium	low	low	high	high	high	high	high
Stock-recruitment	Ricker (1954, 1975); Beverton and Holt (1959); Alstad et al. (1994a,b); Alstad (2001)	medium	medium	medium	low	medium	high	high	high	high
Stochastic differential equation	Poole (1974); Goel and Richter-Dyn (1974); May (1974); Ginzburg et al. (1982); Iwasa (1998); Tanaka (1998); Hakoyama and Iwasa (1998)	high	medium	medium	medium	low	medium	medium	medium	medium
Stochastic discrete-time equation	Poole (1974); Capocelli and Ricciardi (1974); Goel and Richter-Dyn (1974); May (1974); Ginzburg et al. (1982); Ferson (1999)	high	high	medium	high	medium	high	high	medium	high
Equilibrium exposure	Hallam et al. (1983a,b)	high	high	high	low	low	medium	medium	low	medium
Bioaccumulation and population growth	Bledsoe and Yamamoto (1996); Spencer and Ferson (1997b); Bledsoe and Megrey (1989)	high	high	high	high	high	medium	medium	low	low

Note   ◆◆◆ - high
    ◆◆ - medium
    ◆ - low

Table 4.3 Applications of Scalar Population Models

Model	Species	Location	Reference
Malthusian growth	*Daphnia*	Laboratory	Coniglio and Baudo (1989)
	Burrowing mayfly (*Hexagenia* spp.)	Lake Erie	Madenjian et al. (1998)
	Pheasant (*Phasianus colchicus*)	Washington State	Einarsen (1945)
	8 marine mammals	Worldwide	Chapman (1981)
	42 species	Worldwide	Fenchel (1974)
Logistic growth	Yeast	Laboratory	Allee et al. (1949)
	Paramecium	Laboratory	Gause (1934)
	Daphnia	Laboratory	van Leeuwen et al. (1986)
	Burrowing mayfly (*Hexagenia* spp.)	Lake Erie	Madenjian et al. (1998)
	Drosophila	Laboratory	Lotka (1924)
	Song sparrow (*Melospiza melodia*)	Mandarte Island	Gotelli (1998)
	Birds	Hypothetical	Chesser et al. (1994)
	Sheep	Tasmania	Davidson (1938)
	Human	United States	Pearl and Reed (1920)
Stock recruitment	*Daphnia* magna	Laboratory	Smith (1963)
	Daphnia magna	Laboratory	Schoener (1973)
	Zooplankton	Germany	Horn and Benndorf (1980)
	Fathead minnow (*Pimephales promelas*)	Midwestern lakes	Waller et al. (1971) Spencer and Ferson (1997b)
	Salmon and herring	Canada	Butterworth (2000)
	Many fisheries spp.	Worldwide	Gulland (1972)
	700 fish data sets	Worldwide	Myers et al. (1999)
	Harp seal (*Phoca groenlandica*)	Northwest Atlantic	Lett et al. (1981)
Stochastic	Rotifers	Laboratory	Snell and Serra (2000)
	Daphnia galeata	Laboratory	Tanaka (1999)
	Invertebrate spp.	Laboratory	Tanaka (1998)
	Crucian carp (*Carassius auratus*)	Japanese lakes	Hakoyama and Iwasa (1998)
	Bluefin tuna (*Thunnus thynnus*)	Pacific Ocean	Matsuda (1997)

CHAPTER 5

Population Models — Life History

Steve Carroll

Life-history models track characteristics of organisms as a function of age or stage. Stages can be defined by size, morphological state, or any classifying variable deemed to be demographically important. The most common characteristics that are tracked in such models are survival rates and fecundities, collectively referred to as *vital rates*. The typical endpoints for life-history models are:

- Population abundance
- Abundances of individual age or stage classes
- Population growth rate or related parameters (e.g., sensitivity, elasticity)

Life-history models are important for several reasons. In many populations, survival probabilities and fecundities vary among age or stage classes, and toxic chemicals may affect the various classes differently. The predictions made on the basis of life-history models generally differ from those made on the basis of simple population models in which individuals are assumed to be identical. Life-history models allow the exploration of age-specific or stage-specific management options.

For the purposes of this review, matrix models were classified into four categories, with the last two representing software implementations of the first two (Table 5.1). These include:

- Deterministic age- or stage-based matrix models (Caswell 2001)
- Stochastic age- or stage-based matrix models (Caswell 2001)
- RAMAS®* Age, Stage, Metapop, or Ecotoxicology (Ferson 1993; Spencer and Ferson 1997b,c; Applied Biomathematics 2000)
- ULM (unified life model) (Legendre and Clobert 1995)

For an in-depth treatment of matrix population models, see Caswell (2001). Density dependence, or the lack thereof, is not explicit in this classification scheme. Thus, within a category, both density-dependent and density-independent models are included.

DETERMINISTIC MATRIX MODELS (AGE OR STAGE BASED)

These models assume that survival rate and fecundity are functions of the age class or stage to which an organism belongs. Age- and stage-structured models have been important tools in natural

---

* RAMAS is a registered trademark of Applied Biomathematics.

Table 5.1 Internet Web Site Resources for Life-History Models

Model Name	Description	Reference	Internet Web Site
Deterministic age- or stage-based matrix models	Life-history matrix models with fixed fecundity and survivorship parameters	Caswell (2001)	http://www.ramas.com/ramas1.htm; http://www.ento.vt.edu/~sharov/PopEcol/lec7/leslie.html; http://www.ets.uidaho.edu/wlf448/Leslie1.htm
Stochastic age- or stage-based matrix models	Life-history matrix models incorporating random variability in fecundity and survivorship parameters	Caswell (2001)	http://www.ramas.com/ramas1.htm
RAMAS Age, Stage, Metapop, or Ecotoxicology	Software for life-history modeling of age- or stage-structured populations	Ferson (1993); Spencer and Ferson (1997b,c); Applied Biomathematics (2000)	http://www.ramas.com/
ULM (unified life model)	Software for life-history modeling of age- or stage-structured populations	Legendre and Clobert (1995)	http://www.snv.jussieu.fr/Bio/ulm/ulm.html http://eco.wiz.unikassel.de/model_db/mdb/ulm.html

resource management, especially for fish, since the 1950s (Barnthouse 1998). The simplest life-history model of this type is the Leslie population projection matrix (Leslie 1945):

$$N(t) = LN(t - 1)$$

where $N(t)$ and $N(t - 1)$ are vectors of the numbers of organisms in each age class ($N_0, \ldots N_k$).

$$L = \begin{bmatrix} s_0f_1 & s_0f_2 & s_2f_3 & \cdots & s_{k-1}f_k & 0 \\ s0 & 0 & 0 & \cdots & 0 & 0 \\ 0 & s_1 & 0 & \cdots & 0 & 0 \\ 0 & 0 & s_2 & \cdots & 0 & 0 \\ 0 & 0 & 0 & \cdots & s_k & 0 \end{bmatrix}$$

where

$$s_k = \text{age-specific survivorship probability}$$
$$f_k = \text{average fecundity at age } k$$

Life stages, when not defined by age, are generally defined by size (height, weight, etc.) or morphological state. Individuals within a given age class or stage are assumed to have identical survival probabilities and fecundities. It is assumed that the vital rates do not vary as a result of random environmental fluctuations or demographic stochasticity. Depending on whether or not density dependence is included, the vital rates may vary (deterministically) as a function of population size. The vital rates, which comprise the population matrix, can be obtained either from laboratory experiments or from field observations.

Realism — HIGH — These models reflect that vital rates generally change with age. In many cases, they incorporate all available demographic data because measures of variability in vital rates are often unavailable. Within this category of models, age- or stage-structured models with density dependence are more realistic than those without it.

Relevance — HIGH — Potential endpoints include expected population size (age specific), lambda (the asymptotic population growth rate), or lambda-based measures such as sensitivity and elasticity. All of these endpoints are relevant to ecotoxicological assessments. Survivorship and fecundity parameters can be adjusted to reflect the effects of toxic chemicals observed in laboratory tests (e.g., Munns et al. 1997).

Flexibility — HIGH — All parameters are species-specific. The number of age classes, as well as the vital rates, can vary among populations. Therefore, different population structures and life histories can be modeled.

Treatment of Uncertainty — LOW — Neither environmental nor demographic stochasticity is incorporated in the basic approach. However, generalizations are being developed to deal with environmental stochasticity (e.g., see below *Stochastic Matrix Models* and Caswell 2001).

Degree of Development and Consistency — HIGH — Several software programs implement this model type, and documentation and technical support are available. Understanding the workings of the model is relatively easy.

Ease of Estimating Parameters — MEDIUM — Parameters can in principle be obtained relatively easily, either from laboratory experiments or from field observations. Alternatively, fecundity and survivorship values may be approximated from standard survivorship and birth rate curves in the literature. However, particularly when laboratory experiments are not appropriate or feasible, assessing the magnitude of chemical effects on each vital rate may be very difficult.

Regulatory Acceptance — HIGH — The model is being used by several regulatory agencies.

Credibility — HIGH — The model type is well known within academia. Many applications of the model exist.

Resource Efficiency — HIGH — Application of the model requires no programming because software is available. In many cases, available data are sufficient. However, when available data are not sufficient, considerable effort may be needed to obtain new data.

STOCHASTIC MATRIX MODELS (AGE OR STAGE BASED)

Like the deterministic age- or stage-based matrix models, these models assume that survival rate and fecundity are functions of the age class or stage in which an organism resides. In contrast with their deterministic counterparts, the stochastic models incorporate environmental or demographic stochasticity or both in the estimates of survival probabilities and fecundities (Caswell 2001). Depending on whether density dependence is included, the vital rates may vary as a function of population size. The vital rates, which comprise the population matrix, can be obtained either from laboratory experiments or field observations.

> **Realism** — HIGH — In addition to reflecting that vital rates generally change with age (or stage), these models recognize that the rates may also vary as a result of random environmental fluctuations and demographic stochasticity. They generally incorporate all available demographic data. Within this category of models, age- or stage-structured models with density dependence are more realistic than those without it.
>
> **Relevance** — HIGH — Possible endpoints include expected population size (age- or stage-specific), risk of decline, risk of extinction, and expected crossing time (the time at which the population is expected to go either above or below a given size). These measures are all relevant to ecotoxicological assessments. As the deterministic version of this model, the survivorship and fecundity values can be adjusted to reflect the effects of toxic chemicals observed in laboratory tests.
>
> **Flexibility** — HIGH — All parameters are species-specific. The number of age classes or stages, the vital rates, and the variation in the vital rates can vary among populations. Therefore, different population structures and life histories can be modeled.
>
> **Treatment of Uncertainty** — HIGH — Both environmental and demographic stochasticity can be incorporated.
>
> **Degree of Development and Consistency** — HIGH — Several software programs implement this model type, and documentation and technical support are available. It is relatively easy to understand the workings of the model.
>
> **Ease of Estimating Parameters** — MEDIUM — Required parameters include average vital rates and their standard deviations. Furthermore, chemical impact on these parameters must be considered (this can be handled, for example, by estimating two matrices — one with toxins and one without toxins). Obtaining estimates of all such parameters can often be quite difficult. However, model parameters are intuitive and can be interpreted biologically.
>
> **Regulatory Acceptance** — HIGH — The model is used by several regulatory agencies.
>
> **Credibility** — HIGH — The model is well known within academia. There are many applications of the model.
>
> **Resource Efficiency** — MEDIUM — Application of the model requires no programming, as software is available. In some cases, data must be collected, but in many cases, available data are sufficient.

RAMAS AGE, STAGE, METAPOP, OR ECOTOXICOLOGY

These four computer programs (Spencer and Ferson 1997b; Ferson 1990; Ferson and Akçakaya 1988; Akçakaya 1998b) are collapsed into one category because they share a common source and they receive the same ratings. Each program applies matrix models for age- or stage-structured populations to estimate population-level parameters such as growth rate and extinction risk.

In RAMAS Ecotoxicology, a model of population dynamics and toxicant kinetics can be constructed and linked to bioassay data. Using this program, a modeler can import data from standard laboratory bioassays, incorporate these data into the parameters of a population model, and perform a risk assessment by analyzing population-level differences between potentially affected and control samples. The user specifies the control survivorship and fecundity for each age class or stage. Density dependence in the form of ceiling, logistic, Ricker, or Beverton–Holt functions can be added for specific age classes or stages. A user simulates toxic effects by selecting and parameterizing dose–response models (Weibull, probit, or logit) linked to survivorship and fecundity values. Parameters can be specified as scalars, intervals, or distributions to account for environmental variability and uncertainty. Monte Carlo simulations are then used to predict future population trajectories and calculate the risk of adverse events such as extinctions or algal blooms. The software checks the validity of the data input and model structure specified by the user.

Realism — HIGH — RAMAS Age models age-structured populations, and RAMAS Stage models stage-structured populations. RAMAS Metapop can model either age-structured or stage-structured populations. RAMAS Ecotoxicology can also model either age-structured or stage-structured populations and can also explicitly model the effects of toxic chemicals. All four programs can include density dependence as well as environmental and demographic stochasticity.

Relevance — HIGH — Possible endpoints include expected population size (stage-specific), lambda, lambda-based measures such as sensitivity and elasticity, risk of decline, risk of extinction, and expected crossing time (the time at which the population is expected to exceed or to decrease to less than a given size). These measures are all potentially useful in ecotoxicological assessments. Effects of toxic chemicals are not explicitly modeled but can be incorporated by adjusting survivorship and fecundity values to reflect toxic effects (e.g., Munns et al. 1997).

Flexibility — HIGH — All parameters are species-specific. The number of age classes or stages, the vital rates, and the variation in the vital rates can vary among populations. Therefore, different population structures and life histories can be modeled. In RAMAS Ecotoxicology, different toxic chemical dynamics and dose–response functions can be modeled.

Treatment of Uncertainty — HIGH — Both environmental and demographic stochasticity can be incorporated.

Degree of Development and Consistency — HIGH — These models are easy to use and easy to apply to different populations. Each program has a detailed user's manual explaining the scientific basis of the model as well as the capabilities of the program. The programs include several internal checks for consistency.

Ease of Estimating Parameters — MEDIUM — Required parameters include vital rates at the very least. In some cases, average vital rates and their standard deviations need to be estimated. Furthermore, the chemical impact on these parameters may need to be considered (e.g., by choosing and parameterizing a dose–response function or by estimating two matrices — one with toxins and one without toxins). Obtaining estimates of all such parameters can often be quite difficult. Thus, ease of parameter estimation varies among cases. However, model parameters are intuitive and can be interpreted biologically.

Regulatory Acceptance — HIGH — These models are being used by several regulatory agencies.

Credibility — HIGH — These models are widely used within academia. Many applications of the models exist.

Resource Efficiency — HIGH — No programming is necessary to use these programs. In some cases, data must be collected, but in many cases, available data are sufficient.

UNIFIED LIFE MODEL (ULM)

ULM (Legendre and Clobert 1995) is a computer program that implements deterministic or stochastic matrix models for analysis of population dynamics. ULM is somewhat similar to RAMAS, except for its interface and varied capabilities. ULM accommodates a wide range of populations with variable life-history characteristics. ULM can model any species life history as a matrix model with or without density dependence, environmental stochasticity, demographic stochasticity (as branching processes), inter- or intra-specific competition, parasitism, and metapopulations. Results are expressed as population trajectories, distributions, growth rate, population stage- or age-structure, generation times, sensitivities to changes in parameters, probability of extinction, and extinction times. The stochastic models within ULM are implemented as Monte Carlo simulations.

Realism — HIGH — ULM models age-structured and stage-structured populations. The number of age classes or stages is variable. Environmental stochasticity is included.

Relevance — HIGH — Possible endpoints include expected population size (age- or stage-specific), lambda, lambda-based measures such as sensitivity and elasticity, and risk of extinction. These measures are all used in ecotoxicological assessments.

Flexibility — HIGH — All parameters are species-specific. The number of age classes or stages, the vital rates, and the variation in the vital rates can vary among populations. Therefore, different population structures and life histories can be modeled.

Treatment of Uncertainty — HIGH — Environmental stochasticity can be incorporated.

Degree of Development and Consistency — MEDIUM — Some programming is required to use the model. This programming must use a language defined by the authors of the program. Legendre and Clobert (1995) explain the use of the model and provide examples.

Ease of Estimating Parameters — HIGH — Parameters can be obtained relatively easily, either from laboratory experiments or field observations.

Regulatory Acceptance — LOW — No information on the model's regulatory acceptance is available. The model is not likely to be used by a regulatory agency at present because programming is required to use the model.

Credibility — MEDIUM — Several (approximately 20) publications apply the model to different cases.

Resource Efficiency — MEDIUM — Some programming is necessary to use this program. In some cases, data must be collected, but in many cases available data are sufficient.

DISCUSSION AND RECOMMENDATIONS

Life-history models (Table 5.2) are well developed and are already being used for ecotoxicological assessment (Levin et al. 1996; Munns et al. 1997; Crutchfield and Ferson 2000) (Table 5.3; also see Barnthouse [1993] for a review of other applications). These models have already been generalized to be spatially explicit within a metapopulation framework (RAMAS Metapop) and to explicitly include the dynamics and effects of toxic chemicals (RAMAS Ecotoxicology). Deterministic and stochastic matrix models and RAMAS Age, Stage, Metapop, or Ecotoxicology are therefore recommended for further evaluation and use in chemical risk assessment. Suggested future developments include development of software that includes both spatially explicit effects and the dynamics/effects of toxic chemicals. Software should be developed for calculating risk-based (probabilistic) sensitivities with respect to vital rates to answer questions such as *How does the probability of decline change as a result of changes in the vital rates?* This method would be an adaptation of the results of Uryasev (1995). This development would allow the conversion of sensitivity-based methods, such as decomposition of the change in lambda, into the language of risk. Such sensitivity-based applications are used to identify age-specific or stage-specific management strategies. The new methods will generalize widely used methods of lambda decomposition (Caswell 2001).

Table 5.2   Evaluation of Population Models — Life History Models

Model	Reference	Evaluation Criteria								
		Realism	Relevance	Flexibility	Treatment of Uncertainty	Degree of Development	Ease of Estimating Parameters	Regulatory Acceptance	Credibility	Resource Efficiency
Deterministic age- or stage-based matrix	Caswell (2001)	◆◆◆	◆◆◆	◆◆◆	◆	◆◆◆	◆◆	◆◆◆	◆◆◆	◆◆◆
Stochastic age- or stage-based matrix	Caswell (2001)	◆◆◆	◆◆◆	◆◆◆	◆◆◆	◆◆◆	◆◆	◆◆◆	◆◆◆	◆◆
RAMAS Age, Stage, Metapop, or Ecotoxicology	Ferson (1993); Spencer and Ferson (1997c); Applied Biomathematics (2000)	◆◆◆	◆◆◆	◆◆◆	◆◆◆	◆◆◆	◆◆	◆◆◆	◆◆◆	◆◆◆
Unified life model (ULM)	Legendre and Clobert (1995)	◆◆◆	◆◆◆	◆◆◆	◆◆◆	◆◆	◆◆◆	◆	◆◆	◆◆

Note: ◆◆◆ - high
◆◆ - medium
◆ - low

Table 5.3   Applications of Life-History Models

Model	Species	Location/Population	Reference
Deterministic age-based matrix	Burrowing mayflies, *Hexagenia* spp. (*H. limbata* and *H. rigida*)	Western Lake Erie	Madenjian et al. (1998)
	Subtidal snail (*Umbonium costatum*)	Hakodate Bay, northern Japan	Noda and Nakao (1996)
	Mysid (*Americamysis bahia*)	Laboratory	Kuhn et al. (2000)
	Pea aphid (*Acyrthosiphon pisum*)	Laboratory	Walthall and Stark (1997)
	Northern sea lions (*Eumetopias jubatus*)	Marmot Island, Alaska	York (1994)
Deterministic age-based matrix and deterministic stage-based matrix	Polychaetes (*Capitella* sp. *I* and *Streblospio benedicti*)	Estuaries and littoral wetlands throughout much of the United States	Levin et al. (1996)
Deterministic stage-based matrix	Pea aphid (*Aeyrthosiphon pisum*)	Laboratory	Stark and Wennergren (1995)
	Soil mite (*Platynothrus peltifer*), isopod (*Porcellio scaber*), and nematode (*Plectus acuminatus*)	Laboratory	Kammenga et al. (2001)
	Bluegill sunfish (*Lepomis macrochirus*)	Generic lake in central Florida	Bartell et al. (2000)
	Brook trout (*Salvelinus fontinalis*)	Southern Appalachian mountain streams	Marschall and Crowder (1996)
	Estuarine fish (*Fundulus heteroclitus*)	New Bedford Harbor, Massachusetts	Munns et al. (1997)
	Loggerhead sea turtles (*Caretta caretta*)	Trawl fisheries of the southeastern United States	Crowder et al. (1994)
	Yellow mud turtles (*Kinosternon flavescens*) and Kemp's ridley sea turtles (*Lepidochelys kempi*)	Texas	Heppell et al. (1996)
	Killer whales (*Orcinus orca*)	Pacific Northwest	Brault and Caswell (1993)
	Savannah grass (*Andropogon brevifolius Schwarz*)	Venezuelan savannas	Canales et al. (1994)
Deterministic stage-based matrix and stochastic stage-based matrix	Snail kite (*Rostrhamus sociabilis*)	Everglades	Beissinger (1995)
	House sparrow (*Passer domesticus*)	India, Pakistan	Slade (1994)
	Steller sea lion (*Eumetopias jubatus*)	Northeast Pacific	Pascual and Adkinson (1994)
	American ginseng (*Panax quinquefolium*) and wild leek (*Allium tricoccum*)	Canada	Nantel et al. (1996)
	Mountain golden heather (*Hudsonia montana*)	North Carolina	Gross et al. (1998)
Stochastic age-based matrix	Brook trout (*Salvelinus fontinalis*)	Montmorency County, Michigan	McFadden et al. (1967)
	Striped bass (*Morone saxatilis*)	Potomac River	Cohen et al. (1983)

Table 5.3   (cont.)

	Hawaiian stilt (*Himantopus mexicanus knudseni*)	Hawaii	Reed et al. (1998)
	Mediterranean monk seal (*Monachus monachus*)	Atlantic Ocean (North Africa) and eastern Mediterranean	Durant and Harwood (1992)
	Roan antelope (*Hippotragus equinus*)	Parc National de l'Akagera in Rwanda	Beudels et al. (1992)
	Asiatic wild ass (*Equus hemionus*)	Negev Desert of southern Israel	Solbreck (1991); Saltz and Rubenstein (1995)
	African elephant (*Loxodonta africana*)	Tsavo National Park, Kenya	Armbruster and Lande (1993)
Stochastic stage-based matrix	Algae (*Ascophyllum nodosum*)	Swedish coast	Aberg (1992)
	Marine bivalve (*Yoldia notabilis*)	Otsuchi Bay, northeastern Japan	Nakaoka (1997)
	Sea whip coral (*Leptogorgia virgulata*)	Northeastern Gulf of Mexico	Gotelli (1991)
	Grey seal (*Halichoerus grypus*) and ringed seal (*Phoca hispida*)	Baltic Sea	Kokko et al. (1997)
	Desert tortoise (*Gopherus agassiz*)	Western Mojave Desert	Doak et al. (1994)
	Red deer (*Cervus elaphus*)	Rum, Western Isles, Scotland	Benton et al. (1995)
	Redwood (*Sequoia sempervirens*)	California, Oregon	Namkoong and Roberds (1974)
RAMAS Age	Bluegill (*Lepomis machrochirus*)	Hyco Reservoir, North Carolina	Crutchfield and Ferson (2000)
	Striped bass (*Morone saxatilis*)	Santee-Cooper system, South Carolina	Bulak et al. (1995)
	Cod	Atlantic	Ginzburg et al. (1990)
	Marbled murrelet (*Brachyramphus marmoratus*)	Oregon	Oregon Department of Fish and Wildlife (1995)
RAMAS Ecotoxicology	Brook trout (*Salvelinus fontinalis*)	Montmorency County, Michigan	Spencer and Ferson (1997b)
	Lesser kestrel (*Falco naumanni*)	Southern Spain	Spencer and Ferson (1997b)
	Moose (*Alces alces gigas*)	Northeastern Alberta	Spencer and Ferson (1997b)
RAMAS Stage	Giant kelp (*Macrocystis pyrifera*)	Southern California coastal waters	Burgman and Gerard (1989)
	Bluegill sunfish (*Lepomis macrochirus*)	North Carolina	Crutchfield and Ferson (2000)
	Threadfin shad (*Dorosoma petenense*)	South Carolina	Barwick et al. (1994)
	Red-cockaded woodpecker (*Picoides borealis*)	Georgia Piedmont	Maguire et al. (1995)
	Bradshaw's lomatinum (*Lomatium bradshawii*)	Western Oregon	Kaye et al. (1994)
	Sentry milk-vetch (*Astragalus cremnophylax*)	Grand Canyon National Park, Arizona	Maschinsky et al. (1997)
ULM — stochastic stage-based matrix	Grey partridge (*Perdix perdix*)	Northwest and southeast France	Bro et al. (2000)

Table 5.3   (cont.)

Model	Species	Location/Population	Reference
ULM — deterministic stage-based matrix	Yellow-legged gull (*Larus cachinnans*)	Medes Islands, northwestern Mediterranean	Bosch et al. (2000)
ULM — stochastic stage-based matrix	Passerine songbirds	User defined	Legendre (1999)
ULM — deterministic stage-based matrix and stochastic stage-based matrix	Snake (*Vipera ursinii ursinii*) and raptor (*Gyps fulvus fulvus*)	User defined	Ferriere et al. (1996)

CHAPTER 6

Population Models — Individual Based

Helen M. Regan

According to Metz and Diekman (1986), there are two main approaches to mathematical modeling of species populations in ecology. The first approach is called *abundance modeling* or *population-state (p-state) modeling*, in which all individual members of a population are regarded as effectively identical. These are aggregated into a single variable, usually population size or average physiological size. This approach assumes that individuals do not behave differently from one another, have the same physiology and genetic makeup, and are spatially homogeneous. It implicitly assumes that every individual has an identical effect on all other individuals in the population, regardless of their proximity to one another. Models that use this approach include the logistic growth model (Gotelli 1998; Ginzburg et al. 1990 and references therein), Malthusian growth models (Gotelli 1998; Ginzburg et al. 1982 and references therein), all other population models discussed above, Lotka–Volterra models for predator–prey relationships (Lotka 1924; Volterra 1926), and all spatially aggregated ecosystem models.

The second approach to population modeling acknowledges that variations among individuals exist. These models typically include a number of individuals defined in terms of their behaviors (procedural rules) and characteristics and an environment in which the interactions take place (Reynolds 2001). Differences among individuals are called *individual states* (or *i-states*) (Metz and Diekman 1986). One treatment of this variation is to create structured models with categories developed on the basis of age, size, growth rates, and other life-history attributes (Metz and Diekman 1986; Caswell 2001). In such models, the entire population is divided into subsets of individuals with similar attributes. These models are sometimes referred to as *i-state distributions*; they include Leslie matrices (Leslie 1945) and Lefkovitch matrices (Lefkovitch 1965), and they do not generally represent spatial heterogeneity (see above, *Population Models, Life-History Models*). An alternative approach to incorporating variation among individuals in a model is to represent each individual separately by explicitly specifying its age, size, spatial location, gender, energy reserves, and so forth (DeAngelis and Rose 1992). These are sometimes referred to as *i-state configurations* but are more commonly referred to as *individual-based models*. These types of models address the local interaction of individuals with other nearby organisms (Caswell and John 1992) and can only be executed practically in computer simulations.

Reynolds (2001) pointed out that individual-based models overlap with cellular automata models. Cellular automata models are similar to spatially explicit, grid-based, immobile, individual-based models. One meaningful difference between these modeling approaches is whether the simulation's inner loop proceeds cell by cell (i.e., in cellular automata models) or individual by individual (i.e., in individual-based models).

In this chapter, we review individual-based models, in which each individual is modeled explicitly. Many of the models reviewed here also explicitly incorporate spatial heterogeneity. Reynolds (2001) lists many more individual-based models and provides links to web sites with information on each model. Swarm, Echo, and XRaptor are software packages or modeling frameworks that facilitate implementation of individual-based models (see Reynolds 2001 for web site links to additional information).

Endpoints for individual-based models include:

- Population abundance
- Abundances of defined classes of individuals within the population (e.g., age or stage classes, sexes)
- Population growth rate
- Distribution and movement of individuals in space
- Individual characteristics such as body weight
- Feeding history of individuals
- Landscape occupancy patterns

We review the following individual-based models (Table 6.1):

- SIMPDEL (spatially explicit individual-based simulation model of Florida panthers and white-tailed deer in the Everglades and Big Cypress landscapes) (Abbott et al. 1995)
- SIMSPAR (spatially explicit, individual-based, object-oriented simulation model for the Cape Sable seaside sparrow in the Everglades and Big Cypress landscapes) (Nott et al. 1998)
- CompMech, which models compensatory mechanisms in fish populations (EPRI 1982, 1996, 2000; Jaworska et al. 1997)
- EcoBeaker, an ecological simulation program (Meir 1997)
- The *Daphnia* model (Gurney et al. 1990)
- CIFSS (California individual-based fish simulation system) (Humboldt State University 1999)
- WESP (workbench for modeling and simulation of the extinction of small populations) and ECOTOOLS (Sonnenschein et al. 1999)
- GAPPS (generalized animal population projection system) (Harris et al. 1986)
- PATCH (program to assist in tracking critical habitat) (Schumaker 1998)
- The NOYELP (northern Yellowstone Park) model (Turner et al. 1994; Pearson et al. 1995)
- The wading bird nesting colony model (Hallam et al. 1996)

The range of individual-based models in the literature is vast. In this review, we include models that represent the types of features that can be accommodated by individual-based models. Note that the "coupled chemical-fate and population growth" models are omitted here because, although they contain an individual-based component, their features seem more appropriate and relevant for the scalar models section. VORTEX is an individual-based model, but because it is primarily a metapopulation model, we address it in a later section on metapopulation models.

SIMPDEL

SIMPDEL is a spatially explicit, individual-based simulation model of Florida panther and white-tailed deer in the Everglades and Big Cypress landscape (Abbott et al. 1995). As a component of the ATLSS program, it is designed to predict the effects of hydrologic and other environmental impacts on the distribution and abundance of these two species. The model simulates the aging, reproduction, foraging, growth, and mortality of individual animals and their interactions in response to the changing landscape and spatial-hydrologic dynamics (Figure 6.1). Although it is a multispecies model and could be considered under the ecosystem model category, it is addressed

Table 6.1 Internet Web Site Resources for Individual-Based Population Models

Model Name	Description	Reference	Internet Web Site
SIMPDEL	Spatially explicit individual-based simulation model of Florida panthers and white-tailed deer in the Everglades and Big Cypress landscapes	Abbott et al. (1995)	http://www.tiem.utk.edu/~gross/atlss_www/simpdel.html http://sofia.usgs.gov/projects/atlss/
SIMSPAR	Spatially explicit individual-based object-oriented simulation model for the Cape Sable seaside sparrow in the Everglades and Big Cypress landscapes	Nott et al. (1998)	http://atlss.org/forum/simspar.overview http://sofia.usgs.gov/projects/atlss/
CompMech	A flexible model of compensatory mechanisms in fish populations, which has been applied to toxic chemical assessments	EPRI (1982, 1996, 2000); Jaworska et al. (1997)	http://www.esd.ornl.gov/programs/COMPMECH/brief.html
EcoBeaker	A spatially explicit ecological simulation program for classroom use	Meir (1997)	http://www.ecobeaker.com/
The *Daphnia* model	A spatially aggregated bioernergetics model for cladoceran zooplankton	Gurney et al. (1990)	http://es.epa.gov/ncer/progress/grants/95/environmental/envnisbe98.html
CIFSS	A spatially explicit individual-based model of fish populations in California	Humboldt State University (1999)	http://weasel.cnrs.humboldt.edu/~simsys/software.html; http://weasel.cnrs.humboldt.edu/~simsys/EcolArch/FormDoc.html
WESP/ECOTOOLS	WESP is a model for simulation of the extinction of small populations; ECOTOOLS is a general programming environment	Sonnenschein et al. (1999)	http://offis.offis.uni-oldenburg.de/projekte/ecotools/
GAPPS	A spatially aggregated model designed primarily for vertebrate populations with low abundance, extended parental care, or complex age structures	Harris et al. (1986)	http://www.consecol.org/Journal/vol4/iss1/art6/
PATCH	A spatially explicit, individual-based model designed primarily to assist in tracking critical habitat for territorial terrestrial vertebrate species	Schumaker (1998)	http://www.epa.gov/wed/pages/models/patch/patchmain.htm
NOYELP	A spatially explicit, individual-based model for small groups of bison and elk in a multihabitat landscape	Turner et al. (1994); Pearson et al. (1995)	NA
The wading bird nesting colony model	A model of the effects of mercury contamination on wading birds	Hallam et al. (1996)	http://citeseer.nj.nec.com/fishwick96extending.html

Note: NA - not available.

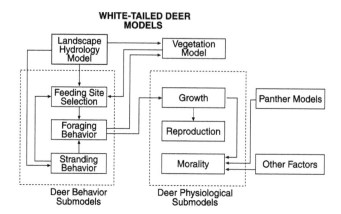

Figure 6.1   Structure of SIMPDEL Deer Model. SIMPDEL Homepage
(http://www.tiem.utk.edu/~gross/atlss_www/html_pictures/deer_model.jpg)
Revised from Comiskey et al. (1997).

here because the individual-based models for white-tailed deer and panther are essentially separate
models linked together, and they form the main components of SIMPDEL.

There are four major interconnected components — hydrology, vegetation, deer, and panthers.
Each component provides input for the other components. Water levels affect the vegetation
distribution and deer movement; deer depend on high-quality vegetation for optimal growth; and
panthers prey on deer. Geographic information system (GIS) maps are used to construct the
landscape topography; the South Florida management model provides spatially explicit hydrology
input; and vegetation is modeled as two classes of forage (high and medium quality), spatially
assigned from satellite images. Vegetation composition is regularly updated in the model in response
to foraging by deer. Deer dispersal is driven by rules that depend on the age of the individual, food
availability, water levels, and the presence of other deer in a spatial cell. Growth in body size is
determined by the individual's current weight, caloric intake, and energy expenditure. Deer mortality
can occur from starvation, natural age-related factors, and panther predation. Similarly, the panther
model attempts to simulate each individual Florida panther on the basis of its behavioral and
physiological characteristics in addition to the interactions between panther and deer.

Realism — HIGH — The model is realistic in that daily food intake, energy expenditure, growth,
reproductive status, fitness, location, foraging success, and other state variables are recorded for
each individual at each time-step. Species-specific rules are adopted to govern each individual's
behavioral activities, such as movement, foraging behavior, and interactions with other individuals.
The model makes extensive use of the output of the landscape, hydrology, and biota models within
the ATLSS program.

Relevance — HIGH — The primary endpoints of SIMPDEL are the abundances and spatial distributions
of deer and Florida panther. SIMPDEL currently does not have the capacity to be simply extended
to include the effects of chemicals. Parameters such as mortality, fecundity, and growth parameters
can be modified by a user in different scenarios to implicitly model chemical effects.

Flexibility — LOW — Like all of the ATLSS components, the explicit spatial nature of SIMPDEL
makes it specific to South Florida wetland ecosystems. Hence, it does not have the flexibility to be
readily applied to other regions or other species.

Treatment of Uncertainty — MEDIUM — Demographic stochasticity is represented in the mortality
and fecundity parameters; natural spatial variation is represented in the landscape and hydrologic
composition. However, mechanisms for dealing with the uncertainty due to measurement error and
ignorance of parameter values appear to be absent.

Degree of Development and Consistency — LOW — The model is implemented as software; although
it has sufficient documentation, it is not easy to apply to a new system because of its specific and

complex nature. The large number of interactions necessitates the use of parallel computing to run the simulations.

Ease of Estimating Parameters — LOW — The level of complexity of this model makes estimation of all the parameters extremely difficult. The large number of processes, interactions, and dependent variables makes it practically impossible to ascertain all parameter values with any great accuracy or realism.

Regulatory Acceptance — HIGH — ATLSS (and hence, SIMPDEL) is being used by the Federal, State, and Tribal Taskforce for the Restoration of the South Florida Ecosystem as the primary ecological tool for assessing water management strategies.

Credibility — LOW — Few publications exist that cover the model's application.

Resource Efficiency — LOW — Owing to the specific nature of the model, much effort is required to apply it in a particular case.

SIMSPAR

SIMSPAR is a component of ATLSS that is designed to predict the impact of alternative hydrologic scenarios on population levels of the Cape Sable seaside sparrow (Nott et al. 1998). The sparrow's range is almost entirely enclosed by the boundaries of the Everglades National Park and Big Cypress Preserve. Breeding occurs during the dry season in late winter and early spring when water levels recede, allowing the sparrows to nest in dense grass tussocks. Areas that are too wet are not suitable breeding habitats.

The model has the same form as other ATLSS components in that the topography, hydrology, and vegetation of the landscape are modeled explicitly. Spatial cells are marked as habitat or not, depending on whether they satisfy a set of criteria. The model follows the state of each individual during a breeding period and records its sex, age, reproductive status, and location. Movement and behavior are governed by a set of rules dependent on the hydrologic conditions in each cell. The establishment of breeding territories, interactions between individuals, the onset of nesting, and the status of eggs and nestlings are tracked daily. The output of the hydrologic model drives breeding activity. Water depth in each cell is recorded daily — sufficient recession in water levels produces nesting habitats, whereas an excessive rise in water levels causes abandonment of nests. Dispersal is determined by habitat suitability, reproductive success in a cell in the previous breeding period, the sex and development of individuals, and the carrying capacity of a cell. Mortality and fecundity parameters are assigned from data obtained in life-history studies.

Realism — HIGH — The model is realistic because daily food intake, growth, reproductive status, location, breeding success, and other state variables are recorded for each individual at each time-step. Species-specific rules are adopted to govern each individual's behavioral activities, such as movement and interactions with other individuals. The individual-based models make extensive use of the output of the landscape, hydrology, and biota models within the ATLSS program.

Relevance — HIGH — The primary endpoints of SIMSPAR are the abundance and spatial distribution of the Cape Sable seaside sparrow. The presence and effect of chemicals are not modeled at all, nor can SIMSPAR include fate, transport, and bio-uptake of chemicals. However, chemical effects can be implicitly incorporated into mortality, fecundity, and growth parameters.

Flexibility — LOW — Like all ATLSS models, SIMSPAR is specific to South Florida ecosystems, and the individual-based component is species-specific. Hence, it does not have the flexibility to be readily applied to other regions or species.

Treatment of Uncertainty — MEDIUM — The model is stochastic in that variability is incorporated into the model via Monte Carlo simulations in which breeding location, mate choice, mating success, clutch size, gender, dispersal, and mortality are assigned randomly in each simulation. Uncertainty due to measurement error and ignorance regarding parameter distributions do not appear to be taken into consideration. Predictions of population sizes under certain hydrologic conditions correlate well with the results of field studies.

Degree of Development and Consistency — LOW — The model is implemented as software; although it has sufficient documentation, it is not easy to apply to a new system because of its specific and complex nature.

Ease of Estimating Parameters — LOW — The level of complexity of this model makes estimation of all the parameters extremely difficult. The large number of processes, interactions, and dependent variables makes it practically impossible to ascertain all parameter values with great accuracy or realism.

Regulatory Acceptance — HIGH — ATLSS (and hence, SIMSPAR) is being used by the Federal, State, and Tribal Taskforce for the Restoration of the South Florida Ecosystem as the primary ecological tool for assessing water management strategies.

Credibility — LOW — Few publications exist that cover the model's application.

Resource Efficiency — LOW — Owing to the specific nature of the model, much effort is required to apply it in a particular case.

COMPMECH

CompMech is a programming strategy devised by EPRI to predict the impact of increased mortality, loss of habitat, and the release of toxicants on fish populations (EPRI 1982, 2000; Jaworska et al. 1997). Models have been developed for a variety of key aquatic species to compare the ecological and financial consequences of alternative operational scenarios implemented by utilities. The results of CompMech models are also used to compare the cumulative effects of utility operations, the impact of fishing, and natural population dynamics on abundant fish populations as opposed to rare and threatened species. The individual-based CompMech formulation is founded on the idea that the population dynamics of abundant species exhibits compensatory mechanisms inherently different from those of rare species. Compensation is described as the tendency of some populations to counterbalance the decreased growth, reproduction, and survival of some individuals within a population by increased growth, reproduction, and survival of the remaining individuals or vice versa (EPRI 1996). In short, compensation mechanisms are density-dependence effects that occur under highly specific environmental and biological conditions. Although the CompMech approach is spatially explicit, it need not be site specific. The abiotic environment can be represented in terms of the physical and chemical characteristics believed to play a dominant role in the population dynamics of the fish (EPRI 1996).

The CompMech approach has been applied to quantify the effects of polychlorinated biphenyls (PCBs) on largemouth bass at a Superfund site and in southeastern U.S. reservoirs (Jaworska et al. 1997). In the reservoirs, a model was constructed for young-of-the-year largemouth bass that simulated the daily development, growth, and survival of largemouth bass from eggs to the end of their first growing season. The effects of PCBs, expressed as increased mortality and reduced growth of individuals, were imposed for a range of exposure concentrations. Outputs of the model included population density, biomass, mean length of fish, mean condition factor, and fraction of eggs surviving. The model was stochastic and used Monte Carlo simulations to propagate variation and uncertainty in model inputs.

Realism — HIGH — CompMech is not a software package that can be manipulated by the user. Rather, it is a modeling method used by EPRI in partnership with the Oak Ridge National Laboratory in which density-dependence mechanisms are modeled in detail. Models can be constructed with as much or as little life-history and spatial detail as desired.

Relevance — HIGH — The primary endpoints of CompMech are fish population abundance and age structure. Utility companies and resource-management agencies have applied the CompMech approach in many situations. Applications include assessments of direct mortality owing to entrainment, impingement, and fishing; instream flow modifications; habitat alteration such as thermal discharges, water-level fluctuations, water diversions, or introduction of exotic species; and effects of toxic chemicals. The CompMech approach has been applied to quantify the effects of PCBs on largemouth bass (Jaworska et al. 1997).

Flexibility — MEDIUM — CompMech (and the models constructed as part of the CompMech program) is not flexible because new models or modifications of established models are required for each combination of species, abiotic environment, and impacts. Once the models are set up, any scenario can be modeled once the life history of component species and their responses to the toxic chemicals of interest are known.

Treatment of Uncertainty — MEDIUM — Demographic stochasticity is incorporated into the models by specifying the probability of survival of each individual and then randomly determining if the individual survives at each time-step. Spatial and temporal variability in the abiotic environment can be modeled explicitly depending on the availability of information and the purpose of the model. Whether measurement error and parameter uncertainty are dealt with is unclear.

Degree of Development and Consistency — LOW — The CompMech program is not implemented as software. Each new case must be modeled separately, either by constructing new models or by substantially modifying existing models.

Ease of Estimating Parameters — LOW — Parameter estimation in an approach that models density dependence in such detail is extremely difficult. Identification and quantification of compensatory mechanisms in fish populations is difficult and depends on the physical, chemical, and behavioral environment of the species (EPRI 1982). The problem becomes more complex when the effects of utility activity are addressed. Complexity is compounded when the effects of chemicals on populations and their consequences for compensation are taken into account.

Regulatory Acceptance — MEDIUM — Utility companies have adopted CompMech programs to assess and manage utility activities. CompMech models have probably not been adopted or used in a regulatory setting.

Credibility — MEDIUM — Several publications cover CompMech models' applications.

Resource Efficiency — LOW — Owing to the specific nature of many CompMech models, much effort is required to apply them in particular cases (other than those for which they were designed).

ECOBEAKER

EcoBeaker is a software package primarily designed for demonstrating ecological processes and interactions to undergraduate students (Meir 1997). It is individual based in that the landscape can be partitioned into grid cells, each of which may fall into a specified habitat type and may be occupied by an individual or a group of individuals. EcoBeaker uses two strategies to model population and community dynamics. The first is to model the changes in each grid cell according to a set of rules determining movement to or from the cell, disturbances to the cell, and the states of surrounding cells. Alternatively, individuals can be modeled by rules governing their movement and behavior. State variables such as size, age, fat content, and the speed with which an organism can move are updated in each time-step. EcoBeaker can simulate single-species population dynamics, predator–prey dynamics, vegetation succession, mark–recapture surveys, sampling designs, and community-level dynamics. However, applications to date have only been made for hypothetical species and for scenarios modeled as undergraduate laboratory exercises.

Realism — MEDIUM — EcoBeaker can incorporate several key factors and processes. It is spatially explicit but does not link GIS-landscape data like some of the other individual-based models.

Relevance — HIGH — The primary endpoints of EcoBeaker are the abundances and spatial distributions of species. EcoBeaker is accompanied by a number of developed models that show key ecological principles. One model explicitly predicts the effects of a chemical on an aquatic trophic cascade, whereas two others (models dealing with biological pest control and diseases on a fish farm) could be applied to populations that are exposed to chemicals. Output includes the spatial composition of species at each time-step as well as the number of individuals of each species and the value of a biodiversity index. Risk curves are not provided.

Flexibility — HIGH — The program is flexible in that species can be added or removed as desired, different movement and action procedures can be selected, disturbances to the landscape can be included, and parameter values such as mortality and fecundity can be changed. Furthermore, amendment of the source code enables users to customize models for their specific purposes.

Treatment of Uncertainty — MEDIUM — Demographic stochasticity is incorporated into the models via transition matrices in which the probability of transition from one state to another (e.g., the probability of mortality by predation) is specified. However, only one simulation can be examined at a time — The software does not summarize results for a number of runs. Spatial environmental variation can be included in the model by setting up a spatially variable habitat across the grid. Methods of dealing with parameter uncertainty are not incorporated.

Degree of Development and Consistency — HIGH — The model is implemented as commercially available software that is easy to apply to a new system. The software has good technical support.

Ease of Estimating Parameters — MEDIUM — As with all individual-based models, the ease with which parameters can be estimated depends largely on the complexity of the model. The more complex and detailed the model is, the more difficult it is to estimate parameters, particularly those related to species interactions.

Regulatory Acceptance — LOW — The model has not been formally adopted by a regulatory agency. No information is available on whether it is used by regulatory agencies.

Credibility — LOW — No known publications cover the application of the model.

Resource Efficiency — MEDIUM — Applying the model to a particular case requires minor changes to the program. In some cases, the source code may need to be modified to suit the application.

DAPHNIA MODEL

The *Daphnia* model described in McCauley et al. (1990) and Gurney et al. (1990) is an individual-based model that is not spatially explicit. The aim of the model is to use available data to predict growth curves, time to maturity, and fecundity parameters as a function of food supply, which can in turn be used in a population model to better understand the dynamic patterns of populations observed in the field. The model was developed on the basis of physiological mechanisms such as food assimilation and energy allocation, body length changes and molts, ability to recover from starvation, maintenance of body length and weight, and detailed dynamics of algal food concentrations. A driving assumption in the model is that food, in the form of assimilated carbon, can be allocated either to reproduction or to the combination of growth and maintenance. Because the available data on *Daphnia* physiology is gathered from various laboratory studies, it is assumed that individuals are raised singly in containers with a food supply that is homogeneous throughout. An individual is characterized by four variables: body length, body carbon content excluding eggs, the mass of assimilated carbon committed to eggs, and the number of eggs in the brood pouch. A system of differential equations is constructed to describe how each variable changes with respect to time. These equations are solved by using numerical methods.

Realism — HIGH — The model is realistic in that it simulates, in great detail, the physiological processes of *Daphnia* for which much is known. The model is detailed enough to capture a wide range of quantitative features of individual *Daphnia* biology but simple enough to use all the available data.

Relevance — HIGH — The endpoints of the *Daphnia* population model are organism growth curves, time to maturity, and fecundity parameters as a function of food supply. The model has great potential for predicting the effects of chemical exposure on populations of *Daphnia* species and possibly other similar species, particularly when food supply is affected. *Daphnia* is important as an experimental subject in toxicological studies of individual feeding, assimilation, growth, reproduction, and mortality.

Flexibility — MEDIUM — The model is moderately flexible. The level of physiological detail means it can only be applied to *Daphnia* species at present. However, with some reparameterization, it could easily be applied to other crustacean species. The structure of the model is currently being modified to apply to *Acaria* species (Bartell 2000, pers. comm.).

Treatment of Uncertainty — LOW — The model is deterministic and does not allow uncertainty and variability to be treated, except for variability in food supply.

Degree of Development and Consistency — LOW — The model is not implemented as software and cannot be applied to a new system without fully understanding the similarities and differences of the new system compared with *Daphnia*.

Ease of Estimating Parameters — HIGH — Because *Daphnia* populations have been studied intensively in the laboratory and in the field, much data exist in the literature with which to assign parameter values.

Regulatory Acceptance — LOW — No information is available on the model's regulatory acceptance. It is not likely to be adopted or used by a regulatory agency.

Credibility — LOW — A few publications cover this model's application.

Resource Efficiency — LOW — Applying the model in a new case would require new data and perhaps new governing equations.

CIFSS

CIFSS is a software system for building, testing, calibrating, and using individual-based models of fish populations and multispecies fish communities. CIFSS is being developed at the Department of Mathematics, Humboldt State University, Arcata, California (Humboldt State University 1999). It aims to provide interfaces and templates so that users can build custom models for specific sites and fish species with simple and realistic behavioral rules. The CIFSS template software uses the Swarm agent-based simulation system (Langton et al. 1999), which uses life-history tables to update state variables at each time-step. Spatially explicit features can be introduced via cells in which environmental variables can change over time through the output of external hydraulic models. The approach focuses on realistic rules for the selection of habitat by individuals in response to food intake and mortality risks. The model assumes that fish move to maximize their probability of survival and chance of growing to sexual maturity (Railsback et al. 1999). Its applications include trout instream flow, a stream cutthroat trout research model for Little Jones Creek, and a Columbia Basin out-migrant survival simulator (Humboldt State University 1999). Each application simulates habitat and environmental variables such as water temperature, stream flow, and food availability; movement rules for individuals that maximize survival; predation; spawning; feeding and growth; and other variables, depending on the level of detail of the model.

Realism — HIGH — The model is realistic in that it is spatially explicit; key factors and processes can be included; and movement rules account for density, predation, food availability, and likelihood of survival.

Relevance — HIGH — The primary endpoints of the CIFSS model are the abundance of a fish population and its spatial distribution. Outputs for the model also include mean body weight and movement patterns. The model does not explicitly address the effects of toxicants on fish populations. However, it does include parameters that can be modified to reflect the impact of chemicals.

Flexibility — MEDIUM — The CIFSS template software is flexible in that it can be applied to any species of fish, a variety of environmental conditions, and a wide range of problems that differ in level of detail and complexity. The authors of CIFSS claim that the template can also be applied to species other than fish. It is not likely that it can be applied to species that do not forage (such as plants).

Treatment of Uncertainty — MEDIUM — The Swarm programming syntax incorporates stochasticity in that probabilities are assigned for events such as mortality and reproduction. Events are then realized by Monte Carlo simulations. Results are not summarized for many simulations. Sensitivity analyses for determining the effects of measurement error and ignorance about parameter values can be done with the CIFSS template software.

Degree of Development and Consistency — HIGH — The model is implemented as software that can be applied to new systems. The user must be familiar with the Swarm programming syntax. However, the software design objectives are to maintain a high level of documentation, to provide graphic and file outputs to allow users to understand the results of simulations, and to allow changes in formulations, species, habitat simulations, and so forth.

Ease of Estimating Parameters — MEDIUM — The ease with which all the parameters can be estimated depends on the level of complexity of the model. As more features are included in the model, the more difficult it is to estimate all the parameters and interactions.

Regulatory Acceptance — MEDIUM — Fish population models have been developed for the U.S. Department of Agriculture (USDA), utility companies, and Columbia Basin Research at the University of Washington by using the CIFFS template. The model has not been formally adopted for regulatory purposes.

Credibility — MEDIUM — CIFSS has a few users and has been covered in several publications.

Resource Efficiency — MEDIUM — Applying the model to a particular case requires changes to the program and recompilation. Many cases require collecting additional data, particularly spatial coordinates.

WESP AND ECOTOOLS

WESP and ECOTOOLS are projects being developed by OFFIS at the University of Oldenburg in Germany (Sonnenschein et al. 1999) in close collaboration with the Centre for Environmental Research at Leipzig–Halle. The focus of these projects is to develop an interactive environment for the modeling of individual-based populations and metapopulations that can be implemented on parallel computers (Lorek and Sonnenschein 1995). ECOTOOLS is a general programming environment equipped with a library, called ECOSIM, which allows the user to customize ecological models by specifying state variables for individuals. Features of the library allow the specification of static and dynamic properties of individuals; rules governing the actions of individuals that may be triggered by external events, interactions with other individuals, or changes in the state variables of the individual itself; specification of dynamically changing environments; updating of the environment owing to temporal events such as catastrophes; and different spatial scales to be incorporated within one environment.

Several well-known models have already been successfully implemented in the ECOTOOLS programming environment. These include schooling fish, flocking birds, storks in the Everglades, social interaction of dragonflies, crowding crows, largemouth bass, predator–prey interactions, and northern codfish. All models are spatially explicit and incorporate movement rules developed on the basis of food availability and density dependence. WESP is a more specific tool designed for small populations that are at risk of extinction. It supports single populations or metapopulations, provides fixed generic models (EXI-DLG for single populations and META-X for metapopulations — a review of META-X is provided in the section on metapopulations), and allows users to customize their own models.

Realism — HIGH — The purpose of these models is to allow the user to customize ecological models by specifying state variables for individuals. Any number of parameters and key processes can be included for any species.

Relevance — HIGH — The primary endpoints of the ECOTOOLS and WESP models are the abundance of a population and its spatial distribution. Both modeling environments can include the effects of chemicals on populations explicitly with functions or variations in mortality and reproductive rates specified by the user.

Flexibility — MEDIUM — Both ECOTOOLS and WESP are flexible in that users can customize models according to desired features. However, the focus is on small populations and metapopulations; hence, many aquatic species, plant species, and invertebrate species may not be easily represented.

Treatment of Uncertainty — HIGH — Monte Carlo simulations can be included to represent environmental and demographic stochasticity, and the faster run times achieved by parallel computing allow sensitivity analyses to be conducted in a manageable time frame. In this way, mechanisms for treating uncertainty can be included.

Degree of Development and Consistency — LOW — Although WESP provides generic models that can be adapted by the relatively simple specification of parameters, both tools require knowledge of programming languages to customize models for a more general setting. WESP requires knowledge of a modeling language called WESP-DL, whereas ECOTOOLS requires knowledge of C++. Implementation of the software requires parallel hardware. Support for the software is patchy in that parts of the documentation appear only in German with no plans to translate them into English.

Ease of Estimating Parameters — MEDIUM — As with all individual-based models, the ease with which parameters can be estimated depends largely on the complexity of the model. The more complex and detailed the model, the more difficult it is to estimate parameters, particularly those related to species interactions.

Regulatory Acceptance — LOW — No information on the model's regulatory acceptance is available. The amount of programming knowledge required to use these models means they are unlikely to be used or formally adopted by a regulatory agency.

Credibility — LOW — Few publications that apply the model to different cases exist.

Resource Efficiency — LOW — Applying the model in a particular case requires programming, testing, and debugging. Additional data may be necessary.

GAPPS

GAPPS allows users to construct individual-based models for single populations (Harris et al. 1986). It is designed primarily for vertebrate populations with low abundance, extended parental care, or complex age structures. It is not spatially explicit and requires rates to be age- and sex-specific. Probabilities of survival, reproductive rates, and other factors are assigned to each individual in each age class. GAPPS can model density-dependent relationships by allowing four user-specified parameters to construct a linear function or the Michaelis–Menton equation that defines the impact of population size on mortality or fecundity rates. GAPPS can also be used to model the genetics of a small population in a simple Mendelian fashion, to study the effects of harvesting strategies, and to simulate immigration.

Realism — LOW — GAPPS cannot incorporate the level of detail shown in the other models reviewed here. The mechanism for simulating the genetics of a population is much simpler and cruder than that used in VORTEX (see reviews of VORTEX in the section on metapopulation models). The model is not spatially explicit. The effects of catastrophes or toxicity on a population can only be modeled implicitly through the harvesting option.

Relevance — HIGH — The primary endpoint of GAPPS is the abundance of a population. The effects of chemicals on populations are not explicitly incorporated into models. Parameters such as survival and reproductive parameters can be modified to analyze the impact of toxicants on a population.

Flexibility — MEDIUM — GAPPS has limited flexibility in that it can only be used to model vertebrate species with state variables dependent on age structure. Population structures dependent on size, for instance, cannot be modeled explicitly. However, the model can be applied relatively easily to vertebrate species with an age structure and low fecundity.

Treatment of Uncertainty — MEDIUM — GAPPS includes demographic stochasticity but not environmental variation or parameter uncertainty.

Degree of Development and Consistency — MEDIUM — GAPPS is implemented as software that runs under DOS. It does not provide graphic output but does produce tabular output that can be manipulated by the user.

Ease of Estimating Parameters — HIGH — Parameters in GAPPS are relatively easy to estimate given the data typically available — survival rates, reproductive rates, fecundity, and so forth.

Regulatory Acceptance — LOW — The model is unlikely to be currently used by a regulatory agency.

Credibility — LOW — No information is available on model credibility.

Resource Efficiency — MEDIUM — Applying the model to a particular case requires minor changes to the program. Additional data collection may be necessary.

PATCH

PATCH is a spatially explicit, individual-based model designed primarily for territorial terrestrial vertebrate species (Schumaker 1998). It is a single-species, females-only model that pays particular attention to habitat pattern and quality. PATCH uses GIS maps to explicitly model habitats and allows the assignment of weights that measure a habitat's importance to the species considered. It can also incorporate landscape change through time. PATCH is an age- or stage-based model in which estimates of survival and fecundity are supplied in the form of a Leslie or Lefkovitch projection matrix. All vital rates can be linked to habitat quality. Movement behavior can be defined in a number of ways. The movement process may be defined by a pseudorandom walk, or individuals may be assigned a complete knowledge of some portion of the landscape that governs their movement behavior. Territorial individuals may or may not elect to move. Movement behavior is controlled by site fidelity parameters, species' movement ability, the linearity of the searcher's movements, the tendency to move up gradients from low- to high-quality habitats, a search radius, and rules governing whether or not an individual should attempt to obtain the best or closest available site. PATCH provides home-range analysis, the frequency of each habitat type, source-sink analysis, demographic information, observed movement rules, and observed occupancy rates, all of which allow the user to identify, track, and modify clusters of habitats that play a critical role in the maintenance of a population.

Realism — HIGH — The model is fairly realistic because it combines several key processes such as survival, reproduction, dispersal, and local extinction. It incorporates factors such as habitat suitability, age/stage structure, density dependence (territoriality), and environmental fluctuations. The structure is detailed enough to use the most available and relevant demographic data.

Relevance — MEDIUM — The primary endpoints of PATCH are the abundance of a terrestrial vertebrate population and its spatial distribution. The potential effects of chemicals are not addressed explicitly. However, PATCH allows modification of vital rates, habitat quality, and movement rules to implicitly incorporate the impact of toxicants. The results include landscape patterns, territory allocation, movement rates, occupancy rates, expected abundance, and expected variation in abundance. The model does not report viability-related results such as risk of extinction or time to extinction.

Flexibility — MEDIUM — The model is flexible in that new species can be modeled by modifying the life-history tables, importing the relevant GIS map, and specifying new movement rules. The flexibility of PATCH is limited because it is designed primarily for territorial terrestrial vertebrate species. The model has been applied mostly to forest-dwelling birds.

Treatment of Uncertainty — MEDIUM — PATCH is a stochastic model in both a demographic and an environmental sense. The model simulates natural variation only through tables of vital rates. It allows only fully correlated fluctuations. Parameter uncertainty is not dealt with explicitly, but several scenarios can be run and compared.

Degree of Development and Consistency — MEDIUM — It is implemented as software run on UNIX® platforms that is moderately easy to apply to new species. A number of simulations can be performed, the results of which can then be tallied and imported into S-PLUS.

Ease of Estimating Parameters — MEDIUM — Many of the parameters are readily estimated, such as survival rates. The most difficult parameters to estimate are those that govern movement behavior (a major feature which forms the basis for PATCH).

Regulatory Acceptance — HIGH — The model is published by a regulatory agency (EPA).

Credibility — MEDIUM — Because the model is used by EPA, the number of users is likely to be high but not greater than 200.

Resource Efficiency — MEDIUM — Applying the model to a particular case does not require any programming. Many cases require collecting additional data, especially on dispersal behavior.

NOYELP

NOYELP is a spatially explicit, individual-based model for small groups of bison and elk in a multihabitat landscape (Turner et al. 1994; Pearson et al. 1995). The model was designed primarily to investigate the effects of fire scale and pattern on winter foraging dynamics and the survival of elk and bison in the park. The search, movement, and foraging activities of individuals are simulated across daily time-steps for 180 days (the entire winter period for 1 year). The model does not include ungulate reproduction or plant regeneration—it must be initialized at the beginning of each winter period. The landscape is represented as grid cells with a resolution of 1 hectare. Spatial heterogeneity is represented as a series of data layers. Six habitat types are represented in addition to burned and unburned areas and slope, elevation, and aspect parameters. The spatial and temporal heterogeneity of snow conditions is also simulated in the model. Initial forage is assigned to each grid cell and is decremented as ungulates graze. Individual elk and bison are categorized into one of three classes for each species: calf, cow, or bull. Body weights, movement, foraging, energetics, and survival rates are assumed to be identical for all individuals within the same class. Movement occurs in groups, the size of which is flexible. Movement rules are developed on the basis of the availability of forage in grid cells and the forage intake. Energy requirements and expenditure through traveling and maintenance and the subsequent effect on body mass are updated in each time-step. A number of scenarios are tested in the model to determine the effects of winter severity, fire size, and fire pattern on ungulate survival.

Realism — HIGH — The model is realistic in that it is spatially explicit and attempts to simulate detailed vegetation patterns and weather effects in northern Yellowstone Park. Movement rules are developed on the basis of forage availability and energy requirements. The structure of the model allows most available and relevant demographic data to be used. The model is initialized with data from the literature and uses GIS-based landscape data.

Relevance — HIGH — The primary endpoints of NOYELP are the abundances of ungulate populations and their spatial distributions. The outputs of the model are daily and total forage consumed, daily and total survival for each ungulate category, and mean biomass remaining in each vegetation class. Although the effects of chemicals on the ungulate population are not modeled explicitly, several features can be used to model chemical effects, such as vegetation availability, change in body mass, survival, and dispersal.

Flexibility — MEDIUM — The model is specific to elk and bison in northern Yellowstone Park. Hence, the model itself cannot be applied to systems other than those for which it was originally developed, although features such as movement rules and energy equations can be utilized in alternative landscapes (although perhaps not for alternative species).

Treatment of Uncertainty — HIGH — The model is stochastic, and the results of Monte Carlo simulations are summarized statistically for a number of replicates. Sensitivity and uncertainty analyses were done by varying parameters of interest and examining the effects on the model outputs.

Degree of Development and Consistency — LOW — The model is not implemented as software that can be readily used in new applications.

Ease of Estimating Parameters — MEDIUM — The model relies on several parameters for which data are available (e.g., snow depth, ungulate densities, and survival rates). However, movement is notoriously difficult to simulate with great accuracy and must be determined on the basis of assumptions for which no data exist.

Regulatory Acceptance — MEDIUM — The U.S. National Park Service and U.S. Forest Service supported development of the model. The model is not likely to be adopted or used in a regulatory setting for toxic chemicals, although the results may be used for resource management.

Credibility — LOW — The model is known in academia. No information is available on the number of users, but, like most individual-based models, it is likely to be low.

Resource Efficiency — LOW — Because the model is spatially explicit and specific to elk and bison in northern Yellowstone Park, much time and effort would be required to apply it to another case.

WADING BIRD NESTING COLONY

The wading bird nesting colony model is designed to study the effects of chemical contamination on avian populations (Hallam et al. 1996) and is a modification of the wading bird model constructed by Wolff (1994). It consists of an individual-based model for a single population coupled with a chemical exposure model and an effects model. The chemical of concern is mercury, and the species of interest is the wood stork. The landscape is divided into cells in which water depth, prey availability, and elevation are defined. Energy requirements and the number of nestlings requiring food determine foraging activity by mature wood storks. Feeding sites are chosen either randomly or by flock location. If the prey level at a site is less than a certain threshold, the bird leaves the cell to search for another cell. Birds are given a time limit of 6 hours per day for foraging activity. Time-steps are 15 minutes. Two conditions for nesting exist: enough forage must exist within a suitable distance from the nesting site, and females must meet certain energy requirements. Results of the model address the population effect of next-generation survival and cannot be projected for future generations.

To include the effects of mercury in the model, a concentration is assigned to fish in each cell. The total body concentration of mercury in wood storks at any time is modeled by a linear, one-compartment accumulation model dependent on the amount of intake, the rate of mercury elimination, and the size of the bird. At the end of each day, the mercury concentration in birds is updated by using the accumulation equation and the previous level of body concentration. The physical effects of mercury contamination that are modeled are hindrances of flight (reduced speed), of foraging (decreased food caught and increased time spent catching food), and of the ability to search for feeding sites (the probability of initiating a search for food decreases). Decreased appetite, premature cessation of egg production, decreased number of eggs produced, reduced egg viability and survival of hatchlings, and retarded progression of nestlings to young birds are all represented in the model as the effects of mercury contamination.

Realism — HIGH — The model is realistic in that it is spatially explicit and attempts to simulate the effects of mercury contamination on the wood stork in the Florida Everglades. It includes exposure pathways and the effects of mercury. The underlying hydrological model varies in time. Movement rules are developed on the basis of forage availability and energy requirements.

Relevance — HIGH — The outputs of the model are population size through time, spatial distribution, and feeding history for adults, subadults, and nestlings. The effects of mercury are modeled explicitly.

Flexibility — MEDIUM — The model is specific to wading birds in the Florida Everglades. Hence, the model itself cannot be applied to systems other than those for which it was originally developed. However, the chemical exposure model and the effects model can be applied to alternative avian species.

Treatment of Uncertainty — MEDIUM — The model is stochastic in that probabilities are used to determine some events. Five simulations are done, and the results are summarized with standard deviations reported. Sensitivity analyses are not done, even though parameters such as initial mercury concentration are varied to determine the effect on the population.

Degree of Development and Consistency — LOW — The model is not implemented as software that can be readily used in new applications.

Ease of Estimating Parameters — MEDIUM — The model relies heavily on parameters pertaining to mercury toxicology, for which wading bird data are limited.

Regulatory Acceptance — MEDIUM — The model is not likely to be adopted or used in a regulatory setting for toxic chemicals, although the results may be used for resource management.

Credibility — LOW — The model is known in academia. No information is available on the number of users, but, like most individual-based models, it is likely to be low.

Resource Efficiency — LOW — Because the model is spatially explicit and specific to mercury contamination in wood storks in the Florida Everglades, much time and effort would be required to apply it to another case.

DISCUSSION AND RECOMMENDATIONS

The individual-based models reviewed here (Table 6.2) all have utility at some level. The level of detail incorporated into these models provides a high to medium level of realism. All of the models have a high degree of relevance because they include parameters such as survival, reproduction, and dispersal rates that can implicitly incorporate the effects of chemicals. Most of the models are not very flexible because they were designed in great detail for specific populations and habitats. Except for the *Daphnia* model, all of the models are stochastic. Stochastic models in which sensitivity analyses can be done relatively quickly (i.e., WESP and ECOTOOLS) deal with uncertainty very well, whereas those in which such analyses are time consuming and impractical have a medium uncertainty rating.

The degree of development varies across the range of models (Table 6.2). Although all the ATLSS components that are individual-based models exist as software (i.e., SIMPDEL and SIM-SPAR), they are not easy to apply to new systems because they are specific and complex. The models that exist as individual-based software modeling packages (i.e., EcoBeaker, CIFSS, GAPPS, and PATCH) have a medium to high rating for degree of development, depending on the ease with which they can be applied to new systems. The exceptions to this were WESP and ECOTOOLS. These modeling environments require knowledge of a modeling language called WESP-DL and access to parallel computers. Most of the models are data intensive and have low to medium ratings for ease of parameter estimation. GAPPS relies on an age-based structure and uses fewer parameters than many of the other individual-based models. For this reason, it has a high rating for ease of parameter estimation. Most of the models have a low level of credibility. This level is not surprising in light of their specificity. A model designed for a specific population would be expected to appear in relatively few publications. Similar considerations apply for the resource efficiency criterion.

Models designed for a specific species and habitat cannot easily be used for alternative habitats and populations. This aspect limits the utility of individual-based models in population assessments. Many individual-based models require parameters at such a detailed level that the necessary data are often not available. Examples of this are models with complex movement rules and interactions with other individuals. A further hindrance to individual-based models is the amount of time required to run simulations. The vast number of interactions and parameter updates in many of these models requires substantial computing resources. Some models, such as SIMPDEL, WESP, and ECOTOOLS, have overcome this by using parallel computing. Although parallel computing does address lengthy computation times, it makes the model less accessible than models that can be run on personal computers (PCs). A further drawback of individual-based models is that the high level of detail relies on many assumptions that usually are not verified.

Except for the *Daphnia* model and GAPPS, all of the individual-based models reviewed here are spatially explicit. In such models, the spatial landscape in which an organism lives is heterogeneous. The movement and behavior of organisms are then tracked through space and time. In some models, such as in the components of ATLSS, CIFSS, PATCH, and NOYELP, the spatial habitat changes in time. Although spatially explicit models incorporate realistic features of ecosystems and landscapes, the trade-off is that they require data in the form of GIS maps or explicit spatial coordinates that are specific to a region. This requirement makes their flexibility and resource efficiency lower than that of more generic models that implicitly incorporate spatial considerations in survival and dispersal rates. One key advantage of spatially explicit models, particularly for ecotoxicological applications, is that an affected site can be spatially represented in detail in the model and linked to habitat suitability parameters if the necessary data are available. In this way, an individual's foraging behavior, dispersal ability, and chances of survival and reproductive success can depend on its proximity to the affected site.

Owing to the specific nature of many individual-based models and their varied utility, we do not offer model recommendations. Rather, we focus on features that may be useful for applications in ecological risk assessment of toxic chemicals. The CIFSS are the most flexible and well developed

Table 6.2   Evaluation of Population Models — Individual-Based Models

Model	Reference	Evaluation Criteria								
		Realism	Relevance	Flexibility	Treatment of Uncertainty	Degree of Development	Ease of Estimating Parameters	Regulatory Acceptance	Credibility	Resource Efficiency
SIMPDEL	Abbott et al. (1995)	♦♦♦	♦♦♦	♦	♦♦♦	♦	♦	♦♦♦		♦
SIMSPAR	Nott et al. (1998)	♦♦♦	♦♦♦	♦	♦♦♦	♦	♦	♦♦♦	♦♦	♦
CompMech	EPRI (1982, 1996); Jager et al. (1999)	♦♦♦	♦♦♦	♦♦	♦♦♦	♦	♦	♦♦	♦♦♦	♦
EcoBeaker	Meir (1997)	♦♦	♦♦♦	♦♦♦	♦♦	♦♦♦	♦♦	♦	♦	♦♦
Daphnia	Gurney et al. (1990); McCauley et al. (1990)	♦♦♦	♦♦♦	♦♦	♦	♦♦	♦♦♦	♦	♦♦	♦
CIFSS	Langton et al. (1999); Railsback et al. (1999); Railsback and Harvey (in prep.); Humboldt State University (1999)	♦♦♦	♦♦♦	♦♦	♦♦	♦♦♦	♦	♦	♦♦♦	♦♦
WESP and ECOTOOLS	Lorek and Sonnenschein (1995); Sonnenschein et al. (1999)	♦♦♦	♦♦♦	♦♦	♦♦♦	♦	♦	♦	♦	♦
GAPPS	Harris et al. (1986)	♦	♦♦♦	♦♦♦	♦♦♦	♦♦♦	♦♦♦	♦	♦	♦♦
PATCH	Schumaker (1998)	♦♦♦	♦♦	♦♦♦	♦♦♦	♦♦♦		♦♦♦	♦♦	♦♦
NOYELP	Turner et al. (1994); Pearson et al. (1995)	♦♦♦	♦♦♦	♦♦	♦♦♦	♦		♦♦♦	♦	♦
Wading bird nesting colony	Hallam et al. (1996)	♦♦♦	♦♦♦	♦♦	♦♦	♦	♦	♦♦	♦	♦

Note: ♦♦♦ - high   ♦♦ - medium   ♦ - low

of the spatially explicit models. It allows users to construct their own models if they have some knowledge of the Swarm programming environment. Documentation and technical support for both Swarm and CIFSS are very good. Although CIFSS was primarily designed for aquatic systems, it could be adapted to other systems. CIFSS is most useful when the movement behavior of individuals is of interest.

Some of the models have already been used in ecotoxicological risk assessments (Table 6.3). These include CompMech, EcoBeaker (although only in a hypothetical, illustrative sense), the *Daphnia* model, and the wading bird nesting colony model. The ecotoxicological features of these models could be adapted to alternative scenarios. We recommend that individual-based models be used when the data exist to parameterize the input variables and when investigating more detailed features of populations than risks of decline (e.g., predicted movement and foraging behavior). Individual-based models are usually too complex and their simulations are too time consuming for them to be used as screening tools.

One often unconsidered potential use of individual-based models is that for which the *Daphnia* model was designed. Detailed individual-based models may have greater utility in estimating parameters such as annual vital rates in particular environments, which can then be incorporated into less detailed population models. We feel that this use is one worth further investigation, particularly in assessments of chemical impacts on populations for which data and parameter estimates are scarce. The level of detail possible in many of the individual-based models may also allow the inclusion of dose–response equations to model the effects of chemicals. This use is also worth further investigation.

Table 6.3 Applications of Individual-Based Population Models

Model	Species	Location/Population	Spatially Explicit?	Reference
SIMPDEL	White-tailed deer (*Odocoileus virginianus seminolus*) and Florida panther (*Felis concolor*) (proposed)	Everglades and Big Cypress Swamp, Florida	Yes	Abbott et al. (1995)
SIMSPAR	Cape Sable seaside sparrow (*Ammodramus maritimus mirabilis*)	Everglades and Big Cypress Swamp, Florida	Yes	Nott et al. (1998)
CompMech	Largemouth bass (*Micropterus salmoides*)	Southeastern reservoirs, United States	Yes	Jaworska et al. (1997)
	White sturgeon (*Acipenser transmontanus*)	Snake River, United States	Yes	Jager et al. (1999)
	Winter flounder (*Pseudopleuronectes americanus*)	Various, United States	Yes	Chambers et al. (1995); Rose et al. (1996); Tyler et al. (1997)
	Smallmouth bass (*Micropterus dolomieu*)	Various, Canada	Yes	DeAngelis et al. (1991, 1993); Jager et al. (1993)
	Striped bass (*Morone saxatilis*)	Santee-Cooper River, South Carolina, and Sacramento River — San Joaquin Delta, California	Yes	Rose et al. (1993)
	Yellow perch (*Perca flavescens*), walleye (*Stizostedion vitreum*)	Oneida Lake, United States	Yes	Rose et al. (1996)
EcoBeaker	No known applications			Meir (1997)
Daphnia	*Daphnia pulex*	Laboratory	No	McCauley et al. (1990); Gurney et al. (1990)
CIFSS	Rainbow trout (*Salmo gairdneri*) and brown trout (*Salmo trutta*)	Tule River, California	Yes	Railsback et al. (in prep.)
	Cutthroat trout (*Oncorhynchus clarki clarki*)	Little Jones Creek, California	Yes	Harvey (1998); Harvey et al. (1999)
	Chinook salmon (*Oncorhynchus tschawytscha*)	Snake River, Washington	Yes	Zabel et al. (1998)
WESP & ECOTOOLS	Wading birds	Everglades, Florida	Yes	Kohrs (1996)
	Crown of thorns (*Acanthaster planci*)	Various, Pacific Ocean	Yes	OFFIS (1999)
GAPPS	Rhinoceros	Africa	No	Dobson et al. (1992)
	Scimitar-horned oryx (*Oryx dammah*)	Niger	No	Dixon et al. (1991)
	Addax (*Addax nasomaculatus*)	Niger	No	Dixon et al. (1991)
NOYELP	Bison and elk (*Cervus elaphus*)	Northern Yellowstone Park, Wyoming	Yes	Turner et al. (1994); Pearson et al. (1995)
Wading bird nesting colony	Wood stork (*Mycteria americana*)	Everglades, Florida	Yes	Hallam et al. (1996)
Lumbricid pesticide exposure	Lumbricid species (*Lumbricus rubellus* and *L. terrestris*)	Laboratory	No	Baveco and de Roos (1996)
Agrochemical exposure	General avian species	Midwestern United States	No	Dixon et al. (1999)
McKendrick–von Foerster *Daphnia*	*Daphnia*	Hypothetical	No	Hallam et al. (1990)
	Daphnia	Generic lake in central Florida	No	Bartell et al. (2000)

CHAPTER 7

Population Models — Metapopulations

H. Resit Akçakaya and Helen M. Regan

A metapopulation is a set of populations of the same species in the same general geographic area with a potential for migration among them (Figure 7.1). Often there is movement of individuals (dispersal) among the different populations of a metapopulation, which may lead to recolonization of empty (extinct) habitat patches. Metapopulation dynamics are important for several reasons. Many species live in naturally heterogeneous landscapes or in habitats that are fragmented by human activities, thereby leading to a series of populations of the same species distributed in space. The dynamics of a spatially structured metapopulation are complicated by the interaction (e.g., exchange of individuals) among populations within the metapopulation. The extinction risk of a metapopulation can be estimated only when all populations are modeled together in a metapopulation model because partially correlated fluctuations in environmental variables lead to complex dependencies in their extinction risks. Existence of a species in multiple populations exposes it to additional threats, such as further fragmentation, isolation, and dispersal barriers, in addition to the threats that affect each of its populations. Finally, the existence of a species in multiple populations allows additional conservation and management options, such as reserve design, habitat corridors, reintroduction, and translocation. Endpoints for metapopulation modeling include the following:

- Metapopulation persistence
- Metapopulation occupancy, local occupancy duration
- Expected abundance, expected variation in abundance
- Movement rates, occupancy rates
- Spatial patterns of occupancy

Typical data inputs for a metapopulation model include the areas and locations of suitable habitat patches; presence/absence data for the species; carrying capacity; survival, fecundity, and dispersal rates; parameters for describing catastrophes; and the time series of habitat maps. Figure 7.1 illustrates the spatial structure of one example of a metapopulation, the California gnatcatcher.

Metapopulation models reviewed below include (Table 7.1):

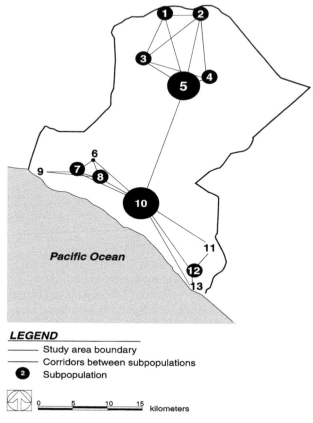

Figure 7.1   Schematic of the spatial structure of a metapopulation. Note: This example of a metapopulation is
for the California gnatcatcher in Orange County. (From Akçakaya and Atwood (1997) A habitat-
based metapopulation model of the California gnatcatcher. *Conserv. Biol.* 11:422–434. With permission of Blackwell Science, Inc.)

- Occupancy-incidence function, a model for determining the occupancy status of habitat patches
  (Hanski 1994; Hanski and Gilpin 1991)
- Occupancy-state transition, a model for predicting the transitions in the status of habitat patches
  (Verboom et al. 1991; Sjögren-Gulve and Ray 1996)
- RAMAS Metapop and RAMAS GIS, structured metapopulation models with population dynamics
  in each patch (Akçakaya 1998a,b; Akçakaya et al. 1995; Kingston 1995)
- VORTEX, an individual-based metapopulation model (Lacy 1993; Lindenmayer et al. 1995)
- ALFISH (ATLSS landscape fish model), a spatial model of fish in the Everglades (DeAngelis et
  al. 1998a; Gaff et al. 2000)
- ALEX (analysis of the likelihood of extinction), a generic metapopulation model (Possingham and
  Davies 1995)
- Meta-X, an occupancy status model currently being developed (UFZ and OFFIS 2000)

In this chapter, we also included individual-based population models that have a strong spatial
component. Thus, in this chapter, the term *spatial models* refers to metapopulation models with
spatial components. For other reviews of metapopulation models, see Gilpin and Hanski (1991),
Breininger et al. (in press), and Akçakaya and Sjögren-Gulve (2000).

We review two occupancy-type models and four software platforms for spatially structured
modeling (RAMAS Metapop and GIS, VORTEX, ALEX, and Meta-X). In an earlier review,
Lindenmayer et al. (1995) compared VORTEX, ALEX, and RAMAS/space (a precursor of RAMAS

Table 7.1 Internet Web Site Resources for Metapopulation Models

Model Name	Description	Reference	Internet Web Site
Occupancy — incidence function	A model for determining the occupancy status of habitat patches	Hanski and Gilpin (1991); Hanski (1994)	www.consci.org/forum/docs2/ch5.pdf
Occupancy — state transition	A model for predicting the transitions in the status of habitat patches	Verboom et al. (1991); Sjögren-Gulve and Ray (1996)	N/A
RAMAS Metapop and RAMAS GIS	Structured metapopulation models with population dynamics in each patch	Akçakaya et al. (1995); Kingston (1995); Akçakaya (1998a,b)	http://www.ramas.com/
VORTEX	An individual-based metapopulation model commonly used in conservation biology	Lacy (1993); Lindenmayer et al. (1995)	http://www.life.umd.edu/classroom/bsci363/assignments/vortex/vortexguide.html http://home.netcom.com/~rlacy/vortex.html
ALFISH	A spatially explicit, individual-based model of fish in the Everglades	DeAngelis et al. (1998a); Gaff et al. (2000)	http://sofia.usgs.gov/projects/atlss/ http://www.tiem.utk.edu/~gross/atlss_www/fish.html
ALEX	A generic metapopulation model applicable to most vertebrates	Possingham and Davies (1995)	http://biology.anu.edu.au/research-groups/ecosys/Alex/ALEX.HTM
Meta-X	An occupancy status model currently being developed	UFZ and OFFIS (2000)	http://www.oesa.ufz.de/meta-x/english/overview.html

Note: N/A - not available

Metapop). In the Discussion and Recommendations section, we identify gaps in the use of these models and potential development directions that will make the models more useful.

OCCUPANCY MODELS — INCIDENCE FUNCTION

The simplest metapopulation approach models the occupancy status of habitat patches (i.e., the presence or absence of the species in the patches) in a geographic region. These models are parameterized by using data on the presence or absence of a species in habitat patches from one or more regional inventories and data on environmental variables. They require that the species has local populations confined to a clearly delimited habitat in a landscape.

Incidence function models (Hanski 1994; Hanski and Gilpin 1991) require data on the areas and geographic locations of suitable habitat patches and on the presence/absence of the species in these patches from *one* complete inventory. A habitat suitability analysis of the species' presence/absence pattern may be required for reliable habitat patch identification and delimitation. On the basis of these data, colonization and extinction probabilities are estimated for each patch by using regression techniques. These estimated probabilities are then used in a simulation to predict metapopulation persistence.

> **Realism — LOW** — Incidence function models incorporate local extinction and recolonization. The factors they incorporate include size of patches and distance among patches. They ignore local population dynamics and do not model fluctuations in the sizes or compositions of the local populations (in terms of sex, age, or stage). They cannot use most available demographic data. The model assumes that the metapopulation is in equilibrium, which is often invalid.
>
> **Relevance — LOW** — The main endpoint is metapopulation persistence. The main model parameters are not directly related to physical or chemical impacts.
>
> **Flexibility — MEDIUM** — The model has been applied to short-lived species (such as butterflies) in terrestrial environments and could be extended to other species. The model has only a few species-specific parameters. The parameters (local extinction risk and recolonization probability) require observations of locally extinct patches (defined as suitable habitats in which the species of interest is absent).
>
> **Treatment of Uncertainty — HIGH** — The model simulates variability in patch occupancy. It can be used in a sensitivity analysis to incorporate uncertainties.
>
> **Degree of Development and Consistency — MEDIUM** — No software programs are available for the model. However, the model structure is simple and would not be too difficult to implement as software.
>
> **Ease of Estimating Parameters — MEDIUM** — Some model parameters (such as local extinction risk) are hard to estimate given typical data sets. Unbiased estimation of parameters requires observations of several occupied and unoccupied patches, and occupancy may be difficult to determine.
>
> **Regulatory Acceptance — MEDIUM** — No information on the model's regulatory acceptance is available. The model is likely to be used by a regulatory agency but is not likely to be formally adopted for use in developing environmental criteria.
>
> **Credibility — MEDIUM** — The model is known in academia; several (10 to 100) publications cover the applications of the model in different cases.
>
> **Resource Efficiency — LOW** — Applying the model to a particular case requires some programming, testing, and debugging. In most cases, available data are not sufficient, and additional data collection is necessary.

OCCUPANCY MODELS — STATE TRANSITION

State transition models (e.g., Verboom et al. 1991; Sjögren-Gulve and Ray 1996) are conceptually related to the incidence function models discussed previously. They also require presence/absence data but from two or more yearly inventories instead of a single snapshot. They predict the transitions in the status of patches (vacant to occupied as a result of colonization, and occupied to extinct as a result of local extinction).

Realism — LOW — These models incorporate the processes of local extinction and recolonization. The factors they incorporate include the size of patches and the distance among patches. They ignore local population dynamics and do not model fluctuations in the sizes or compositions of the local populations (in terms of sex, age, or stage). They cannot use most available demographic data.

Relevance — LOW — The main endpoint is metapopulation persistence. The main model parameters are not directly related to physical or chemical impacts.

Flexibility — MEDIUM — The model has only a few species-specific parameters. The parameters (local extinction risk and recolonization probability) require observations of local extinction and recolonization events. Thus, the model is only applied to short-lived species.

Treatment of Uncertainty — HIGH — The model simulates variability in patch occupancy. It can be used in a sensitivity analysis to incorporate uncertainties.

Degree of Development and Consistency — MEDIUM — No software programs are available for the approach. However, the model structure is simple and would not be too difficult to implement as software.

Ease of Estimating Parameters — MEDIUM — Some model parameters (such as local extinction risk) are hard to estimate given typical data sets. Unbiased estimation of parameters requires several observations of local extinction and colonization events.

Regulatory Acceptance — MEDIUM — No information on the model's regulatory acceptance is available. The model is likely to be used by a regulatory agency but is not likely to be formally adopted for use in developing environmental criteria.

Credibility — MEDIUM — The model is known in academia; several (10 to 100) publications cover applications of the model in different cases.

Resource Efficiency — LOW — Applying the model to a particular case requires some programming, testing, and debugging. In most cases, available data are not sufficient, and additional data collection is necessary.

RAMAS METAPOP AND RAMAS GIS

RAMAS Metapop and RAMAS GIS implement structured metapopulation models (Akçakaya 1998a,b). Structured metapopulation models incorporate the local dynamics of populations in each habitat patch. The main advantage of structured metapopulation models is their flexibility. They can incorporate several biological factors and can represent spatial structure in various ways; they have been applied to a variety of organisms (Akçakaya 2000). Another advantage is that, despite their flexibility, structured models were developed on the basis of a number of common techniques or frameworks that allow them to be implemented as generic programs (such as RAMAS Metapop). This common framework becomes advantageous when models and viability analyses are needed for a large number of species and when time and resource limitations preclude detailed modeling and programming for each species. A third advantage is that structured demographic modeling allows careful risk assessment for species with very few local populations (occupancy models require a larger number), even if no extinctions have occurred.

The user can run these metapopulation models to predict the risk of species extinction, time to extinction, expected occupancy rates, and metapopulation abundance. The programs allow comparison of results from different simulations by superimposing graphs of risk curves, time-to-extinction distributions, trajectory summary, metapopulation occupancy, and other outputs.

Landscape information imported into RAMAS Metapop/GIS may include GIS-generated maps of vegetation cover, land use, temperature, precipitation, or some other aspect of the habitat that is important for the species (Figure 7.2). RAMAS Metapop/GIS then combines the information in all these map layers into a map of habitat suitability expressed as a user-defined habitat suitability function.

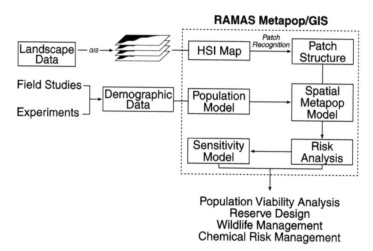

Figure 7.2   Structure of RAMAS Metapop/GIS. (From Applied Biomathematics
(http://www.ramas.com/ramas.htm). With permission.)

RAMAS Metapop/GIS uses a *patch-recognition algorithm* and identifies areas of high suitability as a patch where a subpopulation may survive. The carrying capacity and other population-specific parameters of this patch (survival, fecundity, maximum growth rate) can be calculated as a user-defined function of total habitat suitability, average habitat suitability, core area, patch perimeter (edge), and other habitat characteristics (from the GIS maps) for that patch. RAMAS Metapop/GIS then displays the spatial structure of the metapopulation superimposed with a color-coded map of habitat suitability and other geographical features specified by the user. RAMAS Metapop/GIS can also import a user-defined time series of habitat maps, from which the program creates time series of population-specific parameters for input into the metapopulation model.

Realism — HIGH — RAMAS Metapop considers processes such as survival, reproduction, dispersal, local extinction, habitat loss, habitat growth, demographic stochasticity, catastrophes, recolonization, harvest, translocation, and reintroduction. It incorporates factors such as age structure, stage structure, environmental fluctuations, density dependence, and correlated environments. It does not incorporate genetics. RAMAS GIS includes RAMAS Metapop; it links GIS-based landscape data to the habitat requirements of the species and incorporates wildlife–habitat relationships. The structure of the models allows all available and relevant demographic data to be used. Model assumptions are generally realistic.

Relevance — HIGH — The results include risk of extinction, risk of decline, time to extinction, time to decline, metapopulation occupancy, local occupancy duration, expected abundance, and expected variation in abundance. The model has several parameters (carrying capacities, survival, fecundity, dispersal rates, maximum growth rates, catastrophes) that can be used to model implicitly the effects of habitat degradation, toxicity, hunting/fishing, thermal pollution, timber harvest, and so forth.

Flexibility — HIGH — The model has a large number of parameters that can be modified by the user. However, the model is not applicable to species with complex social interactions. The model has been applied to plants, invertebrates, mammals, fishes, birds, reptiles, and amphibians in terrestrial, marine, and freshwater environments.

Treatment of Uncertainty — HIGH — The model simulates the natural variability in population parameters such as survival, fecundity, and dispersal. It allows correlated fluctuations. RAMAS Metapopulation/GIS has a sensitivity analysis feature that allows multiple simulations with automatically changed input parameters.

Degree of Development and Consistency — HIGH — The model is implemented as user-friendly software that is easy to apply to new cases. It has a detailed user's manual that describes the use of the program and gives background information about metapopulation dynamics and parameter estimation. The program includes several internal checks for consistency. It also checks each user input value for the correct range and for consistency with other input parameters.

Ease of Estimating Parameters — MEDIUM — The model requires more data than occupancy models — for example, stage-specific survival rates and fecundities and their temporal and spatial variation. However, in many cases (especially for vertebrate populations), these types of data may be more readily available than data required for occupancy models (e.g., observations of local extinctions and recolonizations).

Regulatory Acceptance — HIGH — The model is used by several regulatory agencies in different countries (the U.S. Fish and Wildlife Service, EPA, Environment Canada, New South Wales National Parks and Wildlife Service in Australia, and state agencies in the U.S.). The model has been used in listing threatened species and in species recovery plans.

Credibility — HIGH — The model is widely known in academia and has been positively reviewed in the literature (Kingston 1995; Witteman and Gilpin 1995; Boyce 1996). The model has more than 300 users (about 450 if RAMAS/space, a precursor, is included).

Resource Efficiency — HIGH — Applying the model to a particular case does not require any programming. Some cases require additional data collection, but in many cases, available data are sufficient.

VORTEX

VORTEX is an individual-based metapopulation model (Lacy 1993). VORTEX follows a commonly used approach in which the behavior and fate of each individual is modeled in a simulation (DeAngelis and Gross 1992). The behavior and fate (e.g., dispersal, survival, reproduction) of individuals depend on their location, age, size, sex, physiological stage, social status, and other characteristics.

The advantage of individual-based models is that they are even more flexible than structured models and can incorporate such factors as genetics, social structure, and mating systems more easily than other types of models.

One disadvantage of individual-based models is that they are very data intensive. Only a few species have been studied well enough to use all the power of individual-based modeling. Another disadvantage is that their structure often depends on the biology of the particular species modeled. Thus, individual-based models must usually be designed and implemented for each species separately. However, VORTEX (Lacy 1993) has a fixed, age-based structure, even though it was developed on the basis of individual-based modeling techniques.

Realism — HIGH — VORTEX considers processes such as survival, reproduction, dispersal, local extinction, demographic stochasticity, catastrophes, recolonization, and harvest. It incorporates factors such as age structure, environmental fluctuations, density dependence, and genetics. The structure of the models allows all available and relevant demographic data to be used. Model assumptions are realistic.

Relevance — HIGH — The results include risk of extinction, risk of decline, time to extinction, time to decline, expected abundance, and expected variation in abundance. The model has several parameters (e.g., carrying capacity, survival, fecundity, dispersal rates, catastrophes) that can be used to model implicitly the effects of habitat degradation, toxicity, hunting/fishing, thermal pollution, timber harvest, and so forth.

Flexibility — MEDIUM — The model has many parameters that can be modified by the user. The model does not use stage structure, so it may not be appropriate for some species, such as many plants. The model is not practically applicable to highly fecund species (such as many fishes) and to very abundant species. The model does not incorporate habitat relationships.

Treatment of Uncertainty — HIGH — The model simulates natural variability in population parameters such as survival, fecundity, and dispersal. It can be used for sensitivity analysis.

Degree of Development and Consistency — HIGH — The model is implemented as software that is fairly easy to apply to new cases. It has a user's manual that describes the use of the program and gives background information about metapopulation dynamics and parameter estimation.

Ease of Estimating Parameters — MEDIUM — This model requires more data than structured models (such as those related to genetics and mating system).

Regulatory Acceptance — MEDIUM — No information on the model's regulatory acceptance is available. The model is likely to be used by a regulatory agency. The manual has been published by the captive breeding specialist group of the World Conservation Union (IUCN), an international conservation organization.

Credibility — HIGH — The model is widely used. No information is available on the number of users, which is likely to be more than 200.

Resource Efficiency — HIGH — Applying the model to a particular case does not require any programming. Some features (such as density dependence and age-specific fecundity and survival for adults) require writing equations. Some cases require additional data collection, but in many cases, available data are sufficient.

ALFISH

ALFISH (the ATLSS landscape fish model; Gaff et al. 2000) is a component of the across-trophic-level system simulation, an integrated combination of computer simulation modules designed to model the biotic community of the freshwater wetlands of the Everglades and Big Cypress swamp and the abiotic influences on that community (see Chapter 11, Landscape Models). The aim of ALFISH is to predict the effects of hydrologic scenarios on fish densities in the South Florida wetland area and the subsequent food availability to wading birds.

In each cell of ALFISH, the variability in water depth and elevation is characterized statistically. Permanently wet regions such as ponds are modeled explicitly, and all other areas are subject to drying and reflooding. Fish are not modeled individually and do not move between cells. The density of fish is modeled deterministically and depends on the varying water depth in each cell, available food resources from lower trophic levels, and movement between the cells. The model incorporates two distinct functional groups — those fish 7 cm or less in length and those greater than 7 cm. The functional groups are modeled with an age–size structure in which each age class is 30 days. Each age class is divided into six size classes in which length is calculated as a function of age by a von Bertalanffy equation. The time-step is 5 days, during which fish increase in size before moving to the next age class every 30 days. Mature fish produce offspring during their assigned reproductive month, the number and timing of which is determined by parameters specific to each functional group. Mortality has four manifestations: age-dependent background mortality in an uncrowded population; density-dependent mortality from starvation; mortality from predation from other functional groups; and mortality from an inability to disperse to wetter regions after a cell dries out.

Realism — MEDIUM — The model is realistic in the sense that the major influencing factor on fish populations, hydrology, is modeled in great detail. The model includes density dependence within each cell. Population size predictions correlate well with known fish distributions (DeAngelis et al. 1998a).

Relevance — HIGH — The population densities and spatial distributions of fishes are the primary endpoints of ALFISH. The potential effects of chemicals on fish have not yet been modeled in ALFISH, but there are plans to model the effects of mercury. Unlike the more comprehensive model ATLSS, ALFISH does not include simulations of nutrient concentration, bio-uptake, transport, and fate that can be extended to other chemicals. However, mortality, fecundity, and growth parameters can be modified to implicitly incorporate chemical effects.

Flexibility — LOW — The spatial component of the model is specific to the South Florida wetland region, and the stage structure is particular to fish; hence, the model is not flexible enough to implement for other regions or species.

Treatment of Uncertainty — LOW — The fish model is deterministic. Except for spatial variation across the landscape and the variability in pond size within each cell, uncertainty and variability in the model parameters are not dealt with.

Degree of Development and Consistency — LOW — The model is implemented as software. Although this software has sufficient documentation, it is not easy to apply to a new system because of its specific nature.

Ease of Estimating Parameters — MEDIUM — Parameters are relatively easy to estimate because this model is not individual-based, few parameters are used, and data are available for fish.

Regulatory Acceptance — HIGH — ALFISH is part of ATLSS, which is used by the Federal, State, and Tribal Taskforce for the Restoration of the South Florida Ecosystem as the primary ecological tool for assessing water management strategies.

Credibility — LOW — Few publications cover the model's application.

Resource Efficiency — LOW — Owing to the specific nature of the model, much effort is required to apply it in a particular case.

ALEX

ALEX is a generic metapopulation model with features that allow it to be applied to most vertebrates (Possingham and Davies 1995). The main advantages of ALEX include its speed, ability to incorporate a variety of catastrophes, and ability to allow for habitat dynamics. ALEX simulates comparatively large populations very quickly and allows the user to specify a wide variety of environmental processes.

Realism — MEDIUM — ALEX considers processes such as survival, reproduction, dispersal, local extinction, habitat loss, habitat growth, demographic stochasticity, catastrophes, and recolonization. It incorporates factors such as age structure and environmental fluctuations. It does not incorporate genetics. The structure of the model allows most available and relevant demographic data to be used. ALEX allows a simple age structure with only three classes of individuals: newborn, juvenile, and adult. It does not allow stage structure.

Relevance — HIGH — The results include risk of extinction, time to extinction, metapopulation occupancy, expected abundance, and expected variation in abundance. The model has several parameters (e.g., survival, fecundity, dispersal rates, and catastrophes) that can be used to model implicitly the effects of habitat degradation, toxicity, hunting/fishing, thermal pollution, timber harvest, and so forth.

Flexibility — MEDIUM — The model has many parameters that can be modified by the user. The model does not use stage structure, so it may not be appropriate for some species, such as many plants. The model allows only three age classes, so it may not be applicable to species with delayed reproduction. The model has been applied primarily to vertebrates.

Treatment of Uncertainty — HIGH — The model simulates natural variability in population parameters such as survival, fecundity, and dispersal. It does not allow correlated fluctuations. It can be used for sensitivity analysis to incorporate uncertainties.

Degree of Development and Consistency — MEDIUM — The model is implemented as software that is moderately easy to apply to new cases. According to the program documentation, using ALEX requires some limited input from its author, Hugh Possingham.

Ease of Estimating Parameters — MEDIUM — This model requires more data than occupancy models — for example, age-specific survival rates and fecundities and their temporal and spatial variations. However, in many cases (especially for vertebrate populations), these data may be more readily available than the data required for occupancy models (e.g., observations of local extinctions and recolonizations).

Regulatory Acceptance — MEDIUM — No information on the model's regulatory acceptance is available. The model is likely to be used by a regulatory agency, especially in Australia.

Credibility — MEDIUM — No information is available on the number of users, which is likely to be between 20 and 200.

Resource Efficiency — MEDIUM — Applying the model to a particular case does not require any programming. Some cases require additional data collection, but, in many cases, available data are sufficient. ALEX is intended for research and education; its use without the author's permission is discouraged.

META-X

Meta-X is being developed in Germany by Umweltforschungszentrum Leipzig–Halle Sektion Ökosystemanalyse (UFZ) and Oldenburger Forschungs- und Entwicklungsinstitut für Informatik-Werkzeuge und Systeme (OFFIS). We reviewed the beta version of the software.

Meta-X was developed by using the occupancy modeling approach (see above). It models the occupancy status of patches. As opposed to WESP/ECOTOOLS, which is mainly designed for use by model developers, Meta-X incorporates options for selecting available models and is intended for use by ecologists and environmental managers. According to its creators, Meta-X differs from other programs by directly supporting comparative evaluations of alternative environmental management scenarios with respect to their effects on the persistence of the population of concern.

Realism — LOW — Meta-X incorporates local extinction and recolonization. The factors incorporated include location of patches, distance among patches, and correlations among extinction probabilities of patches. It ignores local population dynamics and does not model fluctuations in the size or composition of the local populations (in terms of sex, age, or stage). It cannot use most available demographic data. The model assumes that local extinction risks are constant over time, which is not realistic for increasing or decreasing species populations.

Relevance — LOW — The main endpoints are metapopulation persistence and time to extinction. The main model parameters are not directly related to physical or chemical impacts.

Flexibility — MEDIUM — The model has only a few species-specific parameters. The parameters (local extinction risk and recolonization probability) require observations of local extinctions and are not likely to be applicable to long-lived species. The model assumes that local extinction risks are constant over time and is not applicable to species that are increasing or decreasing in time.

Treatment of Uncertainty — HIGH — The model simulates variability in patch occupancy. It has special features for sensitivity analysis to incorporate uncertainties.

Degree of Development and Consistency — HIGH — The program is simple to use and comes with a manual that includes a tutorial.

Ease of Estimating Parameters — LOW — The model parameters for each population (patch) are probability of local extinction, number of emigrants from the patch per year, and number of immigrants needed to establish (with 50% probability) a new population. These parameters are difficult to estimate given typical data sets.

Regulatory Acceptance — LOW — No information on the model's regulatory acceptance is available. The model has not yet been published and thus is probably not used by a regulatory agency.

Credibility — LOW — No information is available on the number of users, which is likely to be small. The model has not yet been published, and only a beta version has been released.

Resource Efficiency — MEDIUM — Applying the model to a particular case does not require programming. In most cases, available data are not sufficient, and additional data collection is necessary.

DISCUSSION AND RECOMMENDATIONS

Several of the metapopulation models reviewed, in particular RAMAS Metapop/GIS and VORTEX, have high realism, relevance, and flexibility and can be used for risk assessment without additional development (Table 7.2). Metapopulation models have been applied to a variety of species, including representative insects, frogs, birds, and mammals (Table 7.3). Brook et al. (2000) applied several existing models to 21 populations. The results both validated the predictions of these models by comparing their results with observations and showed that models developed by using different software gave similar results when used with the same data sets.

Further development will make these models more suitable for metapopulation-level risk analysis. To make the available metapopulation models more useful for chemical risk assessment, functions could be added such as exposure response curves to model effects of toxic chemicals. Below, we discuss several other potential developments that will enhance metapopulation models.

One potential development is linking a metapopulation model to a GIS-based model that predicts future changes in species habitat. RAMAS Metapop GIS already allows a time series of habitat maps to be input as the basis of the spatial structure of the metapopulation. This feature can be further developed to allow a direct link with a landscape-level model that predicts, for example, the changes in the vegetation structure that result from succession and disturbances. The same approach can be used to link a metapopulation model with a hydrodynamic model that predicts variables such as flow, temperature, and spread of chemicals in an aquatic environment. Such predictions can be used to calculate the population-level parameters (carrying capacity, survival, fecundity, etc.) in each population. The final product would be an integrated model that combines the physical, chemical, and biological aspects of a system in a risk analysis framework.

Another potential development involves adding the features of a structured model to allow sex structure. Matrix-based metapopulation models (such as RAMAS Metapop) can be expanded to model both sexes by (1) allowing the user to input different matrices for males and females and (2) adding options and parameters for a limited set of mating systems. Such development will make these models more applicable to species in which behavioral or physiological characteristics of both sexes may be limiting to population growth.

Most models reviewed in this section were originally developed for terrestrial species, with the notable exception of ALFISH, which was specifically developed for fish populations in the Everglades. Although some other models, such as RAMAS Metapop, have been applied to aquatic organisms, certain additional features would make such models more applicable to species in freshwater and aquatic environments. These features include different types of density dependence and catastrophes affecting dispersal rates (e.g., a rare, large dispersal after a 10-year flood in a river fish metapopulation).

Table 7.2　Evaluation of Population Models — Metapopulation Models

Model	Reference	Evaluation Criteria								
		Realism	Relevance	Flexibility	Treatment of Uncertainty	Degree of Development	Ease of Estimating Parameters	Regulatory Acceptance	Credibility	Resource Efficiency
Occupancy — incidence function	Hanski and Gilpin (1991); Hanski (1994)	♦	♦	♦♦	♦♦♦	♦♦♦	♦♦	♦♦	♦♦	♦
Occupancy — state transition	Verboom et al. (1991); Sjögren-Gulve and Ray (1996)	♦	♦	♦♦	♦♦♦	♦♦♦	♦♦	♦♦	♦♦	♦
RAMAS Metapop and RAMAS GIS	Akçakaya et al. (1995); Kingston (1995); Akçakaya (1998a,b)	♦♦♦	♦♦♦	♦♦♦	♦♦♦	♦♦♦	♦♦♦	♦♦♦	♦♦♦	♦♦♦
VORTEX	Lacy (1993); Lindenmayer et al. (1995)	♦♦♦	♦♦♦	♦♦	♦♦♦	♦♦♦	♦♦	♦♦	♦♦♦	♦♦♦
ALFISH	DeAngelis et al. (1998a,b,c)	♦♦	♦♦♦	♦	♦	♦	♦♦	♦♦♦	♦	♦
ALEX	Possingham and Davies (1995)	♦♦	♦♦♦	♦♦	♦♦♦	♦♦	♦♦	♦♦	♦♦	♦♦
Meta-X	UFZ and OFFIS (2000)	♦	♦	♦♦	♦♦♦	♦♦♦	♦	♦	♦	♦♦

Note:

♦♦♦ - high
♦♦ - medium
♦ - low

Table 7.3 Applications of Metapopulation Models

Model	Species/Ecosystem	Location/Population	Reference
Occupancy — state transition	Pool frog (*Rana lessonae*)	Sweden	Sjögren-Gulve and Ray (1996)
	European nuthatch (*Sitta europaea*)	Denmark	Verboom et al. (1991)
Occupancy — incidence function	Butterfly (*Melitaea cinxia*)	Finland	Hanski et al. (1996)
	Tree frog (*Hyla arborea*)	The Netherlands	Vos et al. (2000)
RAMAS GIS	Bush cricket (*Metrioptera bicolor*)	Southern Sweden	Kindvall (1995, 1999)
	California gnatcatcher (*Polioptila californica*)	Orange County, California	Akçakaya and Atwood (1997)
	Helmeted honeyeater (*Lichenostomus melanops cassidix*)	Victoria, Australia	Akçakaya et al. (1995)
	Florida scrub-jay (*Aphelocoma coerulescens*)	Florida	Root (1998)
	Northern spotted owl (*Strix occidentalis*)	Washington, Oregon, California	Akçakaya and Raphael (1998)
RAMAS Metapop	Zebra mussel	Mississippi River	Akçakaya and Baker (1998)
	Land snail (*Arianta arbustorum*)	Switzerland	Akçakaya and Baur (1996)
	Land snail (*Tasmaphena lamproides*)	Northwest Tasmania	Regan et al. (1999)
	California spotted owl (*Strix occidentalis*)	Southern California	LaHaye et al. (1994)
	Golden-cheeked warbler (*Dendroica chrysoparia*)	Texas	USFWS (1996a)
	Black-capped vireo (*Vireo atricapillus*)	Texas	USFWS (1996b)
	Cottontail rabbit (*Sylvilagus floridanus*)	New England	Litvaitis and Villafuerte (1996)
	Rhizomatous shrub (*Banksia goodii*)	Albany, southwestern Australia	Drechsler et al. (1999)
VORTEX	Lord Howe Island woodhen (*Tricholimnas sylvestris*)	Lord Howe Island, Australia	Brook et al. (1997)
	Leadbeater's possum (*Gymnobelideus leadbeateri*)	Central Highlands, Victoria, Australia	Lindenmayer et al. (1993); Lindenmayer and Lacy (1995a)
	Mountain brushtail possum (*Trichosurus caninus*)	Central Highlands, Victoria, Australia	Lindenmayer and Lacy (1995b); Lacy and Lindenmayer (1995)
ALFISH	Various: small and large fish as food source for higher trophic levels	Everglades and Big Cypress Swamp, Florida	DeAngelis et al. (1998); Gaff et al. (2000)
ALEX	Leadbeater's possum	Southeastern Australia	Lindenmayer and Possingham (1996)
	Greater bilby (*Lagotis macroti*)	Australia	Southgate and Possingham (1995)
	Gliding marsupial (*Petaurus australis*)	Australia	Goldingay and Possingham (1995)
Meta-X	No known applications		
Metapopulation/toxicant model	Hypothetical	Hypothetical	Spromberg et al. (1998)

CHAPTER 8

Ecosystem Models — Food Webs

Steve Carroll

A *food web* is a description of feeding relationships or predator–prey relationships among all or some species in an ecological community. The simplest possible food-web model is a two-species predator–prey model. A food-web model can be as complicated as the modeler chooses, with as many species and feeding relationships as are deemed important. However, as food-web models get more complicated, model uncertainty increases.

Food-web models are important for at least two reasons. First, any given species generally interacts with other species in feeding relationships, either feeding on other species, being fed upon by other species, or both. Second, a receptor of concern in an ecological risk assessment may be exposed to toxic chemicals by ingesting a lower trophic-level species. Therefore, an evaluation of food-web linkages forms the basis for identifying key exposure pathways for bioaccumulative chemicals.

Endpoints for food-web models include:

- Abundances of component species in the food web
- Biomass of component species
- Species richness (i.e., number of species)
- Trophic structure (e.g., food-chain length, dominance)

We review food-web models and computer programs that implement them. For the purposes of this review, predator–prey models were collapsed into one category because considerable argument still exists about how a predator–prey system should be modeled and because the ratings were the same across predator–prey models. We review the following food-web models (Table 8.1):

- Predator–prey models (Lotka 1924; Volterra 1926; Watt 1959; Holling 1959, 1966; Ivlev 1961; Hassell and Varley 1969; Gallopin 1971; DeAngelis et al. 1975; Arditi and Ginzburg 1989)
- Population-dynamic food-chain models (Spencer et al. 1999)
- RAMAS ecosystem (Spencer and Ferson 1997a,c; Spencer et al. 1999)
- Populus (Alstad et al. 1994a,b; Alstad 2001)
- Ecotox (Bledsoe and Megrey 1989).

1-56670-574-6/01/$0.00+$1.50

Table 8.1 Internet Web Site Resources for Food-Web Models

Model Name	Description	Reference	Internet Web Site
Population-dynamic food-chain models	A combination of predator–prey models (in differential equation form) with models of the dynamics of a toxic chemical	Spencer et al. (1999)	http://www.ramas.com/
Predator–prey models	Models that describe the dynamics between a single predator species and a single prey species	Lotka (1924); Volterra (1926); Watt (1959); Holling (1959, 1966); Ivlev (1961); Hassel and Varley (1969); Gallopin (1971); DeAngelis et al. (1975); Arditi and Ginzburg (1989)	http://www.tiem.utk.edu/~mbeals/ predatorprey.html http://www.cbs.umn.edu/class/spring2000/biol/ 3407/lectures/ pred_prey_theory/pred_prey_theory.html
RAMAS Ecosystem	Software for food-web modeling incorporating the effects of toxic chemicals	Spencer and Ferson (1997a,c); Spencer et al. (1999)	http://www.ramas.com/
Populus	Software for population and food-web modeling	Alstad et al. (1994a,b); Alstad (2001)	http://www.cbs.umn.edu/populus/
Ecotox	Software for food-web modeling incorporating the effects of toxic chemicals	Bledsoe and Megrey (1989)	http://www.wiz.unikassel. de/model_db/mdb/ecotox.html

PREDATOR–PREY MODELS

A predator–prey model is the fundamental ingredient in a food-web model. Numerous predator–prey models exist (Lotka 1924; Volterra 1926; Watt 1959; Holling 1959, 1966; Ivlev 1961; Hassell and Varley 1969; Gallopin 1971; DeAngelis et al. 1975; Arditi and Ginzburg 1989), several of which contradict each other.

The Lotka–Volterra equations (Lotka 1924; Volterra 1926) describe a commonly cited, basic predator–prey model. The rate of change of the prey population is described by:

$$dp/dt = rp - cPp$$

where p is the number (or density) of prey, P is the number (or density) of predators, r is the prey's per capita exponential growth rate, c is a constant expressing the efficiency of predation, and t is time. The rate of change of the predator population is described by:

$$dP/dt = acPp - mP$$

where P is the number (or density) of predators, p is the number (or density) of prey, a is the efficiency of conversion of food to growth, c is a constant expressing the efficiency of predation, m is a constant representing the mortality rate of the predator, and t is time. In this simple model, each population is limited by the other. In the absence of predators, the prey population increases exponentially (by the Malthusian growth rate, r). In the absence of prey, the predator population decreases exponentially (by the mortality rate, m). There is a single nonzero equilibrium point (calculated by setting the two equations above to zero and solving for P and p), where

$$P = r/c$$

and

$$p = m/ac$$

On a phase diagram (P plotted against p), these equations define isoclines for the predator and prey populations, respectively. Figure 8.1 shows these isoclines and regions of positive and negative growth for the predator and prey populations. When predator abundance is below some critical level, the prey population increases. When predator abundance is above the critical level, the prey population decreases. Similarly, when prey abundance is above some critical level, the predator population increases. When prey abundance is below some critical level, the predator population decreases. Other than the case represented by the single equilibrium point where the isoclines cross (Figure 8.1), the population sizes of predator and prey oscillate at an amplitude whose magnitude depends on the initial conditions. Two examples of trajectories for oscillating populations are shown in Figure 8.1.

Because the Lotka–Volterra model does not include competition within the predator and prey populations, they are very unrealistic. Adding a self-limitation term to the prey equation (as in the logistic growth model) dampens the oscillations. Jørgensen et al. (1996) and other authors cited describe predator–prey models that incorporate self-limiting terms as well as functional responses of predators to prey density (Holling 1959, 1966). However, no consensus exists on how to model the simple two-species predator–prey system. The kind of model used may depend on the biology of the two species and the dynamics of their interaction.

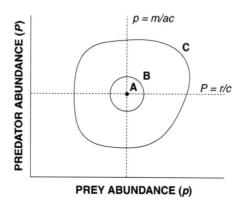

A = Single stationary equilibrium point when the
      abundances of predator and prey are non-zero
B = Oscillatory system with small amplitude
C = Oscillatory system with large amplitude
a = Efficiency of conversion of food to growth of the predator
c = Efficiency of predation
m = Mortality rate of predator population
r = Per capita exponential growth rate of prey population

Figure 8.1   Predator–prey relationships in the Lotka–Volterra model.

Realism — MEDIUM — For any given predator–prey model, the extent to which the model assumptions are realistic with respect to the ecology of the system is unclear.

Relevance — MEDIUM — Potential endpoints include the population size of the predator and prey only. These basic predator–prey models do not include functions for explicit modeling of toxic chemical effects.

Flexibility — LOW — For any given predator–prey model, the user is constrained to accept the assumptions of that particular model when little consensus exists with respect to such assumptions. Species with different life histories can be modeled, but population structure is not explicit in the model.

Treatment of Uncertainty — LOW — Uncertainty is not incorporated in these predator–prey models.

Degree of Development and Consistency — MEDIUM — Several software programs implement these models. However, it is relatively difficult to understand the workings of the model. A lack of consistency characterizes predator–prey models.

Ease of Estimating Parameters — MEDIUM — Obtaining parameter estimates for two species simultaneously is relatively difficult. In particular, feeding rates are difficult to obtain. However, model parameters are generally intuitive and can be interpreted biologically.

Regulatory Acceptance — LOW — To our knowledge, these models are not used by any regulatory agency, except as part of more complex ecosystem and landscape models.

Credibility — HIGH — Predator–prey models are well known within academia. Many applications of the models have been made.

Resource Efficiency — MEDIUM — Application of these models requires no programming because software is available. In most cases, additional data must be collected.

POPULATION-DYNAMIC FOOD-CHAIN MODELS

These models are constructed by using predator–prey models (in differential equation form) with equations modeling the dynamics of a toxic chemical. Population-dynamic food-chain models (Spencer et al. 1999, 2001) can use as building blocks any of the various predator–prey models (e.g., Lotka–Volterra, Holling type II, or ratio-dependent). These models are constrained to be food chains in which each predator has only one prey and each prey has only one predator.

Realism — MEDIUM — These models incorporate processes that are known to be important, such as predation, trophic transfer of toxic chemicals, and sorption (in the case of aquatic organisms). However, they necessarily incorporate at least one formulation of a predator–prey model, although little consensus exists on which, if any, predator–prey model is the most appropriate.

Relevance — HIGH — Potential endpoints include the expected population size of all of the species involved, the concentration of chemicals in all species, and the concentration of chemicals in the environment. All of these endpoints are useful in an ecotoxicological assessment. Chemical toxicity is modeled explicitly.

Flexibility — HIGH — Parameters are species-specific and predator–prey system-specific. Any predator–prey model can be incorporated, and different dose–response functions can be used.

Treatment of Uncertainty — HIGH — The models incorporate both measurement of uncertainty and natural variability in population parameters and interaction parameters (e.g., feeding rate).

Degree of Development and Consistency — HIGH — A well-developed software program (RAMAS Ecosystem) implements this type of model.

Ease of Estimating Parameters — MEDIUM — Several parameters must be estimated, including growth rate of the prey, death rate of the predator, feeding rate, and parameters regarding toxic chemical dynamics. Obtaining all of these parameters can be difficult. However, model parameters are intuitive and can be easily interpreted.

Regulatory Acceptance — LOW — To our knowledge, the model is not used by any regulatory agency.

Credibility — LOW — The model type is relatively novel. Few applications of the model have been made.

Resource Efficiency — MEDIUM — Application of the model requires no programming because software is available. However, in most cases, the available data are not sufficient; parameters may have to estimated in an *ad hoc* fashion.

RAMAS ECOSYSTEM

RAMAS Ecosystem is a population-dynamic trophic model that directly incorporates the effects of toxic chemicals (Spencer and Ferson 1997a, c). For example, the user can define the initial concentration of the toxic chemical, its input rate into the environment, its loss rate, the organism's uptake rate of the toxic chemical, the organism's elimination rate, and a dose–response curve that specifies mortality over a range of toxic chemical doses. RAMAS Ecosystem contains three different models of the predator–prey interaction: the classical Lotka–Volterra model, the Holling type II model, and the ratio-dependent model. The user can build a food web (or a simple food chain), any species of which can be directly affected by a toxic chemical. One can also investigate indirect effects of toxic chemicals by, for example, allowing only a prey species to be directly affected by a toxic chemical and noting the effects on its predator. Elements of RAMAS Ecosystem are similar to RAMAS Ecotoxicology (see Chapter 5, Population Models — Life-History Models, RAMAS Age, Stage, Metapop, or Ecotoxicology).

Realism — MEDIUM — Because applying this program depends on using a specific predator–prey model, the extent to which the model assumptions are realistic with respect to the ecology of the system is unclear.

Relevance — HIGH — Potential endpoints include the expected population size of any population in the model, risk of decline, risk of extinction, and expected crossing time (the time at which the population is expected to either exceed or to decrease to less than a given size). All of these endpoints are potentially useful in ecotoxicological assessments. The model easily accommodates modeling of toxic chemical effects.

Flexibility — HIGH — The number of species and trophic interactions are user defined. One of three predator–prey models and three dose–response functions can be chosen.

Treatment of Uncertainty — HIGH — Both ignorance and natural variability can be incorporated.

Degree of Development and Consistency — HIGH — No programming is required to use the model. The program is user friendly and has a graphic interface. A detailed, clearly written user's manual complements the program.

Ease of Estimating Parameters — MEDIUM — Obtaining parameter estimates for several species simultaneously is relatively difficult. In particular, feeding rates are difficult to obtain. However, model parameters are generally intuitive and can be interpreted biologically.

Regulatory Acceptance — LOW — To our knowledge, the model is not used by regulatory agencies.

Credibility — LOW — Few applications of this program have been made.

Resource Efficiency — HIGH — Application of the model requires no programming. In some cases, data must be collected; in other cases, available data are sufficient.

POPULUS

Populus models a wide variety of ecological interactions by using either differential or difference equations (Alstad et al. 1994a,b; Alstad 2001). Using Populus, one can model a predator–prey system, a food chain, or a general food web. For the predator–prey subprogram, two predefined predator–prey models are available (including the Lotka–Volterra model). One can also use the multiple-species subprogram and define a different predator–prey model. However, the program does not explicitly model the effects of toxic chemicals (i.e., no inputs related to toxic chemical concentration, uptake rates by organisms, etc. are provided). To do so, one would have to run the model with parameters measured without toxic chemicals and compare the results with those from a run of the model using parameters measured with the toxic chemicals. The difference between the results of the two runs would represent the predicted (or simulated) effect of the toxic chemicals.

Realism — MEDIUM — Because this program depends on a specific predator–prey model, the extent to which the model assumptions are realistic with respect to the ecology of the system is unclear.

Relevance — MEDIUM — Possible endpoints include the expected population size of any species in the model. Although the model does not explicitly address toxic chemical effects, several parameters in the model could be adjusted to implicitly model toxicity.

Flexibility — HIGH — The number of species and trophic interactions are user defined. The user can define use of any predator–prey model.

Treatment of Uncertainty — LOW — No treatment of uncertainty is included in this program.

Degree of Development and Consistency — HIGH — The program is easy to use, and graphic outputs are easily obtained. Each model component is well explained in a help file.

Ease of Estimating Parameters — MEDIUM — Obtaining parameter estimates for several species simultaneously is relatively difficult. In particular, feeding rates are difficult to obtain. However, model parameters are generally intuitive and can be interpreted biologically.

Regulatory Acceptance — LOW — To our knowledge, the model is not used by any regulatory agency.

Credibility — LOW — Few applications of this model exist.

Resource Efficiency — MEDIUM — Application of the model may require programming differential equations for a predator–prey model. In some cases, data must be collected; in other cases, available data are sufficient.

ECOTOX

Ecotox is a DOS-based program that explicitly models the effects of toxic chemicals within a food web or a food chain (Bledsoe and Yamamoto 1996). The software implements a bioenergetic food-web model, which is outlined in Bledsoe and Megrey (1989). The model is quite complicated and exclusively uses the predator–prey model of Holling (1966). Using the software requires that the user create a somewhat complicated input file or modify an existing one in a separate word-processing program; a manual accompanying the software describes how to do so.

Ecotox uses differential equations to simulate the dynamics of the energetics (weight as carbon content) and age structure of species populations. Influences on the dynamics of populations result from changes in fecundity linked to food availability and in mortality linked to predation or nutritional status of individuals (i.e., healthy or starved). Multiple toxic chemicals may be transferred through the food web as a result of dietary, dermal, and respiratory exposure, and Ecotox tracks the body burden of contaminants at each trophic level. The program simulates direct mortality due to acute and chronic exposures and reduction in foraging and reproductive rates. Effects of multiple toxicants are linearly additive by default, but nonlinear interactions may be simulated by adding appropriate mechanisms.

Realism — MEDIUM — Because this program depends on a specific predator–prey model, the extent to which the model assumptions are realistic with respect to the ecology of the system is unclear.

Relevance — MEDIUM — Possible endpoints include the expected population size of any species in the model. Although the model does not explicitly address toxic chemical effects, several parameters in the model could be adjusted to implicitly model toxicity.

Flexibility — HIGH — The number of species and trophic interactions are user defined. The dynamics and effects of toxic chemicals are explicitly modeled.

Treatment of Uncertainty — LOW — No treatment of uncertainty is incorporated.

Degree of Development and Consistency — MEDIUM — Use of this model requires some low-level programming. Documentation explaining how to do so is sufficient.

Ease of Estimating Parameters — MEDIUM — Obtaining parameter estimates for several species simultaneously is relatively difficult. In particular, feeding rates are difficult to obtain. However, model parameters are generally intuitive and can be interpreted biologically.

Regulatory Acceptance — LOW — To our knowledge, the model is not used by any regulatory agency.

Credibility — LOW — Few applications of this model exist.

Resource Efficiency — MEDIUM — Some programming is necessary to use this program. In some cases, data must be collected; in other cases, available data are sufficient.

DISCUSSION AND RECOMMENDATIONS

Although the concept of the food web has proved very useful in basic ecology as well as ecological risk assessment, a general theory of food-web structure is not well developed (for a discussion of this topic, see Lawton 1999). Modeling of complex food webs by using individual species is therefore extremely difficult. Despite ongoing debates among ecologists about how to model a predator–prey system, considerable progress has been made in modeling the predator–prey system since the first models in the early 1900s. Thus, simple predator–prey models can be applied to ecological risk assessment problems. Extension to a food chain or a relatively simple food web is also practical. Indeed, food-web models have already been applied in basic ecological research and chemical risk assessments (Tables 8.2 and 8.3).

On the basis of our evaluation (Table 8.2), we recommend further testing and development of the RAMAS Ecosystem and the Populus food-web models. Adding the capability to model food webs in a spatially explicit approach would enhance both of these models. Incorporation of spatial heterogeneity is known to be important in population dynamics and is also important in the dynamics of toxic chemicals. For example, a toxic chemical may have different concentrations at different locations and thus affect populations differently across space. The capability to model the effects of multiple toxic chemicals should be added to both RAMAS Ecosystem and Populus.

Table 8.2 Evaluation of Ecosystem Models — Food-Web Models

Model	Reference	Realism	Relevance	Flexibility	Treatment of Uncertainty	Degree of Development	Ease of Estimating Parameters	Regulatory Acceptance	Credibility	Resource Efficiency
								Evaluation Criteria		
Predator–Prey	Lotka (1924); Volterra (1926); Watt (1959); Holling (1959, 1966); Ivlev (1961); Hassell and Varley (1969); Gallopin (1971); DeAngelis et al. (1975); Arditi and Ginzburg (1989)	♦♦	♦♦	♦	♦	♦♦	♦♦	♦	♦♦♦	♦♦
Population-dynamic food-chain	Spencer et al. (1999)	♦♦	♦♦♦	♦♦♦	♦♦	♦♦	♦♦	♦	♦	♦♦
RAMAS Ecosystem	Spencer and Ferson (1997a,c); Spencer et al. (1999)	♦♦	♦♦♦	♦♦♦	♦♦	♦♦	♦♦	♦	♦	♦♦♦
Populus	Alstad et al. (1994a,b); Alstad (2001)	♦♦	♦♦	♦♦♦	♦	♦♦	♦♦	♦	♦	♦♦
Ecotox	Bledsoe and Megrey (1989)	♦	♦♦	♦♦♦	♦	♦	♦♦	♦	♦	♦♦

Note: ♦♦♦ - high   ♦♦ - medium   ♦ - low

Table 8.3 Applications of Food-Web Models

Model	Species	Location/Population	Reference
Predator–prey	*Didinium* and *Paramecium*	Laboratory	Harrison (1995)
	Algae and *Daphnia*	Laboratory	Spencer and Ferson (1997c)
	Zooplankton and cyanobacteria	Freshwater (data compiled from other studies)	Gragnani et al. (1999)
	Plankton	Theoretical	Jenkinson and Wyatt (1992) Tikhonova et al. (2000)
	Copidium campylum and *Alcaligenes faecalis*	Laboratory	Sudo et al. (1975)
	Spiders	Knox County, Tennessee	Riechert et al. (1999)
	Crops, insects	Laboratory	Trumper and Holt (1998)
	Zooplankton and fish	Freshwater (data compiled from other studies)	Ramos-Jiliberto and Gonzalez-Olivares (2000)
	Spotted seatrout (*Cynoscion nebulosus*) and pink shrimp (*Penaeus duorarum*)	Biscayne Bay, Florida	Ault et al. (1999)
	Fish and humans	Lake Erie	Jensen (1991)
	Microtine rodents	Northern Europe	Hanski et al. (1991) Hanski and Korpimaki (1995)
	Boreal rodents	Boreal and arctic regions	Hanski et al. (1993)
	Elephant and trees	Africa	Swart and Duffy (1987); Duffy et al. (1999)
Population-dynamic food-chain	*Dreissena polymorpha* and calanoid copepods	Lake Erie, Mediterranean Sea	Spencer et al. (1999)
	Mytilus edulis	San Francisco Bay	Spencer et al. (2001)
RAMAS Ecosystem	Red-tailed hawk	Illinois	Long et al. (1997) Katona et al. (1997)
Populus	No known applications beyond classroom use		Alstad et al. (1994a,b); Alstad (2001)
Ecotox	Wetland food web	Kesterson Wildlife Refuge site in the San Joaquin Valley	Bledsoe and Yamamoto (in preparation)
	Walleye pollock (*Theragra chalcogramma*) and Pacific halibut (*Hippoglossus stenolepis*)	Alaska/Pacific shorelines	Bledsoe and Megrey (1989)

CHAPTER 9

Ecosystem Models — Aquatic

Steven M. Bartell

Aquatic ecosystem models are defined here as spatially aggregated* models that represent biotic and abiotic structures in combination with physical, chemical, biological, and ecological processes in rivers, lakes, reservoirs, estuaries, or coastal ecosystems. Aquatic ecosystem models have a relatively long history of development and have been applied to a variety of freshwater, estuarine, and marine systems. Among the models selected for evaluation, a small subset was originally developed to assess ecological impacts and risks posed by toxic chemicals in aquatic ecosystems (Bartell et al. 1988, 1992, 1999; Hanratty and Stay 1994; Park 1998). Only one model (IFEM**) fully integrates exposure and effects assessments in a probabilistic framework (Bartell et al. 1988).

The development of detailed, dynamic models of aquatic ecosystems represents a relatively recent advance in quantitative ecology compared with other ecological modeling efforts (e.g., scalar population models). Early aquatic system models date at least to Riley et al. (1949) and Riley (1965), who were interested in a quantitative description of plankton dynamics in the western North Atlantic Ocean. By the late 1960s, several aquatic ecological models had been derived, primarily to examine hypotheses concerning the plankton populations growing in a dynamic physical and chemical environment. Patten (1968) noted the development of several hundred models of plankton interactions by the late 1960s. Perhaps the first comprehensive biotic–abiotic mathematical descriptions of the physical, chemical, biological, and ecological aspects of production dynamics in aquatic ecosystems resulted from the International Biological Programme (IBP) (McIntosh 1985). Detailed computer simulation models were constructed for Lake Wingra, a small, eutrophic lake in Madison, Wisconsin (e.g., MacCormick et al. 1975), and for Lake George, New York (Park et al. 1974). Following these earlier models, mathematical and computer simulation models have been developed for nearly all imaginable aquatic ecosystems, including streams, rivers, reservoirs, lakes, the Great Lakes, estuaries, coastal oceans, coral reefs, and open oceans.

Aquatic ecosystem models are as diverse in structure and purpose as the set of underlying motivations for their construction. The early IBP models focused on simulating the detailed

---

* Many aquatic ecosystem models have some spatial structure consisting of a minimal number of large habitat compartments (e.g., dividing a lake into an upper mixed layer called the epilimnion and a lower layer called the hypolimnion). Within these compartments, which in reality may be spatially heterogeneous, the ecosystem model assumes homogeneity and predicts average values for state variables. To distinguish models that were initially designed with much more detailed or "gridded" spatial structure from ecosystem models, we term the former *landscape models* and treat them separately in Chapter 11.

** IFEM and CASM are proprietary products of Steven M. Bartell. Trademark registration is in process.

1-56670-574-6/01/$0.00+$1.50

production of aquatic organisms in relation to eutrophication issues. Extensions of these modeling approaches were developed to simulate the flow of energy and/or the cycling of materials through freshwater and marine systems of interest. Other aquatic models emphasized the implications of herbivore–grazer interactions or predator–prey relationships for describing population dynamics, community structure, and system stability. These aquatic ecosystem models invariably included explicit formulations for the abiotic components of aquatic systems (e.g., nutrient concentrations, sediments, physical mixing), as well as differently structured aquatic food webs. To date, no generalized theory concerning the level of structural detail required for accurate description of aquatic ecosystem dynamics has been developed.

Although diverse in their ecological structure, the aquatic models are commonly formulated as sets of coupled differential (or difference) equations on the basis of mass balance of inputs and outputs. The equations have ranged from simple linear equations with constant coefficients, to linear equations with nonlinear terms, to highly nonlinear equations. The most commonly modeled ecological currency has been biomass, carbon, and energy (e.g., joules). More recent modeling advances have attempted to incorporate some of the individual-oriented models (e.g., fish, zooplankton) into more comprehensive simulations of aquatic ecosystems. Earlier attempts at modeling aquatic ecosystems were quite simple in their spatial structure (e.g., completely mixed water column, "two-box" layering of epilimnion and hypolimnion), although models of larger lakes and estuaries might represent the system with several connected spatial regions. Hydrodynamic models are commonly used to provide spatially or temporally varying inputs (current velocities, mixing rates, water temperature, nutrient loadings) to aquatic ecosystem models. Recently, parallel processing computers have been used to develop and implement more spatially detailed, structured models of aquatic ecosystems (see Chapter 11, Landscape Models — Aquatic and Terrestrial).

The primary endpoints for aquatic ecosystem models include:

- Abundance of individuals within species or trophic guilds
- Biomass
- Productivity
- Food-web endpoints (species richness, trophic structure)

We review the following aquatic ecosystem models (Table 9.1):

- Estuarine
  - Transfer of impacts between trophic levels model, an estuarine model to evaluate indirect effects of power-plant entrainment of plankton (Horwitz 1981)
- Lake
  - AQUATOX (CLEAN), a lake/river model (Park et al. 1974; Park 1998; U.S. EPA 2000a,b,c)
  - ASTER/EOLE (MELODIA), a lake model (Salencon and Thebault 1994)
  - DYNAMO pond model, a solar-algae pond ecosystem model (Wolfe et al. 1986)
  - EcoWin, a lake model (Ferreira 1995; Duarte and Ferreira 1997)
  - LEEM (Lake Erie ecosystem model), a model specifically designed to evaluate management issues for Lake Erie (Koonce and Locci 1995)
  - LERAM (littoral ecosystem risk assessment model), a model of the vegetated nearshore zone of lakes (Hanratty and Stay 1994)
  - CASM (comprehensive aquatic system model), or modified SWACOM (standard water column model), lake/river models (DeAngelis et al. 1989; Bartell et al. 1992, 1999)
  - PC Lake, a model designed for evaluating general trends in lakes (Janse and van Liere 1995)
  - PH-ALA, a lake eutrophication model also known as *the Glumsø Lake model* (Jørgensen 1976; Jørgensen et al. 1981)
  - SALMO (simulation by means of an analytical lake model), a simple model designed to evaluate the effects of eutrophication (Benndorf and Recknagel 1982; Benndorf et al. 1985)
  - SIMPLE (sustainability of intensively managed populations in lake ecosystems), the Lake Ontario fisheries model (Jones et al. 1993)

Table 9.1 Internet Web Site Resources for Aquatic Ecosystem Models

Model Name	Description	Reference	Internet Web Site
Estuarine trophic model	An estuarine ecosystem model with a transfer of impacts between trophic levels	Horwitz (1981)	N/A
AQUATOX	An EPA-supported model directly applicable to assessing the effects of toxic chemicals in lakes, reservoirs, and rivers	Park et al. (1974); Park (1998); U.S. EPA (2000a,b,c)	http://www.epa.gov/waterscience/models/aquatox/
ASTER/EOLE	A hydroelectric reservoir model	Salencon and Thebault (1994)	N/A
DYNAMO	A solar-algae pond ecosystem model	Wolfe et al. (1986)	N/A
EcoWin	A lake model incorporating the effects of toxic chemicals	Ferreira (1995); Duarte and Ferreira (1997)	http://tejo.dcea.fct.unl.pt/ecowin/ http://eco.wiz.uni-kassel.de/model_db/mdb/ecowin.html
LEEM	A comprehensive ecosystem model for Lake Erie	Koonce and Locci (1995)	http://www.ijc.org/boards/letf/letfrept.html http://www.ijc.org/boards/cglr/modsum/heath.html http://www.epa.state.oh.us/oleo/lepf/sg16-95.html
LERAM/CATS-4	LERAM is an ecosystem model for risk assessment of littoral systems; CATS-4 is based on LERAM and incorporates the effects of toxic chemicals in aquatic and terrestrial systems	Hanratty and Stay (1994); Traas et al. (1998)	http://www.epa.gov/earth100/records/leram.html
CASM/Modified SWACOM	Comprehensive aquatic system models incorporating the effects of toxic chemicals	DeAngelis et al. (1989); Bartell et al. (1992, 1999)	http://www.u-hommen.de/Software/software.html
PC Lake	A one-dimensional lake model that can be integrated with CATS-4 to yield a model similar to AQUATOX	Janse and van Liere (1995)	N/A
PH-ALA	A lake eutrophication model used to evaluate wastewater treatment alternatives	Jørgensen (1976); Jørgensen et al. (1981)	http://www.wiz.uni-kassel.de/model_db/mdb/ph-ala.html
SALMO	A simple two-layer model of a lake	Benndorf and Recknagel (1982); Benndorf et al. (1985)	http://spree.wasser.tu-dresden.de/salmo.html
SIMPLE	A model to examine the implications of prey availability for competing piscivorous fish populations, which has been applied to Lake Ontario salmonid fisheries	Jones et al. (1993)	N/A
FLEX/MIMIC	A hierarchical lotic ecosystem model	McIntire and Colby (1978)	http://www.fsl.orst.edu/lter/data/models/strmeco.htm
IFEM	An integrated toxic chemical fates and effects model applied to lakes or rivers	Bartell et al. (1988)	N/A
INTASS	A general ecosystem model applicable to aquatic and terrestrial ecosystems	Emlen et al. (1992)	http://biology.usgs.gov/wfrc/jep.htm

Note: N/A - not available

- River
  - FLEX/MIMIC, a hierarchical lotic ecosystem model (McIntire and Colby 1978)
  - IFEM (integrated fates and effects model), a chemical fate and risk model (Bartell et al. 1988)
- General
  - INTASS (interaction assessment model), a general model applicable to aquatic and terrestrial ecosystems (Emlen et al. 1992)

TRANSFER OF IMPACTS BETWEEN TROPHIC LEVELS

Horwitz (1981) derived a model to examine the direct and indirect effects of entrainment of estuarine plankton in power-plant intake structures on the population dynamics of predators. The model describes carbon and nitrogen flows through 11 highly aggregated compartments representing the estuarine ecosystem of Chesapeake Bay.

The model consists of a set of coupled differential equations that describe the population dynamics of organisms that are entrained and the population dynamics of organisms that feed upon the entrained plankton populations. Horwitz (1981) based the model on a Lotka–Volterra approach with added terms for density dependence similar to those in the logistic model for self-limiting populations. He then extended the simple predator–prey model to food chains of three and four species, with self-limiting terms in the bottom trophic level, the top level, or all levels. The main physical forcing factors are temperature, day length, and isolation. The model simulation is based on daily time-steps.

The model demonstrated a consistent negative effect on the entrained populations. However, greater indirect negative impacts were observed on predators of the entrained populations under certain model scenarios. Thus, Horwitz (1981) concluded that single-species models may fail to incorporate indirect effects that are the main source of the greatest mortality associated with the stressor (in this case entrainment). The model also suggested that shifts in the diet of the predators toward detritus and benthic prey often compensated for the loss of entrained prey populations.

Realism — MEDIUM — The Horwitz (1981) model represents populations of plankton and planktonic predators. However, the model incorporates only a single limiting nutrient and does not comprehensively describe estuarine ecosystems.

Relevance — HIGH — The trophic components and endpoints included in the model are relevant to ecological risk assessment. The examination of direct and indirect effects of stressors (e.g., entrainment) is of high interest in ecological risk assessment. Although the model does not explicitly account for toxic chemical effects, several parameters could be adjusted by the user to implicitly model toxicity.

Flexibility — HIGH — The model structure and governing equations could be generalized to other estuarine ecosystems.

Treatment of Uncertainty — LOW — Horwitz (1981) does not report detailed sensitivity or uncertainty analyses for the model.

Degree of Development and Consistency — MEDIUM — The governing equations for the Horwitz (1981) model are similar to formulations that have been proven useful in estimating population dynamics.

Ease of Estimating Parameters — LOW — The model parameters are relatively few in number, and they can be interpreted biologically and ecologically. However, the necessary data are unlikely to be generally available for most site specific applications in chemical risk assessment.

Regulatory Acceptance — MEDIUM — The Horwitz (1981) model was not developed in response to specific regulatory issues, but the assessment of entrainment mortalities is of interest to some regulatory agencies (e.g., EPA).

Credibility — MEDIUM — The model captures some of the population dynamics of plankton and planktivorous predators. The model has not been widely published or used.

Resource Efficiency — MEDIUM — The limited structure of the Horwitz (1981) model suggests that it could be implemented for specific estuarine ecosystems.

AQUATOX

AQUATOX simulates the combined environmental fate and effects of pollutants, including nutrients, sediments, and organic chemical contaminants in streams, ponds, lakes, and reservoirs (Park 1998; U.S. EPA 2000a,b,c) (Figure 9.1). The model addresses potential impacts of stressors on phytoplankton, periphyton, submersed aquatic vegetation, zooplankton, zoobenthos, and several functionally defined fish populations (i.e., forage, game, and bottom fish). AQUATOX simulates important ecological processes, including food consumption, growth and reproduction, natural mortality, and trophic interactions. In addition to addressing acute and chronic toxicity, AQUATOX integrates the results of an environmental fate evaluation, including nutrient cycling and oxygen dynamics, toxic organic chemical phases and transformations (e.g., partitioning among water, biota, and sediments), and bioaccumulation through gills and the diet.

AQUATOX is a combination of algorithms from ecosystem models (e.g., CLEAN by Park et al. 1974), contaminant fate models (e.g., PEST by Park et al. 1982), and the ecotoxicological component from FGETS (Suárez and Barber 1995). AQUATOX was designed for interactive use and flexibility in application to new scenarios. The model reports changes in population biomass on a daily basis. Required input data include nutrient, sediment, and toxic chemical loadings to the waterbody, general site characteristics, properties of each organic toxicant, and biological characteristics of each plant and animal represented in the model.

AQUATOX consists of a set of coupled differential equations that are integrated using an adaptive time-step Runge–Kutta integration routine. The shape of the modeled aquatic system is approximated using idealized geometrical units to describe a pond, lake, reservoir, or stream. Thermal stratification in lakes and reservoirs is modeled in AQUATOX through the specification of a "two-box" epilimnion and hypolimnion. AQUATOX includes a Monte Carlo simulator to facilitate probabilistic risk estimation for aquatic resources.

Various EPA programs have sponsored the model (U.S. EPA 2000a,b,c), and the most recent versions are available on an EPA web site (http://www.epa.gov/waterscience/ models/aquatox/). EPA recently developed AQUATOX Version 2.00, which represents up to 20 chemicals simultaneously, up to 15 age classes for one fish species and two size classes for all other fish species, and 12 or more linked segments (including river channel reaches, backwater areas, and a stratified pond).* In a review of integrated modeling of eutrophication and organic contaminant fate and effects in aquatic ecosystems, Koelmans et al. (2001) concluded that AQUATOX is the most complete model of its type described in the literature.

> **Realism** — HIGH — AQUATOX is a mechanistic model that accounts for important biotic and abiotic interactions within and between several trophic levels and considers associated feedbacks.
>
> **Relevance** — HIGH — The model was developed as a management tool and designed to study the effects of nutrient enrichment and other perturbations on ecologically relevant components of aquatic ecosystems. AQUATOX includes functions representing the effects of toxic chemicals.
>
> **Flexibility** — HIGH — The format of the model is general enough to allow alternative formulations and applications to various site specific conditions. It is currently being applied to a river system (the Housatonic River in Connecticut).
>
> **Treatment of Uncertainty** — MEDIUM — The AQUATOX code includes Monte Carlo simulation capabilities, although it is unclear whether detailed sensitivity analyses have been performed.
>
> **Degree of Development and Consistency** — HIGH — The model has been programmed to facilitate new applications and scenario development and is available as commercial software with excellent technical support. AQUATOX has been validated with data from at least three water bodies, including a data set on PCB transfer in the food web of Lake Ontario (U.S. EPA 2000c).
>
> **Ease of Estimating Parameters** — LOW — AQUATOX has a relatively large parameter set, which means that extensive data are required to apply the model.

---

* Although AQUATOX was originally developed as an ecosystem model, this implementation could be considered a landscape model.

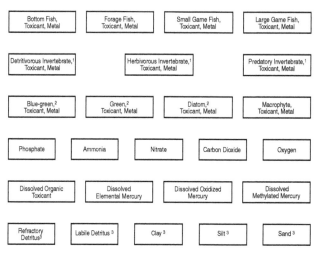

¹ Zooplankton or zoobenthos
² Phytoplankton or periphyton
³ Suspended and sedimented with organic toxicant and metal

Figure 9.1   Compartments (state variables) in AQUATOX. (From Park R.A. et al. 1995. AQUATOX, a general fate and effects model for aquatic ecosystems. Toxic Substances in Water Environments Proceedings, Water Environment Federation, Alexandria, VA. © Water Environment Federation. With permission.)

Regulatory Acceptance — MEDIUM — AQUATOX is used by EPA's Office of Toxic Substances but has no official regulatory acceptance or recommendation.

Credibility — HIGH — The model has been calibrated to a variety of aquatic ecosystems in specific applications. Several published accounts cover AQUATOX applications, and the number of potential users is high given that the model is accessible via the Internet.

Resource Efficiency — HIGH — AQUATOX was programmed for convenient and general application to aquatic ecosystems. The code and a comprehensive user's manual are freely available.

ASTER/EOLE (MELODIA)

Salencon and Thebault (1996) describe the MELODIA model of Pareloup Lake, a hydroelectric generating reservoir in France. ASTER is a biological model (i.e., it incorporates silica, phosphorus, diatoms, and nonsiliceous algae), which was coupled to EOLE, a hydrodynamic and thermal model. Each model is one-dimensional and describes the biological, hydrodynamic, and thermal changes vertically for a water column of specified depth. The two models were coupled to create MELODIA, which was calibrated to data collected in the reservoir. MELODIA was developed to examine lake ecosystem dynamics, particularly spring diatom production in the epilimnion and hypolimnion, in relation to the physical mixing characteristics and onset of stratification in the reservoir. The overall model is specified as set of coupled, partial differential equations. Parameter values were derived from extensive calibration to measurements recorded for Pareloup Lake. The model operates at a daily time scale for simulated periods of up to 5 years. It has been used to evaluate the environmental effects of reservoir management scenarios.

Realism — MEDIUM — ASTER is a multitrophic-level model with several representative species in each level. EOLE is a hydrodynamic, thermal, one-dimensional, vertical model of physical conditions, which assumes horizontal homogeneity. Together, they provide a moderate level of complexity and realism.

Relevance — HIGH — MELODIA was developed to provide a tool for management and decision-making concerning eutrophication of the lake under consideration. The species biomass endpoints are relevant to ecological risk assessment. Although the model does not explicitly account for toxic chemical effects, the user could adjust several parameters to implicitly model toxicity.

Flexibility — LOW — MELODIA was developed specifically for the lake under consideration and is not particularly adaptable to other systems.

Treatment of Uncertainty — LOW — Neither sensitivity analysis nor uncertainty analysis was reported for this model.

Degree of Development and Consistency — MEDIUM — MELODIA was developed by using a modular design that separates the biology, thermal properties, and hydrodynamics of Pareloup Lake into separate submodels that are then linked. The model output compares well with measured data. The model is well documented in the literature.

Ease of Estimating Parameters — MEDIUM — Estimation of approximately 44 parameters is needed to run ASTER and EOLE.

Regulatory Acceptance — MEDIUM — MELODIA was developed with the French Ministry for the Environment but is not likely to be used extensively by regulatory agencies.

Credibility — LOW — The model output captured phytoplankton blooms and collapses but not dynamics and did not capture zooplankton dynamics at all.

Resource Efficiency — MEDIUM — A moderate effort would be required to apply MELODIA to another reservoir. Site specific temperature and hydrodynamics data would also be required.

DYNAMO POND MODEL

Wolfe et al. (1986) describe a model of 2300-L fiberglass ponds used to culture blue tilapia (*Tilapia aurea*). The DYNAMO pond model includes fish, bacteria, algae, carbon dioxide, alkalinity, and dissolved oxygen, as well as nitrate, nitrite, and ammonia as state variables. Exogenous model inputs include values for sunlight, water exchange, aeration, fish stocking density, and fish feeding. The model has realistically simulated the ponds for several 100-day periods. The model was developed to help manage and optimize tilapia production in these small ponds. The model is coded in DYNAMO, a systems modeling platform. The computational time-step is determined by the DYNAMO simulation software in relation to the overall "stiffness" of the model equations. An annotated listing of the model code is appended to the Wolfe et al. (1986) model description.

Realism — MEDIUM — The DYNAMO pond model is based on observations made in a solar-algae pond, which supports a monoculture of fish, and bacteria and algae. Solar fluxes are four to five times higher than in a natural pond of similar depth; so photosynthesis, bacterial metabolism, and chemical activity are higher than normal, resulting in fish densities two to three times higher than those in a natural setting.

Relevance — HIGH — The model was developed to provide insight into the effects of pond management on water quality and the rate of fish growth in relationship to the level of algae present. Although the model does not explicitly account for toxic chemical effects, the user could adjust several parameters to implicitly model toxicity.

Flexibility — MEDIUM — The DYNAMO pond model was developed for specific experimental conditions, which are not found in natural ponds or lakes. The results are perhaps applicable to managed systems.

Treatment of Uncertainty — LOW — The model developers did not perform either sensitivity or uncertainty analysis, but the model could be implemented in such a format.

Degree of Development and Consistency — MEDIUM — The nature of the model structure and equations have been fairly well established in the ecological modeling literature. They were applied to a rather unusual ecological system in this case. The model has been validated for the solar-algae ponds.

Ease of Estimating Parameters — HIGH — The DYNAMO pond model's parameters can be estimated from data that might be expected to be collected from similar aquaculture systems. The parameters are directly interpretable.

Regulatory Acceptance — LOW — The DYNAMO pond model does not appear to have been developed for use by any regulatory agencies.

Credibility — MEDIUM — The authors were able to successfully reproduce maturation of the man-made ecosystem in three separate 100-day simulations. The DYNAMO pond model has not been widely used or documented extensively in the literature.

Resource Efficiency — LOW — Using the model requires knowledge and proficiency in DYNAMO and FORTRAN programming languages and requires resources for placing the model in an uncertainty analysis framework.

ECOWIN

EcoWin is an object-oriented approach to the modeling of aquatic ecosystems, including simulation of the water quality and ecology of rivers, lakes, estuaries, and coastal waters (Ferreira 1995; Duarte and Ferreira 1997). The modeling structure permits specification of physical (advective flows), chemical (nutrients, toxic chemicals), and biological (phytoplankton, zooplankton, benthic plants and animals, fish) components of aquatic systems. The model simulates the dynamics of the specified objects in up to three dimensions for a year by using daily time-steps.

EcoWin consists of a shell, which manages the model input and output, and a set of objects, including state variables and their interactions (those objects that perform the calculations). The basic underlying model structure is that of a compartmental or box-model. EcoWin was developed in Turbo Pascal for an MS-Windows environment and is now available as C++ EcoWin 2000. EcoWin consists of two fundamental groups of objects: one group consists of the ecological components specified in the model, and the second group provides for the interfacing among the various ecological components. EcoWin has thus far been implemented for the Tagus Estuary (Portugal), Carlingford Lough (Ireland), the Northern Adriatic Sea, Sanggou Bay (China), and the Azores Front (North Atlantic).

Realism — HIGH — The programming objects that have been defined in EcoWin to describe aquatic ecosystems emphasize structures and processes that are generally recognized as important for simulating the production dynamics of these systems.

Relevance — HIGH — The ecological model outputs from EcoWin include those endpoints that are routinely included in ecological risk assessments. The model explicitly accounts for toxic chemical effects.

Flexibility — HIGH — EcoWin was purposely designed by using an object-oriented framework to facilitate application to different aquatic ecosystems. It has been used to run zero-dimensional (time-varying only), one-dimensional (varying longitudinally), two-dimensional (varying areally),* and three-dimensional (areal and layered)* models.

Treatment of Uncertainty — LOW — The model as presented does not address uncertainty or perform sensitivity analyses. Such capability might easily be included as another class of objects that could be linked to the overall EcoWin modeling shell.

Degree of Development and Consistency — HIGH — EcoWin has been programmed for highly interactive use. The program operates in a Windows environment and permits parameter inputs through a commercial spreadsheet. The model outputs can be plotted or printed and copied easily into documents by using the Windows clipboard. The model has been developed over a 10-year period.

Ease of Estimating Parameters — HIGH — The ecological process approach to describing aquatic ecosystems provides EcoWin with parameters that have clear interpretations. Parameters are numerous but estimable from typical data available in site specific applications.

Regulatory Acceptance — LOW — The documentation on EcoWin (Ferreira 1995) does not mention a regulatory purpose, use, or recommendation.

---

* With sufficient spatial detail, such implementations of EcoWin would be considered landscape models.

Credibility — LOW — Fewer than ten published accounts of EcoWin applications exist. The number of EcoWin users is unknown but presumably small.

Resource Efficiency — HIGH — EcoWin is essentially an aquatic ecosystem modeling platform that has been designed and implemented to facilitate site specific applications. No new programming would be required for new applications of the model.

LEEM

Koonce and Locci (1995) describe a model developed to examine changes in Lake Erie fish species that might result from various combinations of nutrient loading, introduction of zebra mussels, and different fish management actions. LEEM accounts for changes in the biomass of 16 major game fish and forage fish species in Lake Erie in relation to nutrient loading, food-web dynamics, and human activities. The model also describes the accumulation of toxic contaminants by modeled biota.

LEEM is a component model consisting of population submodels run in parallel and linked by informational constraints (Sturtevant and Heath 1995). The model divides Lake Erie into three distinct basins. Within each basin, the model simulates the dynamics of nutrients, primary producers, zebra mussels, zooplankton, zoobenthos, and fish. Primary production is simulated for macrophytes, edible phytoplankton, inedible phytoplankton, edible benthic algae, and inedible benthic algae, each of which is distributed appropriately throughout the lake. Steady-state approximations between phosphorus loading and primary production are used. In addition to the implicit feedbacks through grazing of primary production, release of phosphates from grazers such as zebra mussels was incorporated (Sturtevant and Heath 1995). At the upper trophic levels, LEEM models 16 major game fish and forage fish species in Lake Erie on an annual basis by taking into consideration large-scale spatial and temporal heterogeneities. Each fish species is modeled as an age-structured population, accounting for well-known physiological and behavioral characteristics of each taxon.

The model has been programmed in Visual Basic and accommodates input and output through a user-friendly interface to Excel spreadsheets. LEEM simulates user-specified scenarios over periods of multiple years using an annual time-step. Koonce and Locci (1995) provide a detailed description of the state variables, model parameters, and computer code for LEEM.

Realism — MEDIUM — The model is a multitrophic-level model with several representative species at each level. The model addresses important biological feedback mechanisms, such as the implications of direct uptake and trophic transfer of bioaccumulative chemicals (e.g., PCBs) on long-term distribution of contaminants and interactions among nutrient loading and water clarity in determining changes in benthic community structure.

Relevance — HIGH — The model was developed to examine the effects of biological and chemical stressors on the Lake Erie ecosystem. The endpoints and the stressors modeled are very relevant to ecological risk assessment of toxic chemicals.

Flexibility — LOW — The model was developed specifically to address ecological problems in the Lake Erie ecosystem and is not easily adapted to other systems.

Treatment of Uncertainty — LOW — The capability to perform sensitivity and uncertainty analyses has not been incorporated into the model.

Degree of Development and Consistency — HIGH — The model has been developed by using commercially available software that runs in a combined spreadsheet and Visual Basic package. A run-time application and source code are available from the authors. Extensive testing and calibration with historical data sets show that LEEM successfully modeled fish populations and predicted the results of potential management efforts (Sturtevant and Heath 1995).

Ease of Estimating Parameters — MEDIUM — The parameter set did not appear unwieldy; the exact number of parameters was not determined. However, the nature of the listed parameters suggests that they have a clear ecological interpretation and might be estimated from data and information available for lakes.

Regulatory Acceptance — MEDIUM — LEEM was initiated by the International Joint Commission to aid in anticipating the effects of declining nutrient loading, invasion of zebra mussels, loading of toxic organic contaminants on fish populations, and other management issues of concern to the Lake Erie Task Force. LEEM was intended to serve as a framework for addressing these issues and as a tool for Lake Erie managers to evaluate possible management strategies.

Credibility — LOW — The authors state that work needs to be done to increase the credibility of the model.

Resource Efficiency — LOW — If the software is used to run the model, one must be able to run Visual Basic and Excel. An uncertainty analysis and sensitivity analysis framework is needed to run the model in a risk assessment context.

LERAM

Hanratty and Stay (1994) and Hanratty and Liber (1996) present an adaptation of CASM (see next model section) called LERAM to describe the impacts of pesticides on littoral zone ecosystems. LERAM is a compartmental model that simulates changes in the biomass of bacterioplankton and in multiple populations of phytoplankton, zooplankton, macrophytes, benthic invertebrates, and fish. LERAM also simulates daily changes in dissolved inorganic nitrogen, phosphorus, and silica, as well as dissolved oxygen.

The daily changes in the biomass of LERAM-modeled populations are determined by coupled bioenergetics-based differential equations. Primary production in the model is determined by daily values of incident light intensity, water temperature, and nutrient availability. LERAM uses the same sublethal toxic stress method used in CASM. LERAM has been implemented for chlorpyrifos and diflubenzuron. Comparisons of model output with empirical observations have proven that LERAM realistically simulates the effects of pesticides on littoral ecosystems. LERAM has also been programmed using difference equations in a Monte Carlo simulation for probabilistic risk estimation and sensitivity/uncertainty analysis.

Traas and colleagues (1998) describe a model called CATS-4 (contaminants in aquatic and terrestrial ecosystems-4) that is very similar to LERAM. Both CATS-4 and LERAM are bioenergetics-based models, but they differ somewhat in the details of the parameterization of physiological processes and in the way the effects of toxic chemicals are modeled. In LERAM (and in CASM, both of which are based on SWACOM), toxicity is expressed as a general stress syndrome, which is a linear extrapolation from a chemical's LC50 if we assume that all bioenergetic processes (e.g., growth, respiration) are affected. In CATS-4, Traas et al. (1998) used entire concentration–effect functions obtained from the results of 48-hour laboratory toxicity tests with mortality as the endpoint. Traas et al. (1998) propose that addition of the mortality due to chlorpyrifos in the model is sufficient and that other bioenergetic parameters remain unaffected by the insecticide.

Realism — HIGH — LERAM models a littoral zone of a generic aquatic system. The model aggregates species in various trophic categories (e.g., all diatoms without distinction).

Relevance — HIGH — Model endpoints include the biomass of all ecologically relevant components of a littoral ecosystem. LERAM has been used as a risk assessment tool to examine the effects of insecticides on an enclosed littoral zone ecosystem.

Flexibility — HIGH — LERAM was developed as a general framework for assessing pesticide effects on aquatic systems. Thus, it is acceptable for evaluating the effects of organic chemicals on a variety of aquatic systems.

Treatment of Uncertainty — HIGH — LERAM incorporates the capability for both sensitivity and uncertainty analyses.

Degree of Development and Consistency — HIGH — LERAM has been implemented as software and includes a self-contained Monte Carlo FORTRAN program. It has flexible, user-specified files for data input.

Ease of Estimating Parameters — MEDIUM — Derived from CASM, LERAM has a large number (~150) of parameters that must be estimated. All of these parameters have clear biological or toxicological meaning and can be estimated from laboratory or field data. However, such extensive data might not be routinely available for aquatic ecosystems.

Regulatory Acceptance — MEDIUM — The model was developed with EPA for pesticide assessments. However, LERAM has no formal regulatory standing or recommendation from EPA.

Credibility — LOW — The output from LERAM reasonably estimates the corresponding observations from the field, but the authors recommend some improvements in the model. The model can generally capture the effects of stressors on the littoral system under study. However, few published accounts of this model exist, and the current number of users is unknown but presumed to be fewer than 20.

Resource Efficiency — MEDIUM — LERAM exists as a FORTRAN code that can be run on UNIX workstations or PCs with commercially available FORTRAN software. Given the parameter estimations needed, and because the model has already been analyzed in a probabilistic framework, a moderate number of resources would be required to run the model.

CASM, A MODIFIED SWACOM

CASM consists of a graphic user interface coupled with a biological and ecological modeling framework that describes the growth of populations of aquatic plants and animals in surface water and sediments of rivers, lakes, and reservoirs. CASM extends the capabilities of SWACOM by including multiple populations of aquatic organisms characteristic of the littoral and benthic communities (DeAngelis et al. 1989; Bartell et al. 1992, 1999) (Figure 1.5). (See also the description of LERAM, which derives directly from CASM.) CASM includes multiple nutrients and can simulate time-varying concentrations of toxic chemicals. Like SWACOM, CASM was designed to provide risk managers with a tool for assessing the impacts and ecological risks posed by chemicals in aquatic ecosystems (Bartell et al. 1999).

CASM is implemented as a set of coupled differential equations based on a bioenergetics description of population dynamics. The model uses a daily time-step to simulate production dynamics on an annual time scale (although multiple-year simulations are possible). Like many other aquatic ecosystem models, CASM calculates the biomass of primary producers by using equations describing physiological processes such as photosynthesis, grazing, nonpredatory death, respiration, and so on. For consumer populations, consumption, egestion, nonpredatory death, respiration, and other processes are considered. The impacts (risks) posed by toxic chemicals can be measured at the population, community, or ecosystem levels in CASM. CASM has also been programmed as a set of coupled difference equations using FORTRAN in a self-contained Monte Carlo simulation for probabilistic risk estimation and numerical sensitivity and uncertainty analyses.

CASM has been implemented for a variety of rivers, lakes, and reservoirs. In a recent application, environmental data and possible exposure scenarios were used to estimate site specific ecological risks posed by organic pollutants, metals, and herbicides in Quebec aquatic ecosystems (Bartell et al. 1998).

Realism — HIGH — CASM considers important biological interactions and associated feedbacks at several trophic levels including trophic-level overlap of species, which allows functional redundancy to be tested at the system level.

Relevance — HIGH — CASM was developed to address questions concerning the resilience of food webs in relation to nutrient inputs, and CASM 2.0 has been used to evaluate both direct and indirect toxic effects in aquatic ecosystems.

Flexibility — HIGH — CASM is easily adaptable to new situations.

Treatment of Uncertainty — HIGH — CASM was developed in a self-contained, general Monte Carlo framework. Both sensitivity and uncertainty analyses have been performed on the model.

Degree of Development and Consistency — MEDIUM — CASM is a self-contained FORTRAN program that has been made available to the general scientific research community. The model has been applied numerous times. However, no Internet web site exists for the model.

Ease of Estimating Parameters — MEDIUM — More than 100 parameters need to be estimated for this model. However, the nature of the parameters increases the likelihood that they can be estimated from site specific information or from the general technical literature.

Regulatory Acceptance — LOW — CASM was not developed for use by any regulatory agencies. Agencies appear to be supporting development of alternative models (AQUATOX was supported by EPA; CATS-5 was supported by agencies in the Netherlands).

Credibility — MEDIUM — Results from applications of the model to northern lakes in Quebec (CASM 2.0) compared fairly well on average with data reported in the literature. CASM has been calibrated to Lakes Biwa and Suwa in Japan. References to CASM are increasing in the open, peer-reviewed literature, and the number of CASM users is probably greater than 20.

Resource Efficiency — MEDIUM — Essentially no new programming would be required to apply CASM to new site specific applications. Site specificity takes the form of the physical and chemical driving variables (e.g., light, water temperature) and the food-web description. Substantial resources may be required to obtain site specific values for the high number of CASM parameters.

PC LAKE

PC Lake is a one-dimensional model that describes the biota and physical–chemical conditions in the water column and upper sediment layer of a lake (Janse and van Liere 1995). The model was initially developed to simulate chlorophyll *a*, transparency, phytoplankton, and submerged macrophytes. Multiple species of phytoplankton can be included in the model. Inputs include lake hydrology, nutrient loading, lake depth, lake size, and sediment characteristics. The food web also includes zooplankton, zoobenthos, whitefish, and predatory fish.

PC Lake includes procedures for calibration, uncertainty analysis, and probabilistic assessments. The model was intended to predict general trends rather than site specific results for individual lakes. PC Lake has been used to evaluate different restoration scenarios in several lakes in the Netherlands. The model was developed by the same group of researchers who applied the CATS-4 model to assess insecticide effects in aquatic and terrestrial systems (Traas et al. 1998) (see also the previous discussion of the similarity of CATS-4 to LERAM). PC Lake and CATS-4 can be integrated to yield a model very similar to AQUATOX (van Leeuwen 2000, pers. comm.)

Realism — MEDIUM — PC Lake includes much of the ecosystem structure that might be of concern in assessing ecological risks posed by chemicals in lakes.

Relevance — LOW — PC Lake was designed more for evaluating general trends than for site specific assessment. Although the model does not explicitly account for toxic chemical effects, the user could adjust several parameters to implicitly model toxicity.

Flexibility — LOW — PC Lake was not developed for site specific applications and relies on calibration to extensive data collections. Although the model might be generally applicable, it does not provide much in the way of site specific outputs.

Treatment of Uncertainty — LOW — The documentation for PC Lake does not address uncertainty (Janse and van Liere 1995). However, the model could be incorporated into a probabilistic framework.

Degree of Development and Consistency — MEDIUM — The PC Lake model was programmed for convenient implementation on PCs. Janse and van Liere (1995) did not mention error checking for input values or the availability of user documentation. Nevertheless, the overall simplicity of the model structure, equations, and number of parameters indicates that the PC Lake model has been programmed for general applications.

Ease of Estimating Parameters — HIGH — The model parameters have biological interpretations. The values might be readily estimated from the kinds of data often available for lakes with nutrient enrichment problems.

Regulatory Acceptance — LOW — No evidence exists that PC Lake was developed for regulatory use.

Credibility — LOW — PC Lake has been calibrated in general to data from 20 lakes in the Netherlands. However, the model underpredicts chlorophyll in lakes with short residence times and overpredicts chlorophyll in lakes with longer retention times.

Resource Efficiency — MEDIUM — The model has not been developed expressly for ease in site specific applications. However, its simple structure suggests that it could be implemented with reasonable programming effort.

PH-ALA

Jørgensen (1976) and Jørgensen et al. (1981) present a lake model consisting of 17 state variables that describe the temporal changes in concentrations of carbon, nitrogen, phosphorus, detritus, phytoplankton biomass, and zooplankton biomass. The model also considers nutrient loading, rate of phosphorus release from the bottom sediment, light extinction, and water temperature in determining the modeled values of the 17 state variables.

PH-ALA consists of a set of coupled, linear differential equations with several nonlinear terms (e.g., nutrient limitation, light attenuation). The model provides daily outputs for its state variables, typically over an annual time period. Input variables include wind velocity and direction over long periods, initial temperature distribution in the lake, daily hours of sunshine, nutrient loading, pollutant discharge location, and mass loading. A procedure was developed to estimate a unique set of model parameter values for model calibration. The model was used to simulate the results of wastewater treatment alternatives on phytoplankton, zooplankton, and nutrients in Glumsø Lake, Denmark.

Realism — HIGH — Multiple trophic levels were included in PH-ALA, and its results for phytoplankton dynamics were compared with results of a model that used Monod's kinetics only.

Relevance — HIGH — PH-ALA incorporates all ecologically relevant components of a lake ecosystem. The model was developed for application to wastewater treatment facilities and focuses on which nutrients need to be controlled to reduce the adverse effects of eutrophication. Although the model does not explicitly account for toxic chemical effects, the user could adjust several parameters to implicitly model toxicity.

Flexibility — MEDIUM — PH-ALA was developed for a shallow, morphologically simple lake with excessive algal growth and a quick turnover time.

Treatment of Uncertainty — LOW — Jørgensen (1976) and Jørgensen et al. (1981) performed neither sensitivity analysis nor uncertainty analysis, but PH-ALA could be placed in such a framework.

Degree of Development and Consistency — LOW — The model structure and equations are generally accepted as useful in describing the production dynamics of aquatic populations. However, no description of the software used to run the model is provided in Jørgensen (1976) and Jørgensen et al. (1981). No mention of a user's manual or evaluation of user-specified model input values is made.

Ease of Estimating Parameters — MEDIUM — An estimated 38 parameters are needed to run the model; most parameter values are generic (not site specific).

Regulatory Acceptance — LOW — To our knowledge, PH-ALA was not developed for use by a regulatory agency.

Credibility — LOW — The model has been the topic of several open literature publications, primarily for a European audience. The number of users is not indicated.

Resource Efficiency — MEDIUM — Virtually no information is provided on the actual computer implementation of PH-ALA. The overall model structure (e.g., governing equations, kinds of parameters, external driving variables) suggests that the model would require minimal reprogramming but some site specific parameter estimation in developing applications to new systems.

SALMO

SALMO was developed as a general model to examine the effects of phosphorus enrichment on phytoplankton and zooplankton by using a simple, two-layer physical description of an aquatic ecosystem (Benndorf and Recknagel 1982; Benndorf et al. 1985). The model simulates daily values of phytoplankton and zooplankton in relation to user-specified manipulations of phosphorus enrichment or imposition of fish predation on zooplankton or both. In its original version, SALMO included three state variables: phytoplankton, zooplankton, and orthophosphate (Benndorf and Recknagel 1982). In the most recent version, SALMO has six state variables: hypolimnetic oxygen, dissolved orthophosphate, total dissolved inorganic nitrogen, phytoplankton, zooplankton, and allochthonous detritus (Benndorf et al. 1985).

SALMO is vertically structured, with one layer representing a mixed epilimnion and the other layer representing a mixed hypolimnion. Sediment–water interactions are incorporated by empirical relationships for nutrient release from sediments and sediment oxygen demand. The modeling approach emphasizes structural simplicity, as evidenced by the small number of state variables, complemented by complexity in the specification of the ecological control mechanisms (e.g., differential temperature dependence of photosynthesis and photorespiration, multiple resource kinetics in control of phytoplankton photosynthesis and zooplankton grazing). Fish are not included as a state variable, but their ecological role is implicitly modeled by the inclusion of a mortality rate for zooplankton.

Values of model parameters have been derived from laboratory experiments, field observations, or the technical literature; parameter estimation does not rely on extensive calibration. The model has been implemented for several lakes and reservoirs of various depths and degrees of eutrophication.

Realism — MEDIUM — SALMO includes a pelagic zone, multitrophic levels with several representative species at each level, and important feedback mechanisms. The model includes interspecific competition for nutrients, nutrient remineralization by zooplankton, sedimentation of particulate phosphorus, and subsequent release of phosphorus into the water column.

Relevance — HIGH — SALMO was developed specifically to evaluate water-quality issues and was applied in several cases, including mesotrophic, oligotrophic, shallow hypereutrophic, and deep hypereutrophic lakes. Although the model does not explicitly account for toxic chemical effects, the user could adjust several parameters to implicitly model toxicity.

Flexibility — MEDIUM — Benndorf and Recknagel (1982) applied SALMO to four different lakes of varying classifications and developed the model as a general water-quality analysis tool. SALMO has been applied to more than 20 lakes and reservoirs, but most of these applications were not published (Benndorf 2000, pers. comm.). The model is not intended for application to shallow lakes (less than approximately 5 m mean depth) because of the dominating role of macrophytes (not included in SALMO) and/or of the sediment–water interactions (included only in simplified form) in these lakes.

Treatment of Uncertainty — HIGH — Benndorf and Recknagel (1982) presented a sensitivity analysis for SALMO. Petzoldt and Recknagel (1992) performed Monte Carlo analysis on the model.

Degree of Development and Consistency — MEDIUM — The model was developed to minimize the number of state variables while maintaining realistic descriptions of the ecological control mechanisms that regulate the production of aquatic populations. No mention was made of the software used to develop or execute the model. Several detailed accounts of SALMO were published in the technical literature, and a user's manual in German is available (Benndorf 2000, pers. comm.).

Ease of Estimating Parameters — HIGH — The parameters required to run the model have fairly clear meaning and should be estimable from available site specific data

Regulatory Acceptance — LOW — The model has been used to address several nutrient management alternatives and in-lake measures such as artificial mixing and biomanipulation (Benndorf 2000, pers. comm.). However, no mention of any regulatory use, acceptance, or recommendation of SALMO was made.

Credibility — MEDIUM — SALMO has been the subject of an increasing number of publications from the Dresden University Water Resources Department, and the model is receiving additional attention from the European ecological modeling community. The number of current users is not known but could easily exceed 20.

Resource Efficiency — LOW — Benndorf and Recknagel (1982) and Benndorf et al. (1985) do not provide much information concerning the computer software implementation of SALMO or any information that might be used to determine how much programming would be required for SALMO's application to additional aquatic systems.

SIMPLE

Jones and colleagues (1993) constructed SIMPLE to examine the implications of prey availability for competing piscivorous fish populations, including economically valuable salmonid species, in the Great Lakes. SIMPLE treats the lake as a single homogeneous unit and uses an annual time scale to simulate interactions among five piscivorous salmonids and their prey species (planktivores and invertebrates). The salmonids included chinook salmon (*Oncorhynchus tshawytscha*), coho salmon (*Oncorhynchus kisutch*), lake trout (*Salvelinus namaycush*), rainbow trout (*Oncorhynchus mykiss*), and brown trout (*Salmo trutta*). The prey included alewife (*Alosa pseudoharengus*), rainbow smelt (*Osmerus mordax*), and slimy sculpin (*Cottus cognatus*).

In SIMPLE, the predicted dynamics of an age-structured population of fishes are characterized in terms of changes in numbers and weights of fish in each age class. Bioenergetics are integrated with the age-class descriptions to simulate the processes of fish growth and losses of forage fish species (e.g., alewife) to predators. The model also predicts the diet composition of the five salmonid species of interest in relation to the availability of alternative prey, including different sizes of alewife, smelt, and sculpin. Recruitment of piscivores is driven by stocking, whereas stock-recruitment relationships are used to calculate recruitment of planktivores. The model was developed to evaluate the consequences of various fisheries management strategies, especially changes in stocking and harvest rates for the five piscivorous salmonid species common in the Great Lakes. The model has been applied to Lake Ontario and Lake Michigan.

Realism — MEDIUM — SIMPLE accounts for important variables for predators, but the approach is questionable for prey. The integration of bioenergetics and age-structured modeling approaches provides a realistic description of different processes that determine the growth dynamics of the fish species of interest.

Relevance — HIGH — The model focuses on state variables and processes that are important to commercial fisheries. SIMPLE does not explicitly include the effects of toxic chemicals, but several variables could be modified to incorporate such effects implicitly.

Flexibility — LOW — SIMPLE was developed specifically for Lake Ontario and its managed ecosystems.

Treatment of Uncertainty — LOW — There is no indication that sensitivity analyses have been done with the existing version of SIMPLE. However, the model could be placed in a Monte Carlo framework with additional programming effort.

Degree of Development and Consistency — LOW — The software implementation of SIMPLE was not described in detail. However, previous models produced by these authors were developed in BASIC or a combination of commercial spreadsheets and BASIC. No indication of error checking of input parameter values was found. SIMPLE has been calibrated but not validated.

Ease of Estimating Parameters — MEDIUM — The number of parameter values required by SIMPLE does not appear unwieldy, although the exact number could not be determined from the reference. The nature of the parameters suggests that values could be estimated from available data for at least some well-studied aquatic ecosystems.

Regulatory Acceptance — LOW — SIMPLE was not developed with any regulatory mandate or agency.

Credibility — LOW — SIMPLE has been published, but references to it are few, and the number of actual users of this model is presumed to be fewer than 20.

Resource Efficiency — LOW — Application of SIMPLE to a particular case study would require considerable programming, testing, and debugging. The necessary fish data would probably not be routinely available in most applications.

FLEX/MIMIC

FLEX/MIMIC is a hierarchical model that describes the ecological dynamics of streams of the northwestern U.S. (McIntire and Colby 1978). The model addresses key ecological processes, including periphyton production, grazing, shredding, collecting, invertebrate predation, vertebrate predation, and detrital conditioning. Using FLEX/MIMIC software, the model yields time-varying integrations of these processes (e.g., shredding, collecting, predation) over an annual period. The user specifies the physical characteristics of the simulated stream in terms of channel width, depth, cross-sectional area, current velocity, suspended load, and discharge. The model operates on a daily time-step.

The key ecological processes are represented by bioenergetics equations. The herbivory version of the model is an update that tracks successional changes and production dynamics of the periphyton assemblage as well as the response of grazers to corresponding changes in food quality and quantity (http://www.fsl.orst.edu/lter/data/models/strmeco.htm). This version also describes the effects of grazing on successional changes within the periphyton assemblage. The riparian version of the model is an update that describes effects of vegetation canopy structure in the riparian zone on process dynamics in the stream. In this version, photosynthesis and partitioning of primary production among the three algal functional groups (diatoms, cyanobacteria, and chlorophytes) are modeled at an hourly resolution instead of a daily time-step.

The FLEX/MIMIC stream ecosystem model has been used to examine the impacts of clear-cut logging on changes in stream structure and function. The original version was primarily a research tool used to generate hypotheses, to synthesize the results of field and laboratory research, and to set priorities for future research. Subsequently, the model was used for teaching purposes and for generating new hypotheses about primary production and grazing in lotic ecosystems.

Realism — MEDIUM — FLEX/MIMIC focuses on benthic processes in small streams with a limited number of trophic levels. The ecological processes included in the model provide a realistic characterization of the growth of aquatic invertebrate populations.

Relevance — HIGH — FLEX/MIMIC was developed to understand the dynamics of small, flowing-water ecosystems and associated subsystems. The model was used to examine the effects of clear-cut logging on a stream. The biotic groups and endpoints in the model are relevant to chemical risk assessments. Although the model does not explicitly account for toxic chemical effects, the user could adjust several parameters to implicitly model toxicity.

Flexibility — HIGH — The model was constructed in a general computational framework to be adapted to any low-order stream.

Treatment of Uncertainty — LOW — Sensitivity and uncertainty analyses were not reported for the FLEX/MIMIC model, but the model could be placed in such a framework.

Degree of Development and Consistency — MEDIUM — FLEX/MIMIC has been coded in a flexible, hierarchical scheme, but this might be outdated in relation to currently available software. It is unclear whether the software or a user's manual remains available.

Ease of Estimating Parameters — LOW — Much information is needed about the stream being studied, and many parameters require estimation from regression formulas on the basis of site specific data. The parameters have fairly clear biological interpretations.

Regulatory Acceptance — LOW — FLEX/MIMIC has not been used by any regulatory agency.

Credibility — LOW — FLEX/MIMIC was described in fewer than ten publications. It does not appear to be generally used within the current community of stream ecologists. The focus on growth processes translates into difficulties in obtaining field measurements appropriate for comparison with model outputs.

Resource Efficiency — MEDIUM — Despite the model's general framework for small lotic ecosystems, using FLEX/MIMIC would require considerable effort to estimate parameters for application to particular case studies. Recompilation would almost certainly be required, although programming changes would probably entail minimal investment.

IFEM

IFEM integrates environmental fate processes, bioaccumulation, and toxicity information to describe the ecological effects of polycyclic aromatic hydrocarbons (PAHs) in lotic ecosystems (Bartell et al. 1988). The biota modeled in IFEM include phytoplankton, periphyton, macrophytes, zooplankton, benthic insects, larger benthic invertebrates, benthic detritivorous fish, and pelagic omnivorous fish (Figure 9.2). Sublethal toxic effects are modeled in relation to dynamic body burdens and reflect differential growth characteristics, bioaccumulation, and the sensitivity of modeled aquatic populations to toxic chemicals.

IFEM is a compartmental model that is a combination of the toxic effects model SWACOM (Bartell et al. 1992; O'Neill et al. 1982) and the dynamic fate model FOAM (Bartell et al. 1981). Coupled differential equations define the daily production dynamics of the modeled populations and the time-varying concentration of the dissolved and sorbed contaminant for a year. The food web in IFEM consists of 11 functional groups.

Population biomass of the primary producers changes in relation to daily values of light, temperature, nutrients, and toxic chemicals. Consumer biomass changes daily as determined by population-specific bioenergetics and grazing or predator–prey interactions. Sublethal effects on growth processes are determined using exposure–response relationships based on LC50 data from toxicity tests and time-varying body burdens calculated from uptake, degradation, and depuration. Parameters that determine the modeled dissolution, photolytic degradation, volatilization, sorption, and bioaccumulation of PAHs in IFEM can be estimated from QSARs. The model has been programmed in FORTRAN by using difference equations in a Monte Carlo framework for probabilistic risk estimation and numerical sensitivity/uncertainty analyses.

Using IFEM to assess the effects of naphthalene in a stream, Bartell et al. (1988) demonstrated that ecosystem modeling allows prediction of certain ecological effects that could not have been predicted directly from laboratory toxicity data. For example, different levels of effects were found for modeled receptors with similar sensitivity expressed for individual-level endpoints in laboratory toxicity tests. The most severe effect on macrophyte growth was observed at the lower naphthalene loading rates. The authors attributed these findings to the variation in toxicokinetics and sensitivity to naphthalene among species, to differential population growth rates, and to trophic interactions.

Realism — HIGH — IFEM was constructed to describe aquatic populations across multiple trophic levels in a generic framework for streams and rivers. The model includes physical, chemical, biological, and ecological processes that realistically describe the transport, fate, and toxic effects of contaminants in streams and rivers.

Relevance — HIGH — IFEM was designed specifically for estimating ecological risks posed by PAHs in streams and rivers.

Flexibility — HIGH — Thus far, IFEM has been used only to examine the effects of PAHs on a hypothetical stream. However, the model can be applied to essentially any lotic ecosystem. The site specificity is determined by the values of user-specified input files.

Treatment of Uncertainty — HIGH — IFEM was developed in a Monte Carlo framework for probabilistic risk assessment and numeric sensitivity/uncertainty analyses.

Degree of Development and Consistency — MEDIUM — IFEM was developed as a FORTRAN program including Monte Carlo capabilities. However, the model has no user's manual. Many model parameters are estimated from regression equations. Thus, the interpolation provides a certain degree

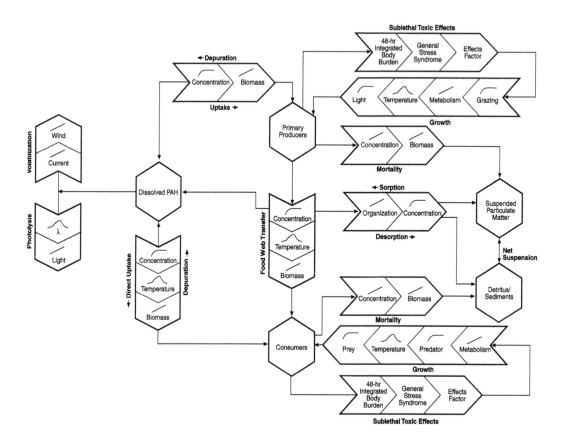

Figure 9.2 Compartments, processes, and pathways for chemical transport and bioaccumulation in IFEM. Note: Hexagons identify state variables. (From Bartell et al. (1988). An integrated fates and effects model for estimation of risk in aquatic systems. pp. 261–274. In *Aquatic Toxicology and Hazard Assessment: 10th Volume*, ASTMSTP 971. American Society for Testing and Materials, Philadelphia. With permission.)

of error checking of parameter values. Bartell et al. (1988) concluded that the exposure–response functions in IFEM could be improved by including toxicity-dependent toxicokinetic parameters.

Ease of Estimating Parameters — MEDIUM — Fewer than 100 parameters are needed for the model. The parameters have clear biological, ecological, or toxicological interpretations. Several key parameters are estimated by using QSARs derived explicitly for PAHs.

Regulatory Acceptance — LOW — IFEM was not developed for use by any regulatory agencies.

Credibility — LOW — IFEM has only been implemented once, and that was for a hypothetical situation. The model results have not been compared with any field data.

Resource Efficiency — MEDIUM — Minimal, if any, new programming would be required to apply the model to any particular case study. The site specificity is determined by the nature of the initial biomass and parameter values required to run the model. These parameters would, in many cases, be available for lotic systems that have been studied sufficiently to warrant site specific applications.

INTASS

The interaction assessment model (INTASS, Emlen et al. 1989, 1992) is a new approach to constructing quantitative expressions for fitness of interacting populations within a biological community. INTASS is a linear (or nonlinear; Emlen, unpublished) model with empirically derived coefficients that express the impact of environmental variables and of the population densities of conspecifics and other populations on the fitness of the target animal populations. INTASS does this by establishing for each defined receptor subpopulation a sphere of response. This sphere of response represents the volume of habitat within which an animal is affected by and responds to environmental variables.

INTASS was developed on the assumption that the environment is a mosaic of microhabitat patches representing potential spheres of response and that the realized spheres of response are a subset of the microhabitat patches actually occupied by the animals at a given time. On the basis of evolutionary theory, Emlen et al. (1992) assert that the animals will move among the patches such that fitness is equivalent among occupied patches. Fitness is defined as the population growth parameter (r). Within any given sphere of response, fitness is modeled as a function of population density and a set of environmental variables. The coefficients for the environmental variables are estimated from empirical observations by minimizing the mean squared error between modeled results and observed results. If more than one microhabitat patch is monitored for a short time period, the patch-specific population growth parameters, the global density-feedback parameter, and a constant in the model can be estimated. This process permits the testing of hypotheses about local disturbances by evaluating the differences in magnitude and shape of density-dependence curves (i.e., r vs. population density) in the presence or absence of a defined stress.

Realism — MEDIUM — INTASS is a general approach to quantifying interactions within and between populations, including the effects of density dependence and multiple environmental variables. The model was developed on the basis of evolutionary and population theory, but the coefficients of the model were parameterized on the basis of empirical data. The greatest strength of INTASS is its ability to quantify effects of disturbances (including toxic chemicals) on population fitness independent of population dynamics. The assumption of homogeneity of fitness among the spheres of response may not be true if significant time lags exist between changes in the environment and perception by the individuals within the population.

Relevance — HIGH — The fitness endpoint of INTASS and the potential applications of the model are highly relevant to chemical risk assessment. The method for quantifying impacts on density dependence and thus overall fitness could be applied directly to assess changes in an ecosystem's status related to chemical stress.

Flexibility — HIGH — INTASS is a very flexible model that may be applied to any ecological receptor for which a subpopulation may be identified. Its method of parameterization, which depends on the minimization of the mean squared error, allows for consistent determination of the coefficients upon the state variables of interest. The model has been applied to fish, land snails, and desert plants.

Treatment of Uncertainty — MEDIUM — The sample application to American eel (*Anguilla rostrata*) presented in Emlen et al. (1992) did not account for uncertainty in the overall analysis. The model structure, particularly with regard to parameterization of the state variable coefficients, requires that the probability density functions be collapsed. However, one could track variability through multiple iterations of the model.

Degree of Development and Consistency — MEDIUM — INTASS is not available as a software package. However, Emlen et al. (1989, 1992) provide sufficient detail to allow programming and application of the model.

Ease of Estimating Parameters — MEDIUM — Because the coefficients of the model are empirical and do not represent some mechanistic functions directly, the parameterization of INTASS is highly dependent on the complexity and size of the habitat considered.

Regulatory Acceptance — LOW — Although the authors have proposed this approach as a substitute for the habitat suitability index models, INTASS has no regulatory status and has not been applied within a regulatory compliance context. The model uses an approach that is unfamiliar to most environmental managers.

Credibility — LOW — The approach used in INTASS has been critically reviewed as part of the peer-review process for publication, but the model has not been distributed widely or achieved general acceptance within the scientific community.

Resource Efficiency — MEDIUM — In assessing impacts resulting from physical disturbance or stress, INTASS would provide a reasonably simple method of quantifying risk. Compared with mechanistic ecosystem models, INTASS would require estimation or calibration of relatively few parameters. However, its lack of availability as software and the requirements for empirical site specific data mean that a relatively high level of effort is required to execute this model.

DISCUSSION AND RECOMMENDATIONS

The continued development, application, and evaluation of aquatic ecosystem models during the past several decades suggest that such models have become increasingly accepted among ecological researchers (e.g., Tables 9.2 and 9.3). Model validation (e.g., comparison of model predictions with observations of real ecosystems) remains an important aspect of this approach for ecological risk assessment. In addition to model performance issues, the feasibility of model implementation for site specific risk assessments is an issue (e.g., Bartell et al. 1998); the resources required for site specific parameter estimation typically increase as a function of model complexity.

On the basis of our evaluation of aquatic ecosystem models (Table 9.2), three models are recommended for more detailed evaluation and application in selected case studies. These models are AQUATOX, CASM, and IFEM. These models were recommended because they are highly developed approaches for estimating ecological risks posed by toxic chemicals in aquatic ecosystems. AQUATOX and CASM have been applied to a variety of case studies and site specific risk assessments. Although not widely applied, IFEM is especially appealing because it combines chemical fate, bioaccumulation, ecological effects, and probabilistic risk estimation within a single modeling framework. Comparisons between model predictions and validation data have shown that these models are capable of predicting ecologically relevant endpoints reasonably well. The models are also programmed in Monte Carlo frameworks that permit probabilistic risk estimation and numerical sensitivity and uncertainty analyses.

Interestingly, the structure and process-level formulations of two of the recommended models (AQUATOX, CASM) are derived from the earlier IBP models. AQUATOX represents a recent manifestation of a series of models derived originally from CLEAN (Park et al. 1974). The modified SWACOM/CASM models are very similar to CLEAN in their basic process formulations but were also derived from the Lake Wingra model (MacCormick et al. 1975).

Koelmans et al. (2001) reviewed integrated models of eutrophication and organic contaminant fate and effects in aquatic ecosystems, including AQUATOX (Park et al. 1974; Park 1998; U.S. EPA 2000a, b, c), CATS-5 (Traas et al. 1998), GBMBS* (the Green Bay mass balance study; Bierman et al. 1992), IFEM (Bartell et al. 1988), an adapted version of QWASI** (quantitative water, air, and sediment interaction; Mackay et al. 1983, adapted by Wania 1996), and Ashley's HOCB*** (hydrophilic organic compound bioaccumulation model) (Ashley 1998). Koelmans et al. (2001) summarized the features of these models and concluded that these tools are invaluable for focusing attention on feedback mechanisms that are often overlooked, for identifying important processes in aquatic systems, for formulating counterintuitive hypotheses about ecosystem function, and for assessing the short-term risk of acutely toxic chemicals. Koelmans et al. (2001) considered AQUATOX the most complete model among those reviewed. These authors also believe that the value of integrated models in predicting long-term effects of contaminant exposure is limited by key limitations in food-web modeling rather than in the representation of contaminant fate.

---

* The GBMBS model includes food-chain accumulation of organic chemicals but does not model toxicity.
** QWASI does not include a food web, bioaccumulation, or effects of toxic chemicals.
*** Ashley's HOCB combines a model of the biological fate of toxic organic chemicals with a detailed model of the carbon cycle but does not include nutrient cycling and toxic effects.

Table 9.2 Evaluation of Ecosystem Models — Aquatic Ecosystem Models

Model	Reference	Realism	Relevance	Flexibility	Uncertainty Analysis	Degree of Development	Ease of Estimating Parameters	Regulatory Acceptance	Credibility	Resource Efficiency
Marine and Estuarine										
Transfer of impacts between trophic levels	Horwitz (1981)	◆◆	◆◆◆	◆◆◆	◆	◆◆	◆	◆◆	◆◆	◆◆
Lake										
AQUATOX (CLEAN)	Park et al (1974); Park (1998); U.S. EPA (2000a,b,c)	◆◆◆	◆◆◆	◆◆◆	◆◆	◆◆◆	◆	◆◆	◆◆◆	◆◆◆
ASTER/EOLE	Thebault and Salencon (1983); Salencon and Thebault (1994)	◆◆	◆◆◆	◆	◆	◆◆◆	◆◆	◆◆	◆	◆◆
DYNAMO pond	Wolfe et al. (1986)	◆◆	◆◆◆	◆◆	◆◆	◆◆◆	◆◆◆	◆	◆◆	◆◆
EcoWin	Ferreira (1995); Duarte and Ferreira (1997)	◆◆	◆◆◆	◆◆◆	◆	◆◆	◆◆◆	◆	◆◆	◆◆◆
LEEM	Koonce and Locci (1995)	◆◆◆	◆◆	◆	◆	◆◆◆	◆◆	◆◆	◆◆	◆
LERAM	Hanratty and Stay (1994); Hanratty and Liber (1996)	◆◆	◆◆	◆◆	◆◆◆	◆◆◆	◆◆	◆◆	◆◆	◆◆
CASM/modified SWACOM	DeAngelis et al. (1989); Bartell et al. (1992, 1999)	◆◆◆	◆◆	◆◆◆	◆◆◆	◆◆	◆◆	◆	◆◆	◆◆
PC Lake	Janse and Van Liere (1995)	◆◆	◆	◆	◆	◆◆	◆◆◆	◆	◆◆	◆◆
PH-ALA	Jørgensen (1976); Jørgensen et al. (1981)	◆◆◆	◆◆◆	◆◆	◆	◆	◆◆	◆	◆	◆◆
SALMO	Benndorf and Recknagel (1982)	◆◆	◆◆◆	◆	◆◆◆	◆◆	◆◆	◆	◆◆◆	◆
SIMPLE	Jones et al. (1993)	◆◆	◆◆◆	◆	◆	◆	◆◆	◆	◆	◆
River										
FLEX/MIMIC	McIntire and Colby (1978)	◆◆	◆◆◆	◆◆◆	◆	◆◆	◆	◆	◆	◆◆
IFEM	Bartell et al. (1988)	◆◆◆	◆◆◆	◆◆◆	◆◆◆	◆◆	◆	◆	◆	◆◆
General										
INTASS	Emlen et al. (1992)	◆◆	◆◆◆	◆◆◆	◆◆	◆◆	◆◆	◆	◆	◆◆

Note:

◆◆◆ - high

◆◆ - medium

◆ - low

Table 9.3   Applications of Aquatic Ecosystem Models

Model	Species/Ecosystem	Location/Population	References
Transfer of impacts between trophic levels	Estuarine plankton and predators	Chesapeake Bay	Horwitz (1981)
AQUATOX, CASM	Lake ecosystems	Generic lake in central Florida	Bartell et al. (2000)
	Lake ecosystems	Lake Biwa	Miyamoto et al. (1997)
	Lake ecosystems	Lake Suwa	Naito et al. (1999)
	Lake ecosystems	Canadian lakes	Bartell et al. (1999)
	River ecosystems	Canadian rivers	Bartell et al. (1999)
	Reservoir ecosystems	Canadian reservoirs	Bartell et al. (1999)
	Ponds	Experimental	Bartell et al. (1992)
ASTER/EOLE	Reservoir ecosystems	Pareloup Lake, France	Salencon and Thebault (1996)
DYNAMO	Blue tilapia (*Tilapia aurea*) food web	Aquaculture ponds	Wolfe et al. (1986)
EcoWin	Aquatic ecosystems	Tagus Estuary Carlingford Lough	Ferreira (1995); Duarte and Ferreira (1997)
LEEM	Great Lakes ecosystems	Lake Erie	Koonce and Locci (1995)
LERAM	Pond littoral zone ecosystems	Minnesota	Hanratty and Stay (1994) Hanratty and Liber (1996)
	Drainage ditch microcosms (CATS-4)*	Netherlands	Traas et al. (1998)
PC Lake	Lake phytoplankton and macrophytes	Netherlands	Janse and van Liere (1995)
PH-ALA	Lake phytoplankton and zooplankton	Glumsø Lake, Denmark	Jørgensen (1976) Jørgensen et al. (1981)
SALMO	Lake phytoplankton and zooplankton	German lakes	Benndorf and Recknagel (1982); Benndorf et al. (1985)
SIMPLE	Lake (salmonid) ecosystem	Lake Ontario Lake Michigan	Jones et al. (1993)
FLEX/MIMIC	Small stream ecosystems	Northwestern U.S.	McIntire and Colby (1978)
IFEM	Stream ecosystems	Experimental	Bartell et al. (1988)
	River ecosystems	Hersey River	Bartell (1986)
INTASS**	Aquatic ecosystems	Deerfield River, Massachusetts, American eel (*Anguilla rostrata*)	Emlen et al. (1992)

*CATS-4 is a modification of LERAM (see discussion of LERAM in text).
** INTASS can also be applied to terrestrial ecosystems.

CHAPTER **10**

Ecosystem Models — Terrestrial

Christopher E. Mackay and Robert A. Pastorok

Terrestrial ecosystem models are defined as mathematical constructs that represent biotic and abiotic components in deserts, forests, grasslands, or other terrestrial environments. Often they include physical, chemical, biological, and ecological processes. They are spatially aggregated such that input parameters and model functions are independent of distance and relative position.

The primary endpoints for terrestrial ecosystem models include:

- Abundance of individuals within species or trophic guilds
- Biomass
- Productivity
- Food-web endpoints (e.g., species richness, trophic structure)

We review the following terrestrial ecosystem models (Table 10.1):

- Desert
    - Desert competition model, a hierarchical model that describes the dynamics of two interacting species of mice in the genus *Dipodomys* (Maurer 1990)
- Forest
    - FVS (forest vegetation simulator), which, like the other forest models listed next, projects forest development through time (USDA 1999)
    - FORCLIM (forest climate model) (Bugmann 1997; Bugmann and Cramer 1998)
    - FORSKA (Lindner et al. 1997, 2000; Lindner 2000)
    - HYBRID (Friend et al. 1993; 1997)
    - ORGANON (Oregon growth analysis and projection) (OSU 1999)
    - SIMA (Kellomäki et al. 1992)
    - TEEM (terrestrial ecosystem energy model) (Shugart et al. 1974)
- Grassland
    - Energy flow for short grass prairie model, an energy flow model for grasslands (Jeffries 1989)
    - SAGE (system analysis of grassland ecosystems), a model of the dynamics of primary producers and consumers (Heasley et al. 1981)
    - SWARD, an air pollution model that predicts the impact of sulfur dioxide on grass species and grazing ruminants (White 1984)

Table 10.1 Internet Web Site Resources for Terrestrial Ecosystem Models

Model Name	Description	Reference	Internet Web Site
Hierarchical model of *Dipodomys*	A model of the dynamics of two interacting species of mice	Maurer (1990)	N/A
FVS	A U.S. Department of Agriculture model for projecting forest development through time	USDA (1999)	http://www.fs.fed.us/fmsc/fvs.htm http://eco.wiz.uni-kassel.de/model_db/mdb/fvs.html
FORCLIM	A model for projecting forest development through time with stress functions that could be easily modified for toxic chemical effects	Bugmann (1997); Bugmann and Cramer (1998)	http://www.pik-potsdam.de/cp/chief/forclim.htm
FORSKA	A model for projecting forest development through time without functions to account for harvesting	Lindner et al. (1997, 2000); Lindner (2000)	http://www.pik-potsdam.de/cp/chief/forska.htm
HYBRID	A model for projecting forest development through time which incorporates a spatially aggregated version of the forest landscape model ZELIG	Friend et al. (1993; 1997)	http://www.wiz.uni-kassel.de/model_db/mdb/hybrid.html
ORGANON	A model for projecting forest development through time developed specifically for several habitats in Oregon	OSU (1999b)	www.cof.orst.edu/cof/fr/research/organon/ http://eco.wiz.uni-kassel.de/model_db/mdb/organon.html
SIMA	A model for projecting forest development through time	Kellomäki et al. (1992)	http://gis.joensuu.fi/research/silmu/juttu4.html
TEEM	A forest ecosystem model used in assessing energy use impacts	Shugart et al. (1974)	http://www.esd.ornl.gov/people/o'neill_bob/index.html
Energy Flow for Short Grass Prairie Model	An energy flow model specific to short grass prairies that describes interactions between primary producers and nesting sparrows	Jeffries (1989)	N/A
SAGE	An air pollution model for predicting the impact of sulfur dioxide on grass species and grazing ruminants	Heasley et al. (1981)	http://eco.wiz.uni-kassel.de/model_db/mdb/sage.html
SWARD	A model of the dynamic equilibrium between primary producers and consumers within a grassland ecosystem	White (1984)	N/A
Multi-timescale community dynamics	A model of species turnover in bird communities	Russell et al. (1995)	N/A
Nested species subset analysis	A model for analyzing patterns of nestedness of species subsets in a biological community	Cook and Quinn (1998)	N/A
SPUR	A model for simulating interactions among soils, plants, and grazing ungulates on rangeland	Hanson et al. (1988)	http://vernon.tamu.edu/taes/rlem/brochure.htm http://eco.wiz.uni-kassel.de/model_db/mdb/spur.html

Note: N/A - not available

- Rangeland
  - SPUR (simulating production and utilization of rangeland), a multispecies plant growth and grazer model (Hanson et al. 1988).
- Island
  - Multi-timescale community dynamics models, models of species turnover in bird communities (Russell et al. 1995)
  - Nested species subset analysis, a model for analyzing patterns of nestedness for subsets of species (Cook and Quinn 1998)

In addition to the models listed above, INTASS (Emlen et al. 1992) is a general ecosystem model that can be applied to terrestrial as well as aquatic systems (see review under Chapter 9, Aquatic Ecosystem Models). MUSE (multistrata spatially explicit ecosystem modeling shell) is a modeling system for Windows that can be used to implement and compare a wide variety of forest models (JABOWA, FORET, FORSKA, and others) (http://biology.anu.edu.au/research-groups/ecosys/muse/MUSE.HTM).

DESERT COMPETITION MODEL

The model of Maurer (1990) describes the dynamics between two species of *Dipodomys* (mice) within a Chihuahua desert scrub ecosystem. It was specifically designed to evaluate the role of extrinsic vs. intrinsic factors in the regulation of population dynamics and community structure. The availability of food is the primary extrinsic factor. Intrinsic factors include species recruitment, foraging efficiency, and reproductive rates. It is a bioenergetic model in which the two species of *Dipodomys* feed on a single homogeneous seed source. The model is parameterized on the basis of observations derived from a 20-ha plot located on the Cave Creek Bajada, 2 km north of Portal, Arizona.

The model is structured as a multicompartment construct in which species-specific biomass values are the principle state variables. Differential equations describe competition between species on the basis of relative transfer of metabolic energy from the food source into the reproductive functions for each species. The foraging capacity of each population is based on relative recruitment, relative assimilation efficiency, and relative foraging efficiency of individuals.

Realism — MEDIUM — The design of the desert competition model is principally a bioenergetics model, in which feeding and reproduction represent the main biological processes that lead to competition between the two species. The predominant driving variable is the availability of food. No consideration is given to health factors (such as response to toxic chemicals) independent of food availability.

Relevance — MEDIUM — The model's primary utility in ecological risk assessment is in quantifying the results of perturbations directly affecting interspecies competition for food. The availability of other ecosystem models that address more than two species limits the application of this model. No explicit consideration of toxicity is included in the model, but functions could be added relatively easily to account for toxic chemical effects.

Flexibility — LOW — Although the model relies on standard quantification techniques common in many bioenergetic and population models, it is specific to seed-eating rodents.

Treatment of Uncertainty — LOW — Examples provided in Maurer (1990) were parameterized on a deterministic basis with no consideration of uncertainty.

Degree of Development and Consistency — MEDIUM — The model is apparently not available as software. However, details provided in Maurer (1990) are sufficient to reproduce the simulation.

Ease of Estimating Parameters — MEDIUM — General parameter estimation for the model would be reasonably easy to do by using bioenergetic principles. However, achieving accuracy in applications to new cases would require parameterization of the basal, active, and reproductive metabolic rates for the two species and of the availability and caloric content of the food source. Empirical

data would also be required to define the relation between food availability and fecundity of
individuals.

Regulatory Acceptance — LOW — This model has no regulatory status and has not been applied in
a regulatory context.

Credibility — LOW — This model is a multicompartment bioenergetic model with governing differential equations formulated on the basis of standard algorithms. Although this approach is very
common in ecosystem simulations, no other reference was found for applications of this particular
model.

Resource Efficiency — MEDIUM — Parameterization of this model is reasonably simple because of
its reliance on bioenergetic principles. However, application would require programming of the
algorithms in software form.

FVS

FVS is a nationally supported framework for standardized projection of forest growth and yield
(USDA 1999) originally based on the prognosis model (Stage 1973; Wykoff et al. 1982). Geographic
variants have been developed covering the major forestlands in the U.S. The FVS modeling system
has been used extensively for developing silvicultural prescriptions, evaluating management scenarios, updating inventory information, and providing input to forest planning models (USDA 1999 and
references therein). Additional capabilities include forecasting vegetative structure, assessing wildlife
habitat, analyzing fire hazard, determining forest health risk, and monitoring ecological processes.
Using a parallel processing extension to FVS (Crookston and Stage 1991) and the SUPPOSE interface (Crookston 1997), FVS can be used to model the development of each stand with dependence
on characteristics of the surrounding forest and to generate landscape-level statistics.

The FVS model simulates a wide range of forest cover types, species, size classes, and stand
densities. It predicts live tree stocking, growth, yield, and mortality, including stand structural stage
and crown cover statistics. FVS simulates the establishment of seedlings and stump sprouts. FVS
differs from other forest-gap models in that some variants include simulation of the understory
component, including the height and cover of grasses, forbs, and shrubs.

Two advanced features of FVS distinguish it from other forest projection systems. First, FVS
uses growth increment data to adjust growth functions to match measured trends, thereby self-
calibrating the equations to input data. Second, using the event monitor in FVS (Crookston 1990),
modelers can define variables to influence simulation results and report additional output values.
It allows management activities to be scheduled conditionally, on the basis of changing stand
conditions. In addition, yield forecasts and information about the dominant vegetation can be
evaluated relative to functions of towns and regions to predict the impact of human activity upon
the ecosystem and vice versa. The event monitor adds a robustness not found in other projection
models.

A graphical user interface and tabular output have been developed to allow easy interaction
with the FVS model. Linkage to the stand visualization system enables graphical display of stand
conditions (e.g., trees, shrubs, down material, fire dynamics).

Realism — HIGH — FVS is one of the most highly developed forest-gap models. Although the
parameters are generalized for broad forest categories, the integration of comparative empirical data
from the Forest Service allows a high degree of realism. The images rendered through the standard
visualization system provide a realistic representation of silvicultural treatments and management
options.

Relevance — MEDIUM — FVS does not specifically include relationships for considering the impact
of toxic chemicals. However, the ability of FVS to predict forest structure on the basis of competitive
growth parameters makes it potentially useful in ecosystem characterization. Functions for toxic
chemical effects could be added relatively easily. Specific modules within FVS account for the
effects of insect pests and fire.

Flexibility — HIGH — FVS was specifically developed to model all major forest types throughout the continental U.S. It can process a single stand, multiple stands, or an entire landscape in a single run.

Treatment of Uncertainty — HIGH — FVS can be run in either deterministic or stochastic mode.

Degree of Development and Consistency — HIGH — FVS has a history of more than 25 years. It is the only nationally recognized and supported forest growth and yield model maintained by the U.S. Forest Service. FVS is available as a free software package from USDA.

Ease of Estimating Parameters — HIGH — FVS is self-calibrating, given a tree input list that includes either radial-increment core data or diameter measurements at two points in time. Keyword modifiers give users the added ability to adjust model output to observed values. The U.S. Forest Service provides links to regional data sets that make simulation setup easy.

Regulatory Acceptance — HIGH — FVS is a model accepted by USDA for estimating potential stand productivity within the continental U.S.

Credibility — HIGH — FVS has a long history of use. Furthermore, because of its development and use by the U.S. Forest Service, it has become the *de facto* credibility standard for all other commercial forestry models. It has stood the test of time and continues to evolve as forest management issues evolve.

Resource Efficiency — MEDIUM — All forest-gap models provide roughly equal types of outputs. FVS scored medium in resource efficiency because, although it is highly reliant on site specific parameterization (for ecosystem characterization), its availability as a user-friendly software package makes its application relatively easy.

FORCLIM

FORCLIM is a forest model developed for Central Europe but also successfully applied in eastern and northwestern North America (Bugmann 1997; Bugmann and Cramer 1998; http://www.pik-potsdam.de/cp/chief/forclim.htm). It is designed to incorporate simple yet reliable functions of climatic influence on ecological processes. FORCLIM consists of three modules, each of which can be executed independently or in combination with the other modules.

The primary ecological module, FORCLIM-P, simulates the population dynamics of forest trees. Size cohorts are simulated as opposed to individual trees. Usually, functional plant groups are modeled rather than individual species. Maximum tree growth is determined from an exponential growth curve modified by nutrient and light availability, summer temperature, and water availability. Nitrogen is the limiting nutrient for growth. Light availability to the canopy is calculated based on the Beer–Lambert function. The effect of summer temperature on tree growth is calculated by using a parabolic relation between annual summer-degree days and the growth rate of the trees. Water availability is expressed as a function of annual evapotranspiration deficits. Rates of establishment of species during succession are a function of light availability at the forest floor, browsing intensity, and minimum temperature. Tree mortality is modeled empirically on the basis of age-related and stress-induced mortality rates. Changes in stand biomass over time are simulated as a function of environmental conditions and specific stresses (by indirectly applying climatic factors through the stress-induced mortality rate).

The second module of FORCLIM, FORCLIM-E, simulates the soil–water balance within the forest. It is an empirical scheme (often referred to as a bucket model) requiring parameterization of only monthly mean temperatures and monthly precipitation sums. Outputs for this module are realized evapotranspiration rates relative to precipitation and capacity during each month of the iteration. These are then used by FORCLIM-P as inputs to the stress-growth and stress-mortality functions.

The third module, FORCLIM-S, simulates soil nutrient cycling and availability to forest trees. State variables for this module include available and unavailable nitrogen and phosphorus pools. The state variables are moderated by concentration-dependent functions that reflect temperature, water, and soil physical–chemical conditions. FORCLIM also includes unique functions for carbon turnover as a factor modifying the rates of nitrogen and phosphorus turnover.

Realism — HIGH — FORCLIM represents one of the later-generation forest models. Its major strengths in comparison with predecessors such as FORECE (Kienast 1987) include more detailed and realistic simulations of both water and nutrient behavior.

Relevance — MEDIUM — FORCLIM might be useful in ecological risk assessment for the long-term characterization of forest habitats. Both the stress-mortality and stress-growth functions could be modified to account for effects of toxic chemicals. However, related models, such as JABOWA, which is intended for forest landscape simulation, are probably better suited for ecological risk assessment because of their widespread use and inclusion of some stochastic functions (see Chapter 11, Landscape Models — Aquatic and Terrestrial).

Flexibility — MEDIUM — FORCLIM was originally designed to model continental European forests. However, it has been successfully applied to forests in the eastern U.S. Adaptation of the model to new cases requires site specific parameterization of functions for growth and mortality, and particularly drought resistance.

Treatment of Uncertainty — LOW — The model as presented does not track uncertainty or variability.

Degree of Development and Consistency — HIGH — FORCLIM exists as a software package and may be available from the authors. Results of a validation indicate that the model is generally successful in projecting forest dynamics in eastern North America, except toward the dry timberline in the southeastern U.S., where it failed to simulate the dominance of drought-adapted species and reduced aboveground biomass.

Ease of Estimating Parameters — MEDIUM — FORCLIM is a relatively complex model requiring moderate effort for parameterization. When applied to forest ecosystems for which the model is intended, parameterization would be extremely efficient because many of the parameter values used in previous applications could be retained.

Regulatory Acceptance — LOW — FORCLIM has no regulatory status and to our knowledge has not been applied in a regulatory context.

Credibility — HIGH — FORCLIM is the latest version of the JABOWA type of forest-gap model. It has been used by numerous researchers and is therefore considered credible.

Resource Efficiency — HIGH — When applied to the forest ecosystems for which it was designed, the level of effort and cost for implementation of FORCLIM would be relatively low. Data requirements could be fulfilled by readily available sources.

FORSKA

FORSKA was originally developed to model forest dynamics in Scandinavia (Lindner et al. 1997, 2000; Lasch et al. 1999; Lindner 2000). It simulates the growth, regeneration, and mortality of individual trees in small forest patches. It differs from the earlier forest-gap models because it includes a greater range of mechanistic functions to model tree growth. The tree-volume index, a measure of growth, is derived through the integration of the difference between net assimilation rates in the leaves of the crown layer and the cost in terms of production and maintenance of sapwood for the entire tree. Total tree mass was not parameterized but instead was mathematically inferred from the product of the diameter at breast height, the overall height, the bole length, and an empirical scaling factor. This integral function also includes a resource depletion coefficient that models the overall loss of rate-limiting nutrients as related to plot maturation. Outputs from this module are provided in terms of net tree-volume gains.

Another unique aspect of FORSKA is the inclusion of a functional competition subcomponent in the overall growth module. Each individual tree is assigned a height-to-diameter ratio that depends on the net difference in solar radiation intensity between the tops and the bottoms of the crowns. Hence, if a tree is in danger of being overtopped, it will allocate resources to increasing vertical growth and thereby increase its height-to-diameter ratio. Effects of these changes on tree volume are determined by using a series of scaling relationships.

In FORSKA, tree mortality depends on empirical functions specific to the species and age of the tree. No consideration was given to modeling any extraneous factors in the determination of mortality rates.

Realism — HIGH — FORSKA provides a greater amount of detail in its functional relationships compared with other forest-gap models. The consideration of metabolic energy balance and changes in morphology due to competition enhance its realism.

Relevance — MEDIUM — FORSKA can simulate a variety of ecologically relevant endpoints, such as tree biomass, stand biomass and age structure, and species richness. FORSKA contains no functions to account for effects of toxic chemicals. In its original form, FORSKA was intended to model a natural forest system, and no modules were included to account for plot management activities such as harvesting. A more recent version has initialization and management routines to enable the simulation of managed forests (Lindner 1998). Therefore, modeling toxic chemical effects would require modification of functions describing physical perturbations.

Flexibility — MEDIUM — In comparison with other forest models, FORSKA can be more flexibly applied to diverse forest environments. However, it still relies on a considerable number of empirically derived functions that might require restructuring and data intensive reparameterization.

Treatment of Uncertainty — LOW — FORSKA does not track uncertainty or variability within its model structure.

Degree of Development and Consistency — MEDIUM — FORSKA has been validated. The model is available as a software package from the author.

Ease of Estimating Parameters — HIGH — Parameter estimation for FORSKA is easier than for other forest models because the empirically derived growth functions have been replaced with functional relationships whose parameterization may be retained if applied in similar forest environments. However, parameterizing the mortality curves on the basis of site- and species-specific empirical observations is still necessary.

Regulatory Acceptance — LOW — FORSKA appears to have no regulatory status and appears not to have been applied in a regulatory context.

Credibility — HIGH — FORSKA uses standard modeling techniques developed over many years and has been cited and used in other independent forest research programs.

Resource Efficiency — LOW — Application of FORSKA was considered to be less efficient than that of other forest-gap models because of the inclusion of functional growth relationships. This increases the requirement for site specific data.

HYBRID

HYBRID is a multilevel ecosystem model that synthesizes a forest-gap model, an ecosystem process model, and a biophysiological photosynthesis model (Friend et al. 1993; 1997). The model predicts tree growth and species succession, with carbon and water fluxes between the forest and the atmosphere. HYBRID originated from the merger of the forest-gap model ZELIG with the ecosystem process model FOREST-BGC. By combining these models, predictions of responses to environmental change can be made for both the biochemical processes of individual trees and forest community structure. In this forest model, growth of individual trees is simulated as carbon fixation and partitioning. State variables in HYBRID include carbon dioxide (CO_2) partial pressure (both regional as well as across leaf cuticular boundaries), relative humidity, precipitation, air temperature, tree morphological metrics, evapotranspiration and respiration factors, soil–water capacity, and overall carbon storage capacity.

HYBRID is structured as a nested compartment model. At the ecosystem level, it closely resembles a spatially aggregated version of the ZELIG model iterated on an annual time-step basis. However, rather than relying on empirically based growth curves, HYBRID substitutes the FOREST-BGC routines for carbon fixation, respiration, and carbon allocation, which are iterated on a daily time-step basis. Thus, each tree is separately modeled with respect to daily transpiration,

photosynthesis/carbon fixation, and respiration. Each individual tree is assigned a set of fundamental physiological parameters, depending on species and state of development. These are used
to calculate carbon/water dynamics for each individual. However, state variables describing the
light environment are treated at the plot level. The photosynthesis component of FOREST-BGC
has been replaced in HYBRID with a detailed photosynthesis and stomatal conductance model.
The fluxes of water are summed across individuals in each plot for each day and subtracted from
the soil water to derive the dynamics in soil–water potential. The net CO_2 assimilation rates are
summed across days to give annual forest productivity, which in turn provides the growth parameters
for the forest-gap model.

Realism — HIGH — HYBRID is a highly refined forest model that realistically accounts for interactions
from the biochemical level to the ecosystem level.

Relevance — HIGH — The model endpoints, including metrics for forest community structure, are
relevant for chemical risk assessments. HYBRID does not include any relationships for the consideration of physical or toxicological impacts. However, of the forest models reviewed, HYBRID
would be among those easiest to modify to include such effects, particularly for factors affecting
stomatal conductance, water availability, or photosynthesis rates.

Flexibility — HIGH — HYBRID can be applied to all major forest types in the continental U.S. This
model calculates leaf-level photosynthesis and stomatal conductance for any C_3 plant species, with
minimal species-specific parameterization.

Treatment of Uncertainty — LOW — HYBRID is deterministic and thus does not track uncertainty.
Some sensitivity analysis of HYBRID has been done.

Degree of Development and Consistency — MEDIUM — Validation for a white oak forest (Knoxville,
Tennessee) and a lodgepole pine forest (Missoula, Montana) indicates a high level of accuracy,
particularly with regard to predictions of productivity. HYBRID is not available as a commercial
software package, and Friend et al. (1993, 1997) do not provide sufficient detail to replicate the
model structure.

Ease of Estimating Parameters — HIGH — HYBRID includes a large database from which to select
default values, particularly for the biochemical parameters. In many cases, additional data would
not be required for species-specific growth curves.

Regulatory Acceptance — LOW — HYBRID has no regulatory status and appears not to have been
applied within a regulatory context.

Credibility — HIGH — HYBRID has a reasonable history of use, having been developed as a synthesis
of an accepted forest-gap model (ZELIG) and well-developed physiological models.

Resource Efficiency — MEDIUM — All forest-gap models within this category provide roughly equal
types of outputs. HYBRID is reasonably easy to parameterize, but it is not available as software
and therefore would have to be converted into an executable format.

ORGANON

ORGANON is an individual-based forest model that uses a list of trees, each with exact measurements, as input data to predict forest plot productivity (OSU 1999). The user can specify periods
of growth in 5-year increments and management activities such as thinning, fertilizing, and pruning.
For each of the requested activities, the individual trees are modified to reflect the effects of the
management actions. The program produces stand statistics at each step as well as yield information
after the final harvest of the stand. Results include time course of tree diameter, height, and structure
(branching and wood quality), as well as overall stand density and likely species composition.

ORGANON has been developed to model three habitats in Oregon: (1) the mixed conifer
young growth; (2) the Douglas fir (*Pseudotsuga menziesii*), grand fir (*Abies grandis*), white fir
(*Abies concolor*), ponderosa pine (*Pinus ponderosa*), sugar pine (*Pinus lambertiana*), and incense
cedar (*Calocedrus decurrens*) forest; and (3) the young growth Douglas fir forest. The model can
project development in both even-aged and uneven-aged stands ranging from 20 to 120 years of

development. Parameter inputs include coded tree species, trunk diameter, tree height, crown ratio, and radial growth. ORGANON relies on empirically derived, species-specific growth curves (modified for tree density) to project potential tree growth and forest productivity on the basis of these input parameters. The model is iterative but uses only a 5-year time-step.

Realism — HIGH — ORGANON's predictions are based on empirical growth and competition functions. For the ecosystems for which it was developed, the model is highly realistic.

Relevance — MEDIUM — ORGANON is adequate for simulating forest plot composition and productivity. It includes algorithms to account for various types of physical disturbance, but not toxicity. Presumably, the functions for physical disturbance could be modified or additional functions added to account for toxic chemical effects.

Flexibility — LOW — ORGANON was developed specifically for, and is only applicable to, forest types found in the northwestern U.S.

Treatment of Uncertainty — LOW — ORGANON is deterministic and thus does not track uncertainty.

Degree of Development and Consistency — HIGH — ORGANON is available as a software package from the University of Oregon's School of Forestry. ORGANON has been validated in the forest types for which it was designed.

Ease of Estimating Parameters — HIGH — Because it is highly specialized, ORGANON already has the necessary growth functions included in the model structure. Therefore, parameterization is limited to considerations of site specific parameters such as the inclusive type of forest, distribution of current tree size and structure, and projected management practices.

Regulatory Acceptance — LOW — ORGANON has no regulatory status and appears not to have been applied within a regulatory context.

Credibility — HIGH — ORGANON appears to have a reasonable history of use, having been developed over a number of years. Numerous publications cover its development and use.

Resource Efficiency — HIGH — All forest-gap models within this category provide roughly equal types of outputs. ORGANON scored high because of its low parameterization requirements and availability as a user-ready software package.

SIMA

The SIMA ecosystem model is a forest-gap model for depicting community and production processes dynamics in a boreal forest ecosystem between the latitudes 60° and 70° N, and the longitudes 20° and 32° E (Kellomäki et al. 1992). In this model, forest structure and productivity are controlled by temperature, light conditions, and the availability of nitrogen and water. It was intended to model not only the short-term changes associated with the availability of water and nutrients but also long-term changes associated with changes in climate. The model is parameterized for Scotch pine (*Pinus sylvestris*), Norway spruce (*Picea abies*), pendula birch (*Betula pendula*), pubescent birch (*B. pubescens*), aspen (*Populus tremula*), and grey alder (*Alnus incana*) in Finland. Ground-cover vegetation is also considered in the model. The model is run in annual iterations for a forest plot of 100 m².

The model incorporates four environmental subroutines describing site conditions in terms of temperature, moisture, frost, and decomposition. These are generalized over the forest stand as daily temperature sum, total soil moisture, available soil nitrogen, and duration of subzero temperatures. The model's state variables track reproduction, plant growth, and mortality (indirectly). Allowances are made for the inclusion of management activities such as thinning, clear-cutting, and fertilization. The environmental subroutines are linked to the demographic subroutines, which determine tree population dynamics (birth, growth, and death of trees). Using a bootstrap technique, the user can simulate these processes and the subsequent succession that takes place in the forest ecosystem. The probability of an event is a function of the current forest structure and seasonality.

Realism — MEDIUM — SIMA was judged adequate for simulating forest plot productivity. Both the starting point and model relationships depend on empirical observations. Therefore, when applied in a comparable situation, the model should be reasonably realistic.

Relevance — MEDIUM — SIMA calculates forest productivity, species composition, biomass, and other endpoints that are very relevant for ecological risk assessment. However, the model contains no state variables to track effects of physical disturbances (other than management activities) or of toxic chemicals. Presumably, the functions for management actions could be modified or additional functions added to account for toxic chemical effects.

Flexibility — MEDIUM — SIMA provides some flexibility in that it is specific to tree species as opposed to forest type. It relies heavily on empirical relationships. To date, it has been parameterized solely for Finnish boreal forests.

Treatment of Uncertainty — LOW — Although SIMA is run in a probabilistic manner, the presentation in Kellomäki et al. (1992) does not track uncertainty or provide probabilistic density functions as output. The development of probability density functions would require adding a second-order Monte Carlo analysis.

Degree of Development and Consistency — HIGH — SIMA is available as a software package.

Ease of Estimating Parameters — MEDIUM — For applications of SIMA within the context for which it was designed (Finnish boreal forests), parameter estimation would be relatively easy because the model has been calibrated with default parameters and relationships. Application to other species would require a moderate effort to reparameterize the model.

Regulatory Acceptance — LOW — SIMA has no regulatory status and appears not to have been used in a regulatory context.

Credibility — LOW — Although SIMA depends on standard methods used in forest-gap models, no apparent history of use for this particular construct exists.

Resource Efficiency — HIGH — All forest-gap models within this category provide roughly equal types of outputs. SIMA was rated high primarily on the basis of ease of use owing to a high degree of development and ease of parameterization when applied within the context for which it was intended.

TEEM

TEEM is an ecological model for stimulating energy transfers in forests (Shugart et al. 1974). TEEM is a high-resolution construct with a recommended maximum period of simulation equal to 3 years. Model outputs include annual growth of individual trees, overall forest productivity, and relative energy balance between the three identified forest components: primary producers, consumers, and decomposers. The spatial scale for TEEM is the forest stand, which is assumed to have minimal heterogeneity. Although TEEM may be parameterized for any forest type, it is designed specifically to model an eastern deciduous forest.

TEEM consists of three modules that simulate primary producers, consumers, and decomposers. The primary producer module consists of time-dependent differential equations for predicting gross photosynthesis and respiration. *Gross photosynthesis* is defined as a function of water potential, temperature, and physiological time (duration of solar irradiance). *Net photosynthesis* (carbon assimilation) is defined as functional gross photosynthesis minus the sum of maintenance respiration and energy required to complete photosynthesis. Respiration is modeled as an exponential function and is inversely related to temperature. Net photosynthesis is proportional to temperature and solar irradiance and is modeled as an asymptotic function as maximum photosynthetic rates are approached. Growth and development within the primary producer module are modeled as the integration of the productivity algorithms for three classes of plant tissue (leaves, boles, and roots) and for storage (unincorporated carbohydrates).

The consumer module is an energy-balance construct, in which net biomass is modeled as a function of food intake rates and losses resulting from predation, maintenance respiration, and nonpredatory mortality.

The final component, the decomposition module, is founded on a cryptozoan food chain and an abiotic relationship describing the rate of decomposition. The cryptozoan food chain exhibits two levels of control in the model. The first level includes physiological constraints, such as respiration, feeding, and acclimation rates for each decomposer species. The second level includes a set of parameters for population dynamics of the decomposers including reproductive rates, death rates, and differential impacts related to population densities (based on classic Lotka–Volterra dynamics; see Chapter 8, Ecosystem Models — Food Webs, Predator–Prey Models). As with the consumer model, this module is structured as a differential-sums equation to account for feeding rates and bioenergetic losses across all modeled decomposer species.

Decomposition of organic matter in litter and soil is modeled by the abiotic soil relationship. This simulation relies solely on the rate of energy loss from dead material as: (1) dead material that is readily associated with its living source, (2) dead material that is not associated with the living source but is not mixed with a mineral soil, or (3) dead material mixed with a mineral soil and not associated with its living origin.

Realism — HIGH — TEEM is a parameter-intensive model with more than 50 state variables. The model captures all major physiological and ecological interactions that may potentially affect forest productivity.

Relevance — LOW — TEEM is principally a bioenergetics model designed to describe overall forest productivity and energy transfer. Individual constituents may be modeled separately for ecosystem characterization. The model would have to be extensively modified to address toxic chemical effects.

Flexibility — MEDIUM — TEEM provides comparatively more flexibility than other forest-gap models but is still limited by the number of different tree species considered. Specifically, the model is parameterized for an eastern deciduous forest.

Treatment of Uncertainty — LOW — TEEM is a deterministic model and does not track uncertainty.

Degree of Development and Consistency — MEDIUM — TEEM is not immediately available as a software package. However, descriptions provided in Shugart et al. (1974) were complete enough to allow programming and application of the model.

Ease of Estimating Parameters — MEDIUM — TEEM would require intensive parameterization to support the inordinately large number of state variables. However, many default values are provided for the eastern deciduous forest biome, so application of the model in the intended context would be relatively straightforward.

Regulatory Acceptance — LOW — TEEM has no regulatory status and appears not to have been used in any regulatory context.

Credibility — MEDIUM — Although TEEM depends on standard physiological and ecological principles and algorithms, no history of use could be identified for this model.

Resource Efficiency — MEDIUM — All forest-gap models within this category provide roughly equal types of outputs. TEEM was rated medium in resource efficiency because, although it is highly parameter intensive, apparently adequate default values are provided.

SHORT GRASS PRAIRIE MODEL

The Short Grass Prairie Model is an energy-flow model specific to short grass prairies that predicts interactions among primary producers, insect herbivores, and nesting clay-colored sparrows (*Spizella pallida*) (Jeffries 1989). Endpoints for this model are the biomasses for the respective biotic compartments.

The model is structured as a series of discrete difference equations representing multiple compartments with specific relationships controlling energy flows between them. Because predator density is assumed not to directly affect prey density, accumulation modeling (as opposed to collision modeling) is applied in this model. The food web consists of a single primary producer compartment, three separate consumers (representing three species of grasshopper), and five separate life stages within the sparrow population (adults, pre-laying embryos, post-laying embryos, nestlings, and fledglings). Each sparrow life stage is assigned a specific duration on the basis of growth and reproductive cycling within a single summer season. At an appointed time, all energy within a stage is transferred to the next level of development.

Because no import or export of metabolic carbon is considered, energy inputs are limited to net photosynthesis by primary producers. *Net photosynthesis* is defined as the product of the autotrophic assimilation rates and a circadian coefficient to account for diurnal changes in solar radiation. Availability of water is not considered rate-limiting for photosynthesis. Productivity is estimated by using a time-based sum of net assimilation, net storage, and net transfer to consumers. The distribution to consumers is assumed to be proportional to their species-specific relative population sizes. Ingestion rates, basal metabolic rates, and biomass assimilation rates are treated as constants estimated on the basis of empirical observations. Biomass of each of the sparrow life stages is determined from a time-dependent bioenergetics model. Energy transfer from grasshoppers to the sparrow life stages is also considered to be circadian and dependent on an assumed foraging period (12 h).

> **Realism** — LOW — The Short Grass Prairie Model is unrealistic because it reflects a highly limited ecosystem with no consideration of seasonal fluctuations or important environmental constraints, such as drought.
>
> **Relevance** — MEDIUM — The biomass endpoint is relevant for ecological risk assessment, but the model uses are limited to ecological characterization. The model lacks state variables or relationships to represent the effects of physical disturbance or toxic chemicals. Given its limited realism, modifying the model to account for toxic chemical effects is probably not worthwhile.
>
> **Flexibility** — HIGH — The Short Grass Prairie Model could generally be applied to any grassland ecosystem.
>
> **Treatment of Uncertainty** — LOW — Uncertainty and variability were not considered in this model.
>
> **Degree of Development and Consistency** — MEDIUM — This model is not immediately available as a software package. However, Jeffries (1989) provided sufficient detail to permit its programming and application.
>
> **Ease of Estimating Parameters** — HIGH — The model has low data requirements as determined on the basis of its limited number of parameters.
>
> **Regulatory Acceptance** — LOW — The model has no regulatory status and appears not to have been used in a regulatory context.
>
> **Credibility** — LOW — No history of use or references other than Jeffries (1989) could be identified for the Short Grass Prairie Model.
>
> **Resource Efficiency** — HIGH — Although the model is not immediately available as a software package, its application to any short grass prairie would be extremely easy.

SAGE

SAGE is an air pollution model that predicts the effects of sulfur dioxide on a grassland ecosystem (Heasley et al. 1981). Structurally, it is a multiple compartment model that simulates the movement of carbon, nitrogen, and sulfur among soil, grassland plants, and ruminant animals. Within this context, the model examines concentration-dependent impacts of sulfur dioxide on both plant growth and soil–litter composition. These impacts are cascaded through the model to simulate primary production, ruminant production, system sensitivity to secondary perturbations, and overall availability of nutrients. The primary driving variables are solar radiation, air temperature, precipitation, relative humidity, and wind speed. The model simulates the responses of a typical square meter of grassland generalized to the entire ecosystem. The difference equations representing the subsystem processes are iterated as daily time-steps.

SAGE is structured as five modules. The abiotic module consists of two submodules. The waterflow submodule simulates the flow of water through the plant canopy and several layers of soil. The temperature profile submodule simulates the daily solar radiation, maximum canopy air temperature, and soil temperature at 13 separate points within the soil profile. Input parameters to both these submodules include daily rainfall, cloud cover, wind speed, maximum and minimum air temperatures, and relative humidity (all determined at 2 m above soil level).

The primary producer module uses carbon, nitrogen, and sulfur as the principal state variables for grassland plants. Three plant types are modeled concurrently: cool-season grasses, warm-season grasses, and cool-season forbs. Structurally labile forms of carbon, nitrogen, and sulfur are represented separately. Above-ground plant parts are divided into young, actively growing tissue, and nongrowing but photosynthetically active mature tissue. Below-ground plant structures are separated into root crowns or rhizomes/roots. The photosynthesis model can distinguish between C_3 and C_4 plants. Dynamic nutrient uptake is modeled as a Michaelis–Menton relation modified to depend on soil temperature, soil moisture, and relative nutrient availability. Within the plant, available nutrients are allocated to meet maintenance functions (respiration) first, then growth and reproduction. Senescence is modeled as a function of tissue age, temperature, and soil moisture level. Standing dead material is assumed to enter the soil process module as litter fall.

The soil module simulates inorganic nutrient transformations, litter composition, microbial processes, fractionation of soil organic matter, and transport of nutrients between soil layers. The abiotic and biotic nutrients are modeled separately; the abiotic are represented as dynamic state variables (dependent on soil depth), and the biotic as a series of population life-cycle dynamics submodules. Fungi and bacteria are modeled separately, including their responses to sulfur dioxide concentrations.

The ruminant consumer module consists of differential sums equations representing food consumption, metabolic energy requirements, and nitrogen and sulfur requirements for growth and maintenance in a population. Separate hierarchical functions are used to model metabolic demand. The order of priority is (1) energy requirements that have been established to meet maintenance functions (basal metabolic rate, plus activity, plus thermoregulatory requirements); (2) biomass accumulation (growth); and (3) seasonally dependent reproduction and lactation.

The deposition of sulfur dioxide is simulated as a diffusion resistance model. All atmospheric sulfur converted to sulfate is made available as a nutrient within the primary producer module. If sulfur dioxide enters the leaf at a rate greater than that of the oxidation cascade (a rate limited by sulfite conversion to sulfate), then tissue will be damaged, essentially converting it into standing dead biomass in proportion to the sulfite concentration within the leaf. Stomatal resistance and carbon assimilation are also functionally linked to sulfur dioxide concentrations in the leaf.

Realism — HIGH — SAGE is an ecosystem model describing relationships among soils, plants, and grazing ungulates. The results of model validation suggest that the model is realistic for most of the governing processes.

Relevance — HIGH — SAGE specifically describes the effects of air pollutants on grassland productivity and the resulting effects on ruminants. The model could be extended relatively easily to other air pollutants.

Flexibility — HIGH — SAGE is a general model that could be applied to any grassland ecosystem with little structural modification.

Treatment of Uncertainty — LOW — SAGE does not track uncertainty or variability.

Degree of Development and Consistency — LOW — SAGE is not immediately available as a software package. Furthermore, Heasley et al. (1981) do not provide details on the algorithms to allow programming and application of the model. The model has been validated.

Ease of Estimating Parameters — LOW — Parameter estimation for SAGE is extremely difficult because of the large number of physiological parameters that represent higher level interactions.

Regulatory Acceptance — LOW — SAGE has no regulatory status and appears not to have been used in a regulatory context.

Credibility — MEDIUM — SAGE was developed on the basis of recognized environmental and physiological modeling approaches. However, no history of use or further development was evident.

Resource Efficiency — LOW — SAGE was judged to be an extremely difficult program to execute because of its extensive parameterization requirements, its large number of interconnected functions, and its lack of availability as a software package.

MODIFIED SWARD

The Modified SWARD Model is a multispecies model that examines the dynamic equilibrium between primary producers and consumers within a grassland ecosystem (White 1984). The model is designed to simulate the collective effects of grazing animals and insects on a variety of grassland types indigenous to New Zealand. The model is a dynamic (time-dependent), deterministic, compartment model based on a difference equation applied in daily time-steps. The primary producers are considered as an aggregate canopy as opposed to being modeled on an individual basis. Alternately, consumers are considered as individuals and only aggregated after the application of modifying functions. No decomposer or nutrient flow module is included.

Conceptually, the model is separated into two basic components: the primary producer component and the consumer component. Primary productivity is modeled by using an abiotic and a photosynthetic module. The abiotic module is primarily a water flow and temperature profile construct. Within the photosynthetic module, total carbon assimilation is modeled mechanistically as either prime (growth of the year) or old (storage from previous years) and separated into shoot, root, flower head, crown, or stem compartments. Initiation of flowering is regulated by accumulated carbon pool reserves, thus permitting aperiodic flowering. Allocation strategies are species-specific and responsive to current plant carbon balances. Four simultaneous parameterization options are provided to account for differences between growth forms and specific plant phenology. Seedling germination and radical/hypocotyl growth are not modeled, thereby excluding annual plants from consideration in simulations longer than a year.

The multispecies consumer component is structured with inputs and outputs interfaced directly with both the photosynthetic and the abiotic modules. Mechanistic representations of consumption, assimilation, and excretion processes are included in this component. As many as eight consecutive life stages can be represented simultaneously, with differential biomass transfer, reproduction, growth, and mortality rates. This structuring permits the overlapping of many stages at any given time period to match seasonal developmental shifts in one or more parameter values (dietary preference, activity metabolism, stock management changes). The model permits only one reproductive cycle per species per annum. However, multiple cycles can be substituted by representing a species as independent subpopulations with complementary cycles in parallel.

Realism — HIGH — The Modified SWARD Model considers not only the bioenergetics of primary producers but also growth rate-limiting factors, such as nutrient availability and grazing.

Relevance — MEDIUM — Despite its relevant endpoints, the Modified SWARD Model lacks state variables to represent extrinsic stressors such as physical disturbance or toxic chemicals. Such impacts could be simulated through indirect modification of the stock management coefficients for consumers. Overall, the model could yield information on grassland distribution and productivity that would be useful in characterizing an area potentially affected by toxicants.

Flexibility — HIGH — The Modified SWARD Model is general enough to be applied to any grassland ecosystem with little modification.

Treatment of Uncertainty — LOW — Variability and uncertainty are not tracked in this model.

Degree of Development and Consistency — LOW — The Modified SWARD Model is not immediately available as a software package.

Ease of Estimating Parameters — LOW — The model requires a high degree of parameterization with site specific data.

Regulatory Acceptance — LOW — The Modified SWARD Model has no regulatory status and appears not to have been used in a regulatory context.

Credibility — MEDIUM — The model is based on recognized approaches to modeling abiotic variables, primary producers, and consumers.

Resource Efficiency — MEDIUM — Compared with similar models, a moderate level of effort is required to run the Modified SWARD Model.

SPUR

SPUR simulates interactions among soils, plants, and grazing ungulates (Hanson et al. 1988). SPUR is composed of five modules: hydrology, domestic animals, wildlife species, economics, and plant growth. Only the latter module is reviewed here because other modules were not as well developed, and an integrated model combining all modules was not available for review.

The primary production module consists of differential sums equations, which are integrated over time and applied with daily time-steps. Nine species can be modeled simultaneously, and both intra- and interspecific competition are considered in the model. Abiotic variables used in the plant growth model include daily minimum and maximum temperatures of air and soil, precipitation, soil-water potential, daily solar radiation, accumulated wind run, and soil bulk density. The state variables include plant biomass and nitrogen and carbon content of various environmental compartments. The specific compartments within the model include standing green vegetation, live roots, propagules, standing dead vegetation, litter, dead roots, soil organic matter, and soil inorganic nitrogen. Carbon (biomass) accumulation is simulated by a photosynthesis subcomponent, which is a differential sums equation dependent upon solar radiation as well as soil–water potentials. It incorporates processes common to both C_3 and C_4 plants but cannot model plants that utilize crassulacean acid metabolism. Specific endpoints for this model are carbon/nitrogen available in forage to grazing ungulates on an areal basis.

Realism — HIGH — The SPUR plant model is a multicomponent dynamic flow model that describes variations in plant carbon accumulation and availability to grazing animals. It apparently simulates all major factors affecting both.

Relevance — MEDIUM — The specific endpoints for the SPUR plant model are grassland productivity and availability of forage to grazing mammals. It has potential utility in ecological risk assessment for evaluating chemical exposures on the basis of resource availability and forage requirements of these specific receptors. However, it does not address the effects of physical disturbance or toxic chemicals. Substantial modifications would be required to incorporate toxic chemical effects.

Flexibility — HIGH — The SPUR plant model could be applied easily to a wide range of grassland types.

Treatment of Uncertainty — LOW — The SPUR plant model as presented is deterministic and does not track uncertainty or variability.

Degree of Development and Consistency — HIGH — The SPUR plant module is a subcomponent of the overall SPUR Model software package and thus is available as software.

Ease of Estimating Parameters — LOW — The structure of the SPUR plant model is such that the results are highly dependent on detailed and accurate parameterization with empirical data. Default assumptions are available; however, their applicability has not been fully validated.

Regulatory Acceptance — HIGH — The parent program of SPUR was intended to fulfill a mandate of the Soil and Water Conservation Act (1978). SPUR is used by the U.S. Department of the Interior.

Credibility — HIGH — The SPUR plant model has not been extensively used or cited in the scientific literature. However, it was developed on the basis of several recognized plant ecosystem models.

Resource Efficiency — LOW — Because site specific parameterization is required, accurate and defendable results require a high level of empirical support. SPUR is currently available as software, so model implementation would require only nominal effort.

MULTI-TIMESCALE COMMUNITY DYNAMICS MODELS

Multi-timescale models are community dynamic models that describe the temporal aspects of bird community turnover (Russell et al. 1995). *Turnover* is defined as changes in the composition of a biological community as the result of either the immigration or the local extinction of species. Two models are presented. The first is an equilibrium model in which the number of species is assumed constant and therefore the number of immigrations is equal to the number of extinctions. The determination of probability of extinction (and thus the equilibrium turnover) is made on the basis of least-squares curve fitting by using data from 13 islands off the coast of Great Britain and the Republic of Ireland.

The second model is a dynamic model of island biogeography. Community expansion rates are determined by nonlinear regression of the natural logarithm of the number of species against time. In the model, changes in species number are described relative to variations in the probability of extinction. An error component is also included in the dynamic model to account for year-to-year turnovers that were in effect transient events not representative of actual extinction. This error function is quantified by comparing cumulative variation in species number with observed 4-year variation rates.

  Realism — LOW — Multi-timescale models consistently underpredicted observed species turnover
    rates. Therefore, a substantial moderating factor is not accounted for within the structure of these
    models.
  Relevance — MEDIUM — Multi-timescale models might prove useful in the identification of sub-
    populations potentially affected by multiple factors within the context of the regional avian popu-
    lation. However, they could not be directly applied in chemical risk assessments. These models do
    not incorporate parameters or functions that could be easily modified to account for effects of toxic
    chemicals.
  Flexibility — HIGH — These models could theoretically be applied in any situation involving physical,
    spatial, or temporal barriers limiting interactions between two or more components of a wildlife
    population or community.
  Treatment of Uncertainty — LOW — These models are deterministic; although they are parameterized
    on the basis of the probability of extinction, probability density functions associated with these
    estimates are not conserved in the final estimates of turnover.
  Degree of Development and Consistency — MEDIUM — Multi-timescale models are not immediately
    available as a software package. However, Russell et al. (1995) describe the models in sufficient
    detail to permit programming and application. Although the models were parameterized on the basis
    of observations from 13 independent islands, they were found to be inaccurate during a comparison
    of model predictions with field observations.
  Ease of Estimating Parameters — MEDIUM — These models are relatively easy to parameterize
    because the modeled receptors are defined as distinct populations and as such can be treated as
    single autonomous groups.
  Regulatory Acceptance — LOW — These models are not known to have any regulatory status or
    prior applications within a regulatory context.
  Credibility — MEDIUM — Multi-timescale models do not have any substantial recognition in the
    scientific literature. However, they were developed by using algorithms that are well established.
  Resource Efficiency — MEDIUM — These models require moderate effort and cost to implement.
    Because of the lack of a software package, some programming is needed to implement them.

NESTEDNESS ANALYSIS MODEL

The nestedness analysis model is a randomization model specific to nestedness subset analysis (Cook and Quinn 1998). The concept of *nestedness* depends on an ecological process whereby species occupying small or species-poor sites form a proper subset of richer species assemblages. This process implies that individual species have a strong tendency to be present in all assemblages of the same size as or a greater size than the smallest one in which they can occur. Under these conditions, all assemblages of species can be shown to form a nested series in which each is included as a subset of the next largest assemblage in the series.

Several algorithms are used to evaluate nestedness. The principle algorithm presented was RANDOM1. This algorithm is a class of null models designed to test patterns of interspecific association, notably the effects of competition. To parameterize this model, species observations are compared in a matrix to identify exclusive species pairs (pairs of species that never co-occur). The pattern of exclusive species pairs can then be compared with the frequency distribution of values generated by randomization of the differing habitat subsets. If no differences are found between the observed pattern of exclusive species pairs and the results of the random model, then the degree of nestedness is assumed to be zero.

Realism — MEDIUM — In the context of defining species interactions within open populations, the nestedness analysis model is adequate under many circumstances. However, Cook and Quinn (1998) identified some situations (such as predictions associated with bird migrations) for which it is known to underpredict the degree of nestedness.

Relevance — LOW — The model is designed to examine patterns resulting from competitive interactions. It could be parameterized to specifically test for effects of physical disturbance or toxic chemicals. However, the consideration of nestedness is not currently considered a relevant endpoint in ecological risk assessment.

Flexibility — HIGH — The nestedness analysis model represents a generic approach utilizing site specific matrices of observed species occurrence in comparison with a random pattern. Therefore, the structure of the model does not limit its application to a variety of systems.

Treatment of Uncertainty — HIGH — The model, which is founded on probabilistic analyses of deviation from randomness by subset analysis, provides a direct measure of community patterns.

Degree of Development and Consistency — LOW — The nestedness analysis model is not immediately available as a software package. Furthermore, Cook and Quinn (1998) provided insufficient information for programming and application of the model.

Ease of Estimating Parameters — MEDIUM — The receptors in the nestedness analysis model are defined as distinct populations requiring only the identification of exclusive species pairs. Therefore, parameterization is simplified compared with that for other types of ecosystem models.

Regulatory Acceptance — LOW — The nestedness analysis model is not known to have any regulatory status or prior applications in a regulatory context.

Credibility — MEDIUM — The concept of nestedness in itself is a highly contentious issue in ecology. However, if nestedness is considered credible, then the nestedness analysis model, which uses algorithms that are well established, could be considered credible.

Resource Efficiency — HIGH — The nestedness analysis model requires relatively little effort and cost to apply.

DISCUSSION AND RECOMMENDATIONS

Although many terrestrial ecosystem models have been developed for and applied in basic ecological research, none of those reviewed could be applied directly to most ecotoxicological risk assessments (Tables 10.2 and 10.3). Only the grassland model SAGE considered the effects of toxic chemicals (sulfur dioxide). Most of the models have not been accepted in regulatory programs (Table 10.2).

Terrestrial ecosystems are spatially heterogeneous such that the availability of resources, both primary (such as food) and secondary (such as cover), varies greatly with location. Consequently, ecosystem models that are not spatially explicit provide highly uncertain predictions and are not cost-effective for use in ecological risk assessment of toxic chemicals in terrestrial systems. Rather, it would be more efficient to use available population models or landscape models that permit spatially explicit parameterization.

Ecosystem models are best used as heuristic tools for understanding basic ecological processes and identifying sources of uncertainty in predictions. For example, ecologists have used them as descriptive constructs to evaluate the sensitivity of specific ecological parameters. In this way, they

Table 10.2   Evaluation of Ecosystem Models — Terrestrial Abiotic/Biotic Ecosystem Models

Model	Reference	Realism	Relevance	Flexibility	Uncertainty Analysis	Degree of Development	Ease of Estimating Parameters	Regulatory Acceptance	Credibility	Resource Efficiency
Terrestrial										
Desert										
Hierarchical model of *Dipodomy*	Maurer (1990)	♦♦	♦♦	♦	♦	♦♦	♦♦	♦	♦	♦♦
Forest										
FVS	USDA (1999)	♦♦♦	♦♦	♦♦♦	♦♦	♦♦♦	♦♦♦	♦♦♦	♦♦♦	♦♦♦
FORCLIM	Bugmann (1997); Bugmann and Cramer (1998)	♦♦♦	♦♦	♦♦	♦♦	♦♦♦	♦♦♦	♦♦	♦♦♦	♦♦♦
FORSKA	Lindner et al. (1997, 2000); Lindner (2000)	♦♦♦	♦♦	♦♦	♦	♦♦	♦♦♦	♦♦	♦♦	♦
HYBRID	Friend et al. (1993, 1997)	♦♦♦	♦♦♦	♦♦♦	♦	♦♦	♦♦♦	♦	♦♦♦	♦♦
ORGANON	OSU (1999a,b)	♦♦♦	♦♦	♦	♦♦	♦♦♦	♦♦♦	♦♦	♦♦♦	♦♦♦
SIMA	Kellomäki et al. (1992)	♦♦	♦♦	♦♦	♦♦	♦♦♦	♦♦			♦♦♦
TEEM	Shugart et al. (1974)	♦♦♦	♦	♦♦	♦	♦♦	♦♦	♦	♦	♦♦
Grassland										
Energy flow for short grass prairie	Jeffries (1989)	♦	♦♦	♦♦♦	♦	♦♦	♦♦	♦♦	♦	♦♦
SAGE	Heasley et al. (1981)	♦♦♦	♦♦♦	♦♦♦	♦♦	♦♦	♦♦	♦♦	♦♦	♦
SWARD	White (1984)	♦♦♦	♦♦	♦♦♦	♦	♦	♦		♦♦	♦♦
Rangeland										
SPUR (plant submodel)	Hanson et al. (1988)	♦♦♦	♦♦♦	♦♦♦	♦	♦♦	♦	♦♦♦	♦♦♦	♦
Island										
Multi-timescale community	Russell et al. (1995)	♦	♦♦	♦♦♦	♦	♦♦	♦♦	♦♦	♦♦	♦♦
Nested species sub-set analysis	Cook and Quinn (1998)	♦♦	♦	♦♦♦	♦♦♦	♦	♦♦		♦♦	♦♦♦

Note: ♦♦♦ – high
            ♦♦ – medium
            ♦ – low

Table 10.3 Applications of Terrestrial Ecosystem Models

Model	Species/Ecosystem	Location/Population	Reference
Desert[a]			
Hierarchical model Dipodomys	*Dipodomys* spp.	Arizona desert	Maurer (1990)
Forest			
FVS	Temperate zone forest	United States	USDA (1999 and references therein)[b]
FORCLIM	Beech/oak transition forest stands	Europe; North America	Bugmann and Cramer (1998)
FORSKA	Boreal forest under production	Bavaria, Scandinavia, and other European areas	Lindner et al. (1997, 2000); Lasch et al. (1999); Lindner (2000)
HYBRID	Oak forest	Tennessee	Friend et al. (1993, 1997)
ORGANON	Mixed conifers	Oregon	OSU (1999)
SIMA	Forest (type user-defined)	Continental United States	Kellomäki et al. (1992)
TEEM	Forest (type user-defined)	User-defined	Shugart et al. (1974)
Grassland			
Energy flow for short grass prairie	Multiple dominant grass species, grasshoppers, reproductive sparrows	Saskatchewan boreal forest	Jeffries (1989)
SAGE	Multiple grassland species; sulfur dioxide impacts	User-defined	Heasley et al. (1981)
SWARD	Multiple grassland species, sheep	New Zealand; depleted open grassland	White (1984)
Rangeland			
SPUR (plant submodel)	C_3 and C_4 plants (excludes Crassulacean acid-dependent succulents)	Semi-arid rangeland in Texas and Colorado; user-defined	Hanson et al. (1988)
Island			
Multi-timescale community dynamics	User-defined avian species	User-defined	Russell et al. (1995)
Nested species subset analysis	Multiple avian species	California	Cook and Quinn (1998)

[a] INTASS, which is evaluated in Chapter 9, Aquatic Ecosystem Models, has also been applied to desert plants (Emlen et al. 1989, 1992).
[b] FVS has been applied extensively to forests throughout the U.S., including assessments of disturbances by fire and insects (see USDA 1999 for examples).

may be useful in characterizing ecosystems or in identifying key parameters for other models. For risk assessment of toxic chemicals in terrestrial systems, further development of landscape models is likely to prove more useful than the development of existing ecosystem models. Hence, none of the terrestrial ecological models reviewed (Table 10.2) was recommended for further evaluation and development as a tool for chemical risk assessment.

CHAPTER 11

Landscape Models — Aquatic and Terrestrial

Christopher E. Mackay and Robert A. Pastorok

In contrast to ecosystem models, which are spatially aggregated models, *landscape models* are spatially explicit models that may include several types of ecosystems. In landscape models, the values of one or more state variables are dependent upon either distance or relative location. A landscape model may be totally constructed on a spatial basis, such as cellular automata models using a GIS platform. Some ecosystem models can be easily applied in a landscape mode. For example, AQUATOX is currently being applied to the Housatonic River in Connecticut by dividing the model into discrete segments and linking results from each segment to input information for downstream segments (Beach et al. 2000). Thus, models like AQUATOX and CASM were considered in the development of recommendations for landscape models.

Example endpoints for landscape models include:

- Spatial distribution of species
- Abundance of individuals within species or trophic guilds
- Biomass
- Productivity
- Food-web endpoints (e.g., species richness, trophic structure)
- Landscape structure indices (Daniel and Vining 1983; FLEL 2000a,b; Urban 2000)

We review the following landscape models (Table 11.1):

- Marine and Estuarine
  - ERSEM (European regional seas ecosystem model), a model of marine benthic systems (Eben-hoh et al. 1995; Baretta et al. 1995)
  - Barataria Bay ecological model, a model of an estuary (Hopkinson and Day 1977)
- Freshwater and Riparian
  - CEL HYBRID (coupled Eulerian LaGrangian HYBRID), a coupled chemical fate and ecosystem model for lakes and rivers (Nestler and Goodwin 2000)
  - Delaware River Basin model, a segmented river model (Kelly and Spofford 1977)
  - Patuxent River Watershed model, a whole watershed model comprising ecological and economic systems (Voinov et al. 1999a,b; Institute for Ecological Economics 2000)

Table 11.1 Internet Web Site Resources for Aquatic and Terrestrial Landscape Models

Model Name	Description	Reference	Internet Web Site
ERSEM	A marine benthic ecosystem model for European regional seas	Ebenhoh et al. (1995); Baretta et al. (1995)	http://www.ifm.unihamburg.de/~wwwem/res/ersem.html
Barataria Bay ecological model	An early generation model of an estuarine system	Hopkinson and Day (1977)	Updated models at: http://lts2.ocs.lsu.edu/guests/wwwcei/modeling.html
CEL HYBRID	Models that combine population dynamics with detailed fate modeling for toxic chemicals	Nestler and Goodwin (2000)	http://www.wes.army.mil/el/elpubs/genrep.html
Delaware River Basin model	A spatially explicit model of a river system	Kelly and Spofford (1977)	http://www.state.nj.us/drbc/over.htm
Patuxent River watershed model	A watershed model incorporating human interactions	Voinov et al. (1999a,b); Institute for Ecological Economics (2000)	http://iee.umces.edu/PLM/PLM1.html
ATLSS	A landscape modeling system for the Everglades with specific modeling approaches tailored to each trophic level	DeAngelis (1996)	http://www.atlss.org/; http://sofia.usgs.gov/projects/atlss/
Disturbance to wetland vascular plants model	A spatially explicit model for predicting the impacts of hydrologic disturbances on wetland community structure	Ellison and Bedford (1995)	http://www.mtholyoke.edu/offices/comm/profile/ellisoncv.html
LANDIS	A landscape model for describing forest succession over large spatial and temporal scales	Mladenoff et al. (1996); Mladenoff and He (1999)	http://www.nrri.umn.edu/mnbirds/landis/landis.htm
FORMOSAIC	A cellular automata landscape model	Liu and Ashton (1998)	http://www.ctfs.si.edu/newsletters/inside1999/liu1999.htm
FORMIX	A landscape model for a tropical forest	Bossel and Krieger (1991)	http://eco.wiz.unikassel.de/model_db/mdb/formix.html
ZELIG	A forest landscape model with probabilistic mortality functions	Burton and Urban (1990)	http://www-eosdis.ornl.gov/BOREAS/bhs/Models/Zelig.html; http://eco.wiz.uni-kassel.de/model_db/mdb/zelig.html
JABOWA	A highly developed landscape model for mixed species forests	Botkin et al. (1972); West et al. (1981); Botkin (1993a,b)	http://www.naturestudy.org/services/jabowa.htm; http://eco.wiz.uni-kassel.de/model_db/mdb/jabowa.html
Regional landscape model	A model for evaluating the impact of ozone exposure upon forest stands and associated water bodies	Graham et al. (1991)	N/A
Spatial dynamics of species richness model	A model for evaluating the effects of habitat fragmentation on species richness	Wu and Vankat (1991)	N/A
STEPPE	A gap-dynamic model of grassland productivity	Coffin and Lauenroth (1989); Humphries et al. (1996)	http://eco.wiz.unikassel.de/model_db/mdb/steppe.html
Wildlife-urban interface model	A vegetation cover and wildlife habitat utilization model for evaluating the impacts of urban development	Boren et al. (1997)	N/A
SLOSS	A model of nestedness of species assemblages in habitat patches of varying size	Boecklen (1997)	N/A
Island disturbance biogeographic model	A model for evaluating the effects of perturbations on the distribution of species within a series of linked island habitats	Villa et al. (1992)	http://www.uchaswv.edu/courses/bio345-01/biogeo.htm; http://fp.bio.utk.edu/bio250/lab/jamie/island_biogeography.html
Multiscale landscape model	A model of landscape structure based on the probability of species occurrences	Johnson et al. (1999)	http://es.epa.gov/ncerqa_abstracts/grants/97/ecoind/richards.html

Note: N/A - not available

- Wetland
  - ATLSS (across-trophic-level system simulation), a landscape model of the Everglades (see DeAngelis 1996)
  - Disturbance to wetland vascular plants model, a model of wetland plant communities (Ellison and Bedford 1995)
- Forest
  - LANDIS (landscape disturbance and succession), a forest landscape model along with the following four models (Mladenoff et al. 1996; Mladenoff and He 1999)
  - FORMOSAIC (forest mosaic) model (Liu and Ashton 1998)
  - FORMIX (forest mixed) model (Bossel and Krieger 1991)
  - ZELIG (Burton and Urban 1990)
  - JABOWA (Botkin et al. 1972; West et al. 1981; Botkin 1993a, b)
  - Regional landscape model, a model of ozone effects on a forest and associated water bodies (Graham et al. 1991)
  - Spatial dynamics of species richness model, a model to evaluate the effects of habitat fragmentation (Wu and Vankat 1991)
- Grassland
  - STEPPE, a gap-dynamic model of grassland productivity (Coffin and Lauenroth 1989; Humphries et al. 1996)
  - Wildlife-urban interface model, a model to predict the effects of human activities on wildlife (Boren et al. 1997)
- Island
  - SLOSS (single large or several small), a model of distribution of species assemblages in habitat patches (Boecklen 1997)
  - Island disturbance biogeographic model, a model of species distributions within linked island habitats (Villa et al. 1992)
- Multi-scale
  - Multi-scale landscape model, a model of landscape structure based on probability of species occurrences (Johnson et al. 1999).

ERSEM

ERSEM was developed as a comprehensive model of carbon dynamics and major nutrients (nitrogen, phosphorus, silicon) along the coastal shelf of the North Sea (Ebenhoh et al. 1995; Baretta et al. 1995). The model represents the North Sea as a set of "geographical boxes" that describe regional differences in physical, chemical, and biological characteristics in one to three dimensions. The model consists of pelagic, benthic, and transport submodels. The pelagic submodel includes populations of phytoplankton, zooplankton, and fishes representative of the regions. The benthic component of the model is connected to the pelagic production dynamics by the settling of pelagic detritus and sinking diatoms. The benthic submodel emphasizes the biology of the benthic organisms, the functional importance of bioturbation, and the role of nutrient profiles in regulating microbial activity. The biological populations are based on the concept of functional groups with common processes such as food intake, assimilation, respiration, mortality, and nutrient release but with different parameters for each group.

ERSEM has been used to examine the functional dependence of the benthic system on inputs from the pelagic system, the importance of predation as a stability-conferring process in model subsystems, and the importance of detritus recycling in the benthic food web. The kinds of data inputs needed for ERSEM include annual cycles of monthly mean (or median) values together with ranges of variability, time series of river input of dissolved and particulate nutrient loads for all continental rivers, time series of daily water flow across the borders of horizontal compartments, time series of solar irradiance, and time series of boundary conditions for nutrients.

Realism — HIGH — The overall spatial structure and detailed physical, chemical, and biological components of ERSEM suggest that the model provides a realistic description of major features of the North Sea.

Relevance — HIGH — The endpoints for modeled organisms in both the pelagic and benthic submodels are useful for assessing ecological impacts and risks posed by chemical contaminants. Although the model does not explicitly account for toxic chemical effects, several parameters could be adjusted by the user to implicitly model toxicity.

Flexibility — MEDIUM — The modeling framework has been developed for the North Sea. However, the geographical-box model approach might be adapted for other similarly scaled marine systems.

Treatment of Uncertainty — LOW — ERSEM has not been the subject of detailed sensitivity or uncertainty analyses.

Degree of Development and Consistency — MEDIUM — The development of ERSEM as a set of coupled submodels might lend the model to application to other systems. The model has been implemented, and a software version is probably available from the authors.

Ease of Estimating Parameters — MEDIUM — The model has a considerable number of physical, chemical, and biological parameters to estimate. However, the parameters have fairly understandable interpretations that can facilitate estimation.

Regulatory Acceptance — LOW — ERSEM was constructed to evaluate impacts of nutrients introduced to the North Sea. The model has regulatory applicability, but the reference did not specifically mention any U.S. or international regulatory use or acceptance.

Credibility — MEDIUM — Model calibration and model:data comparisons suggest that the model captures some of the key ecological dynamics characteristic of the North Sea. However, few published references to the model exist, and the number of actual users is unknown but presumably fewer than 20.

Resource Efficiency — LOW — The spatial nature of the model, combined with the food-web detail in the pelagic and benthic submodels, suggests that the model would require a major commitment of resources to implement for specific case studies.

BARATARIA BAY MODEL

The Barataria Bay model is an early generation model that describes carbon and nitrogen flows within an open estuarine ecosystem (Hopkinson and Day 1977). Although the state variables are not directly distinguished with regard to space, transfer coefficients representing fluxes between model compartments are distance-dependent. Seven state variables are tracked for carbon (biomass) and nine state variables for nitrogen (rate-limiting nutrient). Living marsh plants are modeled as the dominant species, *Spartina alterniflora*. The nonmarsh plants consist almost exclusively of phytoplankton. Two separate detrital communities were modeled, one in association with a marsh, and one in association with the open marine environment. Both include not only litter material but also associated decomposing organisms such as bacteria and fungi. Both also exhibit similar dynamics because detritus from higher-level marsh plants is transported by tidal action from the marsh into the marine environment. Therefore, differences between the two detrital communities were primarily due to differing relative amounts of plankton, zooplankton, and high-level plant material inputs. A single state variable for marsh fauna accounted for insects, raccoons, muskrats, birds, snails, crabs, and mussels. Similarly, the state variable for marine fauna accounted for all fish.

Transfer relationships between the state variables are based on steady-state kinetics. Estimates of transfer coefficients were calculated as the product of the compartment capacity (e.g., biomass of zooplankton) at equilibrium and the modeled rate of change in capacity.

Realism — LOW — The Barataria Bay model uses a rudimentary approach to modeling landscape effects by embedding the spatial constituents within the underlying algorithm. This embedding was done by spatially defining all of the state variables and thus making the transfer coefficients distance-dependent. Generalized definitions of state variables such as marsh fauna and marine fauna make the model less realistic than similar models. Results from simulations indicate that this aggregation has the greatest effect on the model's overall realism.

Relevance — LOW — The Barataria Bay model primarily describes the dynamic flows of carbon and nitrogen in the estuarine environment. Because food-web components are highly aggregated in this model, it has limited relevance for ecological risk assessment of toxic chemicals.

Flexibility — LOW — The Barataria Bay model is the least flexible of the aquatic landscape models. Its inherent structure defines fixed spatial compartments within the model. Moreover, its steady-state approach to defining the major state variables limits applications.

Treatment of Uncertainty — LOW — Neither uncertainty nor variability was tracked in the execution of this model.

Degree of Development and Consistency — MEDIUM — The model was not validated. Although this model was developed as software, no indication exists as to its availability. However, Hopkinson and Day (1977) provide sufficient details for programming and application of the model.

Ease of Estimating Parameters — LOW — The Barataria Bay model is fairly complex and must be parameterized with empirical data.

Regulatory Acceptance — LOW — To our knowledge, the model does not have any regulatory status and has not been applied in a regulatory context.

Credibility — MEDIUM — The Barataria Bay model depends on very fundamental modeling techniques and contains no mechanistic functions.

Resource Efficiency — HIGH — The model was deemed efficient to implement because, although it is heavily parameterized, the parameters are estimated on the basis of steady-state conditions.

CEL HYBRID

CEL HYBRID is a spatially explicit model for aquatic ecosystems developed by researchers at the U.S. Army Corps of Engineers (Nestler and Goodwin 2000). This model attempts to join the disparate mathematical approaches of population dynamics with chemical fate modeling. The idea is to integrate biological functions and physical processes by using a mixed-modeling framework. The approach includes a semi-Lagrangian model (Priestly 1993) in which physical and chemical processes are modeled on a Eulerian grid and biological organisms are modeled with a separate individual-based model (Figure 11.1). The points of connection between the two systems update times at which localized biomasses representing organisms are integrated (or perhaps appropriately averaged) over the spatial grid. This approach permits the representation of real feedback between the chemistry and the biology. An individual-based population model is a specific example of the broader CEL HYBRID approach to modeling. What individual-based modeling does for population modeling, CEL HYBRID does for ecosystem modeling (Nestler 2001, pers. comm.).

The modeling strategy inherent in CEL HYBRID has subtle problems in maintaining conservation when any sources or sinks are present and a problem with inflation of error when the two time-steps are not identical. It would be useful to somehow enable the individual-based component to handle extremely large numbers of individuals, such as might be necessary for fish in reservoirs. Supercomputing might facilitate this, but the solution might eventually involve hybridizing the individual-based approach with a frequency-based model in which some "individuals" are really exemplars that represent an entire class of similar organisms.

Realism — HIGH — CEL HYBRID could incorporate key population-dynamic and chemical processes, including density dependence, physical transport (for both chemicals and organisms), chemical uptake, bioaccumulation, and toxicant kinetics. Because the model has not been fully articulated, we cannot assess the number of assumptions it requires.

Relevance — HIGH — CEL HYBRID provides output that is directly relevant to the endpoints used in population-level ecotoxicological risk assessment. Several parameters can be used to describe the ecosystem-level impacts of toxic chemicals.

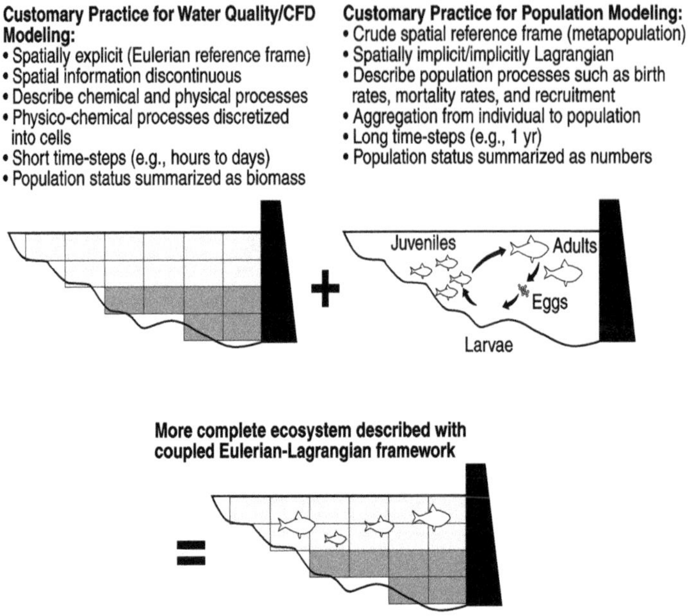

Customary Practice for Water Quality/CFD Modeling:
- Spatially explicit (Eulerian reference frame)
- Spatial information discontinuous
- Describe chemical and physical processes
- Physico-chemical processes discretized into cells
- Short time-steps (e.g., hours to days)
- Population status summarized as biomass

Customary Practice for Population Modeling:
- Crude spatial reference frame (metapopulation)
- Spatially implicit/implicitly Lagrangian
- Describe population processes such as birth rates, mortality rates, and recruitment
- Aggregation from individual to population
- Long time-steps (e.g., 1 yr)
- Population status summarized as numbers

More complete ecosystem described with coupled Eulerian-Lagrangian framework

Figure 11.1   Structure of the CEL HYBRID Model. (From Nestler and Goodwin (2000) Simulating Population Dynamics in an Ecosystem Context Using Coupled Eulerian-Lagrangian Hybrid Models (CEL HYBRID Models). ERDC/EL TR-00-4, U.S. Army Engineer Research and Development Center, Vicksburg, MS.)

Flexibility — HIGH — CEL HYBRID could permit alternate formulations of the dose–response functions. It could also support several different models of population growth. The model should be applicable to a wide variety of organisms in different environments.

Treatment of Uncertainty — LOW — In principle, one could introduce uncertainty and risk analysis into CEL HYBRID by enclosing the model within a Monte Carlo shell. However, the computation costs for this approach are likely to be quite high.

Degree of Development and Consistency — LOW — The inner workings of CEL HYBRID are fairly difficult to understand. The model has not yet been implemented in software. The programming effort needed for this task is considerable. Nevertheless, elementary feasibility and consistency checks would be simple to implement.

Ease of Estimating Parameters — LOW — The effort needed to estimate parameters for CEL HYBRID (once they have been specified) could be substantial.

Regulatory Acceptance — MEDIUM — The model is being developed by scientists at the U.S. Army Corps of Engineers, which is a regulatory agency. Although the model has not yet been used, it will likely be supported and used by the U.S. Army Corps of Engineers in the future.

Credibility — LOW — CEL HYBRID is unknown in academia; few publications describe the approach and, as yet, the model has no applications.

Resource Efficiency — LOW — Applying CEL HYBRID to a particular case would require programming, testing, debugging, and data collection.

DELAWARE RIVER BASIN MODEL

The Delaware River Basin model is a spatially segmented river model designed to evaluate effects of nutrients and toxic chemicals, specifically phenolic compounds (Kelly and Spofford 1977). As a segmented river model, the environmental conditions in the upstream reaches affect conditions in successive downstream reaches. The reaches within the model are treated as homogeneous mixed water bodies with net active water flow serving as the only link between regions. The model is structured as a generalized compartment model using differential equations describing rates of change in state variables. Because the principal application of the model was within a static economic framework, all relationships were designed to describe steady-state conditions.

Biotic compartments within the model are defined as trophic levels to allow evaluation of toxicological impacts on ecologically relevant endpoints such as biomass of primary producers, herbivores (zooplankton), omnivores (fish), and decomposers (bacteria) (Figure 11.2). Abiotic parameters, specifically nitrogen, phosphorus, organic matter, and dissolved oxygen, are included as inputs to functions regulating the rates of transfer of matter or energy among the principal biological state variables. The other state variables, phenolic toxicants and temperature, are included as extrinsic factors affecting the biotic systems.

Aside from primary producers, the definition of the biomass at each trophic level depends on two main processes in each reach. The first is direct input from upstream reaches. The second is accumulation of biomass as a result of ingestion and carbon accumulation. This second input depends on prey availability, predator population size, temperature, and oxygen concentration, as well as the concentration of toxic chemicals. For the most part, functions were empirically derived as either exponential or inverse relationships. Other processes that limited biomass accumulation were respiration, death, excretion, predation, and loss downstream. Rates of predation depend upon the relative population sizes of each predator and prey pair.

To model primary producers, the rate of nutrient uptake is determined on the basis of two concurrent Michaelis–Menton relationships (one for phosphorus and one for nitrogen), both modified by coefficients dependent on the availability of light in the water column. Light availability in turn depends on surface-level radiation, water turbidity, and the water depth profile. Grazing rates are modeled as a function of the abundance of primary producers, the abundance of consumers, and the individual consumers' ingestion rates.

Concentrations of toxic chemicals in biota depend on empirical determinations of uptake and release rates. Release rates are inversely proportional to a concentration-dependent detoxification rate. The derivation of the exposure–response relationship to account for toxicity was not discussed.

Realism — MEDIUM — The Delaware River Basin model simulates transfer of mass, nutrients, and energy between trophic guilds on the basis of spatial locations. The relationships defined in the model appear adequate to account for the main ecological interactions. The assumption of homogeneity within each river reach requires careful differentiation of river reaches under real environmental conditions.

Relevance — HIGH — The Delaware River Basin model is specifically designed to evaluate the effects of toxic chemicals on biomass at various trophic levels (Figure 11.3). The model has been parameterized for phenolic compounds.

Flexibility — HIGH — The model uses a river reach structure and therefore could potentially be applied to other riverine ecosystems.

Treatment of Uncertainty — LOW — Neither uncertainty nor variability is tracked in the structure of the Delaware River Basin model.

Degree of Development and Consistency — MEDIUM — Although the Delaware River Basin model was developed as software, its availability is unclear. However, Kelly and Spofford (1977) provide sufficient details to program and apply the model. No validation of the model was done.

Ease of Estimating Parameters — LOW — The Delaware River Basin model requires separate parameterization for each of the river reach units that compose the landscape. Furthermore, almost all modifying relationships acting upon the biological state variables are empirically derived. Therefore, it is considered to be highly data intensive.

Regulatory Acceptance — MEDIUM — The model was developed as part of the Delaware River Basin Commission's Resources for the Future research program. However, there is no indication in the cited reference or on its Internet web site that it was used within a regulatory context.

Credibility — MEDIUM — The Delaware River Basin model is the product of a history of development of aquatic trophic-interaction models. However, there is no information about its acceptance or future development.

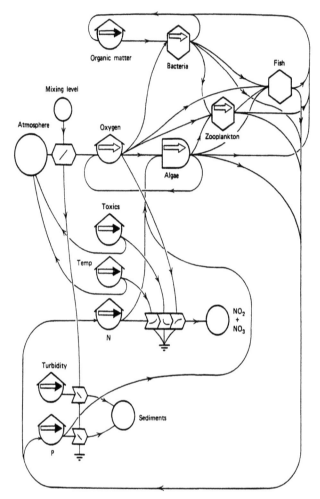

Figure 11.2 Structure of the Delaware River Basin model. (From Kelly and Spofford (1977). Application of an ecosystem model to water quality management: the Delaware estuary. Chapter 18. In *Ecosystem Modeling in Theory and Practice*. C.A.S. Hall and J.W. Day, Jr., (Eds.). John Wiley & Sons, New York. With permission.)

Resource Efficiency — HIGH — The Delaware River Basin model is considered to be among the most efficient of the aquatic landscape models because of the relatively limited number of parameters and comparatively simple structure.

PATUXENT WATERSHED MODEL

Voinov et al. (1999a, b) developed a spatially explicit model of the Patuxent River watershed (see also Institute for Ecological Economics 2000). The major model components include a land-use conversion submodel, a hydrology model, and an ecological model that consists of nutrient, macrophyte, consumer, and detritus submodels. Submodels also have been developed to examine production dynamics in forested and agricultural components of the watershed. The model is used to address questions about the dynamic linkages between land use and the structure and function of terrestrial and aquatic ecosystems, the role of natural and anthropogenic stressors and how their effects change with scale, and the economic effects of alternative management strategies and policies.

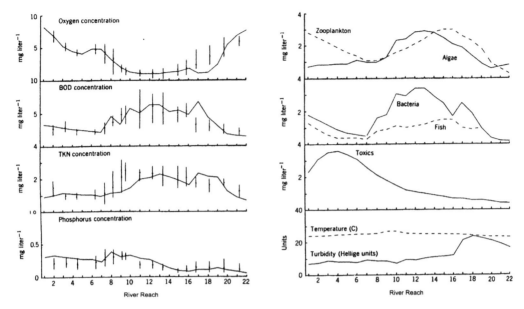

Figure 11.3   Example output of the Delaware River Basin model. Note: Vertical bars show variability for data. (From Kelly and Spofford (1977). Application of an ecosystem model to water quality management: the Delaware estuary. Chapter 18. In *Ecosystem Modeling in Theory and Practice*. C.A.S. Hall and J.W. Day, Jr., (Eds.). John Wiley & Sons, New York. With permission.)

The Patuxent model subdivided the watershed into a set of individual landscape units linked within a GIS, and the submodels are set up for each of these spatial units. The Patuxent model has been implemented in an integrated simulation system called the Spatial Modeling Environment. Spatial scales can be specified as 200 m or 1 km. The different submodel components are calibrated independently at spatial and temporal scales of resolution corresponding to scaled data sets.

Within the Patuxent model, the general ecosystem model (GEM) (Fitz et al. 1996) is designed to simulate a variety of ecosystem types using a fixed structure across a range of scales (Institute for Ecological Economics 2000). GEM predicts the response of macrophyte and algal communities to simulated levels of nutrients, water, and other environmental inputs determined from outputs of algorithms for upland, wetland, and shallow-water habitats. It explicitly incorporates ecological processes that determine water levels, plant production, nutrient cycling associated with organic matter decomposition, consumer dynamics, and fire. Biomass values of producers and consumers, as well as phosphorus and nitrogen, can be simulated on an annual time scale for different land-use categories. GEM is essentially an ecosystem model that can simulate system dynamics for a single homogenous habitat. GEM is replicated throughout the framework of the overall grid-based model using different parameter sets for each habitat to create the landscape-level analysis. The developers used a basic version to simulate the response of sedge and hardwood communities to varying hydrologic regimes and associated water quality.

GEM expresses the dynamics of various ecological processes as the interaction between state variables (biological stocks) and flows of material, energy, and information (Institute for Ecological Economics 2000). Vertical or within-cell dynamics are simulated, and the landscape modeling program processes the results of the unit models. The spatial model calculates the exchange of material between grid cells and simulates temporal changes in water availability, water quality, and landscape structure related to habitat or ecosystem type. For each grid cell, a successional algorithm redefines the habitat/ecosystem type of cells as conditions change and selects parameter sets as necessary. Ecosystem functions and parameters for each grid cell are determined by the cell's land use or habitat designation at the beginning of any simulation time-step. The ecological processes and fluxes are calculated according to that land use and the values of the state variables at that time for the cell. Human activities can affect the system simulation through the land-use designation of a cell or through the ecological processes that occur within a cell conditioned on its land use.

Realism — HIGH — The Patuxent watershed model considers the hydrological, biological, economic, and spatial factors that are important for describing the ecological characteristics of the watershed.

Relevance — HIGH — The ecological populations and endpoints that are represented in the Patuxent watershed model are of concern and are commonly represented in ecological risk assessments. Although the model does not explicitly account for toxic chemical effects, several parameters could be adjusted by the user to implicitly model toxicity.

Flexibility — MEDIUM — The model was developed specifically for the Patuxent watershed. However, the general ecosystem model that provides the main ecological component (GEM) of this overall modeling construct could be applied to other aquatic ecosystems.

Treatment of Uncertainty — MEDIUM — Sensitivity and uncertainty analyses have been done on some parts of the Patuxent watershed model; submodel components could be placed in a Monte Carlo uncertainty analysis framework.

Degree of Development and Consistency — HIGH — The Patuxent watershed model is highly developed and can be accessed on the Internet. It is well documented with examples of applications.

Ease of Estimating Parameters — MEDIUM — Given the spatial detail of the Patuxent watershed model, many parameters for a wide range of physical, chemical, and biological processes are required to run the full model. The parameters in general have clear process-level meaning, and many might be estimated from the data usually available for well-studied watersheds.

Regulatory Acceptance — LOW — The Patuxent watershed model was developed by an educational institution. No reference was made to regulatory acceptance or recommendation.

Credibility — MEDIUM — Results from individual model components were comparable for the most part with observed data for the Patuxent, but no reported results from implementation of the full model were available. The Patuxent watershed model is a modified version of the coastal landscape simulation model developed by Costanza et al. (1990).

Resource Efficiency — LOW — Applications to case studies that did not directly involve the Patuxent watershed would require substantial efforts in parameter estimation. However, major reprogramming efforts probably would not be required.

ATLSS

ATLSS is a multicomponent modeling framework for the Florida Everglades that is constructed in a cellular automata format (DeAngelis 1996). ATLSS is a set of integrated models that simulate the hierarchy of whole-system responses across all trophic levels and across spatial and temporal scales that are ecologically relevant to a large wetland system like the Everglades (Figure 11.4). ATLSS uses different modeling approaches tailored to each trophic level, including differential equations for process models of lower levels and age-structured and individual-based models for higher levels. Much of ATLSS was developed on the basis of empirical data for the Everglades.

In ATLSS, process models are used for modeling lower trophic levels (periphyton and macrophytes, detritus, micro-, meso- and macroinvertebrates), with a series of differential equations defining state variables for biomass of various taxonomic or functional groups. To account for seasonality, the growth and death parameters vary sinusoidally over the year. This allows the system to respond differentially to perturbations occurring during different times of the year. No functions in the process models represent predation losses of plant or macroinvertebrate biomass. Rather, such consumption is considered a separate state variable calculated by modules that describe these higher trophic-level consumers. The amount of material consumed is subtracted from the appropriate state variables in a lower trophic module before its next iteration.

In the detritus model, the generation of detritus is proportional to the death term in the primary productivity module. The disappearance of detritus is proportional to the current stock of detritus modified by a seasonal coefficient. The growth of the invertebrate group is assumed to vary with

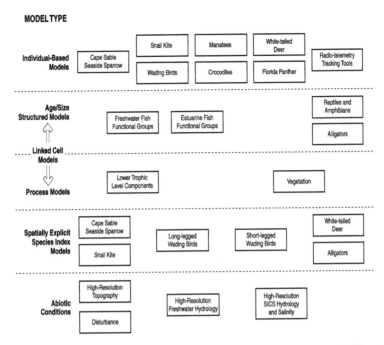

Figure 11.4 Structure of across-trophic-level system simulation (ATLSS) multimodel. (From http://atlss.org/science.forum.html. With permission.)

the product of this group's own populations and the plant or detritus stocks upon which they graze. As with the plant models, death varies with the square of their respective stocks, thereby keeping their populations constrained.

Age-structured population models are used to simulate intermediate trophic levels consisting of five functional groups of macroinvertebrates and fishes (see review of ALFISH in Chapter 7, Population Models — Metapopulations). Each spatial cell within the landscape is assumed to be homogeneous, with a certain carrying capacity for macroinvertebrates and fish, as determined by the process module. Mature individuals of each functional group produce a set number of viable offspring during their reproductive season. Baseline age-dependent mortality is assigned to each functional group on the basis of empirical observations. Rates of predation of larger fish on smaller fish are a function of the ratio of predator biomass to prey biomass.

Individual-based models are used to simulate the top-level consumers (see reviews of SIMP-DEL, SIMSPAR, and the wading bird nesting colony model in Chapter 6, Population Models — Individual-Based Models). For example, the wading bird nesting colony model simulates the activities of reproductive adults just before and throughout the nesting season as well as the activities of offspring. Prey densities are defined by values returned for the state variables in the macroinvertebrate and fish guild models. Each bird decides when to forage and chooses foraging locations on the basis of its knowledge of the system (e.g., knowledge related to prey density in a given cell). The simulation starts near the end of a wet season, when prey are assigned densities across the landscape. The prey in a given cell are assumed to be available to the wading birds in a given cell only when the average water level of the cell is within the bird's foraging depth range. The foraging efficiency of wading birds is a function of the number of prey in the cell. The bird functions are programmed such that birds tend to reside for longer periods in cells with high prey densities.

The nesting adults are described by a set of species-specific algorithms that govern their behaviors from one time interval to the next. This set is structured as a decision tree. The first choice is whether to nest. Nesting begins if the female is able to obtain 20% more than her food requirements for 3 consecutive days. Egg production is asynchronous, and hatching takes place over a period of weeks. Each adult must meet a maintenance energy demand each day. A decision nexus is established at which the adult decides when to bring food back to its nestlings. First, food is allocated to self and, when satisfied, then to offspring. The nestlings compete for food, with a greater proportion of the food taken by the largest nestling. If a nestling receives less than a defined percentage of its cumulative food needs during any 5-day interval, it dies. Likewise, if the parents cannot find enough food to meet their requirements, they will be assigned to the nonnesting state variable, and the nestlings die. Fledging occurs after a prescribed age and once a threshold level of accumulated food has been acquired. If the nestling does not receive this amount of food before a prescribed time, then it dies.

Realism — HIGH — ATLSS is a complex model with detailed algorithms that provide a realistic simulation of the Everglades. The main weakness of ATLSS is its reliance on the assumption of homogeneity of hydroperiod within cells greater than or equal to 4 km^2.

Relevance — MEDIUM — ATLSS can provide a very useful tool for the integrated characterization of wetland communities on both landscape and ecosystem levels. The model endpoints, including species abundances and biomass, species richness, and organism distributions, are ecologically relevant. The current version of ATLSS focuses on the ecological effects of variations in the hydrological regime. Although it does not presently consider the effects of toxic chemicals, it is being modified to account for the effect of mercury on receptors at high trophic levels.

Flexibility — HIGH — ATLSS's true strength is that it accommodates different types of models as modules within its overall structure and allows users to choose an appropriate level of resolution to simulate upper trophic levels. This strength makes the model very flexible. As a spatially explicit approach, the ATLSS framework could potentially be applied to other landscape systems.

Treatment of Uncertainty — LOW — Neither uncertainty nor variability are tracked within the structure of ATLSS. However, individual components, such as SIMPDEL and SIMSPAR, incorporate demographic stochasticity, typically as a Monte Carlo simulation.

Degree of Development and Consistency — HIGH — A software package for ATLSS is available from the developers.

Ease of Estimating Parameters — LOW — ATLSS is a parameter-intensive model. This quality stems from its broad scope, both spatially and ecologically. Applying the ATLSS approach to a new system would require substantial effort.

Regulatory Acceptance — MEDIUM — ATLSS has no regulatory status. However, it has been applied in permitting negotiations involving the Florida Everglades.

Credibility — HIGH — ATLSS has been in development for a number of years and is under constant review and revision.

Resource Efficiency — LOW — Because of the comprehensive nature of ATLSS, it would require a great deal of effort to fulfill its data requirements. Even though the model is available as software, the efficiency of application is considered low.

DISTURBANCE TO WETLAND VASCULAR PLANTS MODEL

Ellison and Bedford (1995) present a spatially explicit model that addresses the impacts of hydrologic disturbances on community structure of wetland vascular plants. The model is a grid-based representation of functionally aggregated species of vascular plants. The functional groups were created by considering plant morphology, life history, and seed dispersal and germination properties for 169 species of plants. The model can incorporate up to 10 functional groups of plants, each of which is defined by a combination of 10 life-history characteristics. The model was used to simulate the changes in vegetation structure of a sedge meadow and a shallow marsh located next to a 1000 mw coal-fired power plant in south-central Wisconsin.

The spatial grid that described the wetland was subdivided into individual elements (e.g., 100 × 100 cells). The structure of the aggregated plant community changed as a function of the vegetation-specific parameters defined for each grid element and the status of the adjacent grid elements (e.g., occupancy status, seed bank, type of species present). The model addresses spatial variation in seed dispersal, plant growth patterns, mortality, and water levels. Competitive interactions between species occupying neighboring cells can be simulated. The simulation year is divided into four seasons, and each species is assigned a specific season in which growth occurs. Fast- and slow-growing species differ in their rates of spread to adjacent grid cells. A gamma distribution is used to describe the distance of seed dispersal. The probability that a plant grows decreases linearly with water depth. The probability of plant death also increases with water depth (as an exponential function) to simulate adverse effects of flooding.

Realism — MEDIUM — The disturbance to wetland vascular plants model describes the plant community as a series of functional groups. This approach limits the realism of the model because species-specific characteristics that may influence the effects of a particular disturbance are not considered.

Relevance — HIGH — Wetland community structure is relevant for many ecological risk assessments. Assessing the potential impacts on plant communities (and other ecological endpoints) from the generation of electric power remains an important area of interest. Although the model does not explicitly account for toxic chemical effects, the user could adjust several parameters to implicitly model toxicity.

Flexibility — MEDIUM — Ellison and Bedford (1995) suggest that the model might be useful for predicting the consequences of anthropogenic disturbances on other freshwater wetlands.

Treatment of Uncertainty — LOW — The authors did a limited sensitivity analysis, but implementing the grid model in an overall uncertainty framework would require substantial effort.

Degree of Development and Consistency — MEDIUM — The model formulations describe vegetation changes in relation to within-cell and between-cell interactions. For future use, the model software would probably require some reprogramming to implement the code on modern computer platforms (Ellison 2000, pers. comm.). Some model validation has been performed.

Ease of Estimating Parameters — MEDIUM — The functional aggregation of the plant species provides for a reasonable number of parameters. However, the number of spatial cells increases the demand for parameter estimation.

Regulatory Acceptance — LOW — The model was not developed for any explicit regulatory application and does not appear accepted or recommended by any regulatory agency.

Credibility — LOW — The model results were only generally in rank-order agreement with 7 years of observed vegetation changes. The model has not been extensively published and no longer appears to be used.

Resource Efficiency — MEDIUM — The model might not require extensive reprogramming for application to particular case studies. However, the spatial detail of the model suggests that the costs of parameter estimation would be considerable.

LANDIS

LANDIS is a spatially explicit model designed to simulate forest landscape change over large area and time domains (Mladenoff et al. 1996; Mladenoff and He 1999). The major modules of LANDIS are forest succession, seed dispersal, wind and fire disturbances, and harvesting.

LANDIS was developed by using an object-oriented modeling approach operating on raster GIS maps. Each cell can be viewed as a spatial object containing unique species, environmental factors, and disturbance and harvesting information. LANDIS simulates tree species as the presence or absence of 10-year age cohorts in each cell, not as individual trees. This approach enables LANDIS to simulate forest succession at varied cell sizes (e.g., 10×10 m or 1000×1000 m). Unlike most other landscape models, LANDIS simulates disturbance and succession dynamics.

During a single iteration, species birth, death, and growth routines are performed on age cohorts, and a random background mortality is simulated. Wind and fire disturbances occur stochastically in terms of the sizes of disturbances, the time intervals between them, and their locations. Environmental factors summarized as various land types set the initial fire disturbance status and fire return intervals. Fuel accumulation derived from the succession module with estimated wind-throw regulates fire severity class. Wind disturbance is less related to environmental factors than is fire. Stand age determines the species' wind susceptibility: the older the individuals are in the stand, the more susceptible it is. The harvesting module uses a strategy similar to that used for disturbances because older trees are more desirable for harvesting.

Realism — HIGH — LANDIS is a realistic model with state variables related to landscape processes such as fire, wind, insect disturbance, succession, and seed dispersal, as well as forest management. For each tree species, the model incorporates life-history characteristics such as longevity, shade tolerance, fire tolerance, seeding distance, and sprouting probability.

Relevance — HIGH — LANDIS is a spatially explicit simulation model that predicts forest landscape change during long time periods (hundreds of years) and for large areas (thousands of hectares). The model can simulate a variety of ecologically relevant endpoints, such as tree species presence and absence, age structure, and species richness. The potential impact of toxic chemicals could be incorporated into the many species-specific life-history traits such as mortality, seed dispersal, and so forth.

Flexibility — HIGH — LANDIS can be applied to different forest landscapes by specifying the life-history characteristics for tree species and the initial conditions.

Treatment of Uncertainty — HIGH — Disturbance events such as fire and wind-throw are simulated stochastically on the basis of mean return intervals and disturbance size.

Degree of Development and Consistency — HIGH — LANDIS is available as a software package that includes a manual.

Ease of Estimating Parameters — MEDIUM — LANDIS would require moderate effort for parameterization to apply to a new forest landscape.

Regulatory Acceptance — LOW — To our knowledge, LANDIS has not been used in a regulatory context.

Credibility — HIGH — LANDIS is a well-known model, and a large number of publications in books and peer-reviewed journals describe the model and its applications.

Resource Efficiency — MEDIUM — Although LANDIS has many parameters, applying it to a new landscape would be relatively straightforward using available species information and GIS data without new programming.

FORMOSAIC

FORMOSAIC simulates natural forest dynamics and the influence of forest management practices (Figure 11.5) (Liu and Ashton 1998). The model has been used to evaluate the long-term impacts of various logging strategies in tropical rainforests of Malaysia. So far, FORMOSAIC has been applied to species groups, not individual tree species. For each species group, the model predicts biomass and number of trees in five distinct canopy layers.

FORMOSAIC is one of the few forest models specifically designed for a cellular automata format with a hierarchical structure consisting of the landscape, the focal forest, the grid cell, and the specific tree. The landscape consists of the focal forest plot and surrounding areas that may or may not be forested. The nature of the surrounding areas directly affects parameters such as tree growth and recruitment. The focal forest is represented as a collection of 100-m^2 cells, each of which contains many individual trees of different species with their own state variables.

FORMOSAIC consists of three modules that simulate tree growth, recruitment, and mortality. Relative growth depends upon size (determined as diameter at breast height), neighborhood shading influences, slope, elevation, and position relative to the closest wet area. Each tree is assumed to have a species-specific maximum size.

The second module simulates recruitment in four guilds: emergent, canopy, understory, and successional species. All mature trees are assumed to have the same probability of reproductive success (i.e., no density dependence in seed production within a grid cell). Recruits in each grid cell come from seeds immigrating from outside of the focal forest, seeds immigrating from other cells within

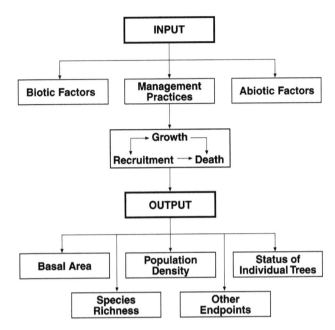

Figure 11.5 Conceptual model of FORMOSAIC. (From *Ecol. Modeling* 106 (2–3), Liu and Ashton, FORMO-SAIC: an individually-based spatially explicit model for simulating forest dynamics in landscape mosaics. pp. 177–200. © 1998, with permission from Elsevier Science.)

the focal forest, or seeds produced within the same grid cell. Sources of recruits (and hence species) are determined by simultaneously modeling all adult trees within the focal forest and then applying empirical seed-distribution functions to determine the probable final location of each progenitor.

The final module of FORMOSAIC tracks mortality. Empirical exponential mortality functions are applied to two size classes of trees: those with a breast-height diameter less than 30 cm, and those with a breast-height diameter greater than or equal to 30 cm. One interesting aspect of the mortality module is that it not only accounts for inherent mortality but also possesses a function to account for damage to surrounding trees as the result of tree fall. The potential damage is quantified as the product of the total number of trees in the affected area and the empirical probability that any inclusive tree would be killed as the result of tree fall.

Realism — HIGH — All growth and mortality coefficients in FORMOSAIC are derived empirically. This approach provides a high degree of realism when applied in an appropriate environment, as was demonstrated in a validation of the model.

Relevance — HIGH — FORMOSAIC can simulate a variety of ecologically relevant endpoints, such as tree biomass, stand biomass, and age structure. FORMOSAIC has no function for modeling effects of toxic chemicals. However, the physical perturbation coefficient could be modified to implicitly model toxicity.

Flexibility — MEDIUM — FORMOSAIC has some flexibility because all of its major functions are determined on the basis of empirically derived growth curves. However, the current model structure restricts application of FORMOSAIC to evergreen rainforests at sites with constant climatic conditions and no droughts throughout the year.

Treatment of Uncertainty — LOW — FORMOSAIC has no inherent mechanism for the conservation of either uncertainty or variability.

Degree of Development and Consistency — MEDIUM — FORMOSAIC was initially coded in C++. Availability of the software was undetermined but assumed to be at the discretion of the authors.

Ease of Estimating Parameters — MEDIUM — Because all the major functions in the model are determined on the basis of empirical observations, this model would be difficult to apply in a forest environment that deviates substantially from the one for which it was intended.

Regulatory Acceptance — LOW — FORMOSAIC has no regulatory status and does not appear to have been used within a regulatory context.

Credibility — HIGH — FORMOSAIC uses standard modeling techniques developed over many model generations and has been cited and used in other independent forest study programs.

Resource Efficiency — MEDIUM — Assuming the availability of software, FORMOSAIC would be extremely easy to implement because almost all of the governing parameters could be conserved. Therefore, relatively few parameter values need to be determined from site specific data.

FORMIX

FORMIX is a forest model intended to represent the growth of a natural tropical forest before and after logging (Bossel and Krieger 1991). Because the dynamics of the forest are determined by canopy cover, the basic geometric unit is a gap, with a functional area on the order of 0.01 to 0.1 ha. Landscape-level processes are modeled in the spatial patterns of forest dynamics resulting from interactions among a large number of neighboring gaps.

In FORMIX, a tropical forest is structured as five tree-canopy layers recognized as functionally distinct developmental stages. These include seedlings, saplings, poles, main canopy, and emergent trees. Although conditions differ for each of these classes, the processes at each stage are identical. These processes include photosynthesis, respiration, shading of lower classes, transfer of trees from lower classes, and others. Seedling recruitment is a function of seed production from mature trees in the main canopy and emergent layer. This variable is modified by parameterized germination rates or planting rates. The model also has a seed dispersal function similar to that described for FORMOSAIC. When the seedlings have attained a threshold height, they enter the sapling stage. Total tree density is calculated by integrating the transition rates (from one class to another) representing tree growth. Maximum potential density is a function of canopy size relative to light availability. Exceedance of this internal constraint results in proportional mortality within the class.

Gross biomass production in FORMIX is modeled as a function of photosynthesis and is determined on the basis of light distribution within the crown (Michaelis–Menton equation). Net biomass accumulation is simulated as carbon fixation through photosynthesis minus energy loss through processes such as respiration, litter loss, and seed production increment.

Stage-specific mortality rates are determined as the product of the total number of trees and a specified mortality rate. The baseline mortality rate is expressed as a loss of biomass and is determined independently of density-dependent mortality functions. Physical perturbation, specifically logging, is accounted for in the main biological state variables as a specified proportion of biomass loss from each of the developmental stages.

One of the unique aspects of FORMIX that separates it from other forest-gap models is the consideration of tree geometric relationships. This aspect is simulated through the use of a geometric packing model based on crown-to-diameter ratios for individual trees relative to the overall tree densities in the plots. The total available leaf area for photosynthesis is derived from these ratios.

Realism — HIGH — FORMIX provides many functions specific to forest structure that increase the realism of the model, particularly with regard to tropical forests.

Relevance — MEDIUM — FORMIX can simulate a variety of ecologically relevant endpoints, such as tree biomass, stand biomass and age structure, and species richness. FORMIX does not possess functions or state variables that could be applied to describe effects of toxic chemicals. However, the model does include a physical perturbation function (described in the harvesting module) that could potentially be modified to this end.

Flexibility — LOW — FORMIX is highly specific to tropical forests. Its basic design was intended to mimic the multilayer canopy structure characteristic of this type of ecosystem.

Treatment of Uncertainty — LOW — Uncertainty and variability are not tracked in the applications of FORMIX. Some aspects of landscape applications are limited in this case because the selection of cellular plots is deterministic.

Degree of Development and Consistency — MEDIUM — Apparently, the model has been applied in a management context. However, no mention was made about the availability of FORMIX as a software package.

Ease of Estimating Parameters — MEDIUM — The application of FORMIX in the environment for which it was intended would require limited parameterization because many of the default values used in the development of the model could be retained. Application to other forest types would require moderate effort for reparameterization.

Regulatory Acceptance — LOW — FORMIX apparently has no regulatory status and does not appear to have been used within any regulatory context.

Credibility — HIGH — FORMIX uses standard simulation techniques developed over many model generations. FORMIX has been cited and used in other independent forest study programs.

Resource Efficiency — MEDIUM — Assuming the availability of the model as a software package, FORMIX would be reasonably easy to apply because most of the parameter values can be conserved. However, no indication exists that such a software package is available.

ZELIG

ZELIG, like most forest-gap models, simulates the annual growth, mortality, and reproduction of individual trees on a series of small plots corresponding to the zone of influence of a single canopy tree (approximately 0.04 to 0.1 ha) (Burton and Urban 1990). The spatially explicit version of ZELIG reviewed here simulates dynamics across a landscape of model plots. The basic approach used to simulate demographic processes within each modeled plot is to begin with maximum potential behavior (i.e., maximum growth rates, in-seeding rates, etc.) and subsequently constrain this potential by resource limitation. Constraints include availability of light, soil quality, temperature range, and soil moisture (Figure 11.6). ZELIG, like FORSKA, models the responses of the trees to differences in solar radiation for layers within each tree's crown. ZELIG does not mechanistically model photosynthesis. Rather, light availability is used as a moderating function, which is converted to a quantified increase in growth by comparison with a species-specific shade tolerance index. This result is then applied as a constraint on the empirically derived growth curves. Growth is modeled not as biomass accumulation but as the height from the ground to the base of the crown. Soil fertility is an empirical parameter. Soil moisture and ambient temperature are simulated by using data on the monthly mean and variances of precipitation and temperature.

The mortality rates in ZELIG are determined probabilistically. Trees are assigned a probability of dying on the basis of the age of the tree by assuming that an individual has a 2% chance of reaching its maximum age. The probability of mortality increases if an individual experiences consecutive years of suppressed growth. Because ambient weather, mortality, and in-seeding are modeled as stochastic processes, output from one model plot represents just one possible trajectory of forest dynamics. Hence, a Monte Carlo approach is used to derive a distribution and an average trajectory of stand dynamics. This generates plot-to-plot variation over the entire landscape such that trends can be described as probabilistic outcomes overlaid on the landscape.

Realism — MEDIUM — ZELIG, like FORET (Forests of Eastern Tennessee), is highly dependent on empirically based functions to describe forest dynamics. Therefore, when applied appropriately, it is accurate and realistic. However, studies that have attempted to modify ZELIG to other circumstances, such as for application to coniferous rather than deciduous forests, have found serious problems related not to parameterization but to the underlying assumptions inherent in its method of simulating crown structures.

Relevance — MEDIUM — ZELIG can simulate the temporal dynamics of a variety of ecologically relevant endpoints, such as tree biomass, stand biomass, age structure, and species richness. ZELIG does not model the effects of either chemical or physical stress. Substantial effort could be required to include functions for physical and chemical effects.

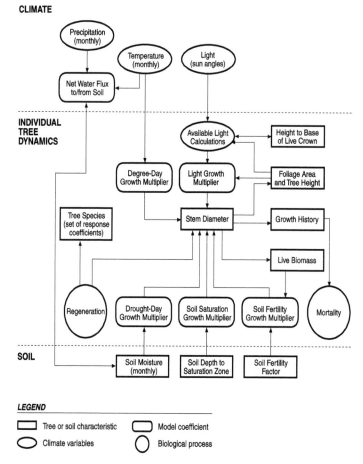

Figure 11.6 Structure of the ZELIG tree simulator model. (From the BOREAS Information System (http://www-eosdis.ornl.gov/BOREAS/bhs/Models/Zelig.html). See also Knox et al. (1997); Weishampel et al. (1999). Redrawn with permission by Robert G. Knox.)

Flexibility — MEDIUM — ZELIG has been shown to be applicable to most deciduous forest types. However, its canopy model has been demonstrated to be inappropriate for coniferous and mixed stands.

Treatment of Uncertainty — HIGH — Variance was tracked in the structure of the model. The potential uncertainty in spatial structure was considered, but this was limited to the mortality module.

Degree of Development and Consistency — HIGH — ZELIG has been validated and is available in numerous software packages in several different versions, including the rastered landscape version.

Ease of Estimating Parameters — HIGH — When ZELIG is applied in the appropriate setting, the parameterization requirements for ZELIG are limited to general environmental conditions. The growth functions, which have been developed for almost every major deciduous tree species, can be conserved.

Regulatory Acceptance — MEDIUM — ZELIG has been considered for use in the fulfillment of regulatory requirements in both Europe and the U.S. with regard to planned forest management. However, it is not known to have any regulatory status.

Credibility — HIGH — ZELIG was developed on the basis of the FORMAN model lineage and has one of the longest periods of development of any of the current forest-gap models.

Resource Efficiency — HIGH — Of all the forest landscape models, ZELIG is probably the easiest to execute. Parameter estimation has been simplified by the large database of species and site specific growth and mortality functions available from numerous sources.

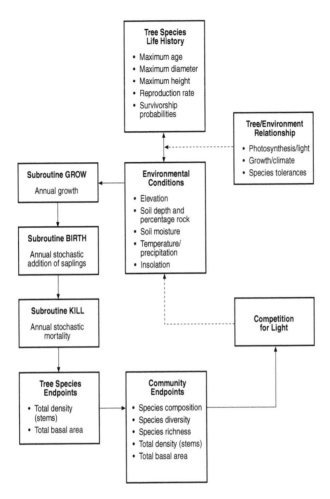

Figure 11.7 Structure of JABOWA.

JABOWA

JABOWA is a generalized model of the reproduction, growth, and mortality of trees in mixed-species forests in response to environmental conditions (Botkin et al. 1972; West et al. 1981; Botkin 1993a,b) (Figure 11.7). It was among the first successful multispecies computer simulations of terrestrial ecosystems and has gone through extensive modification and use during the 30 years since it originated. It is copyrighted in several versions as JABOWA, JABOWA II, and JABOWA 3 (Botkin 2000, pers. comm.).

The model structure includes a landscape consisting of 10 × 10 m grid sections (the default value, which is user-adjustable in the early versions of the model). Model processes affecting growth and mortality take place independently within each grid cell; no interactions take place between adjacent cells. The user determines the kind and number of tree species. Current versions allow for up to 45 species. The species characteristics can be changed either in real time by a user or through initial input files readily accessible to the user. In its original version, the model made use of data from the Hubbard Brook Ecosystem Study in northeastern North America, in which the forest contained ten deciduous species and three coniferous species.

Each tree is assigned a series of state variables that determine the shape of the tree, growth, and mortality. For each species that can grow in the defined environment, the same algorithms that determine growth determine reproduction. These algorithms include the effects of light, soil mois-ture, soil depth, soil water-holding capacity, soil nitrogen, percentage of rocks in the soil, latitude, and snow melt rate. Specific grids are assigned state variables determining the number of saplings within a stand on the basis of shade, elevation, soil type, soil capacity, percentage rock in soil, and monthly temperature and precipitation. Direct competition among individuals is restricted to com-petition for light and is a direct function of relative tree height as modified by a species-specific shade tolerance coefficient. The basic growth relation is a species-specific sigmoidal growth curve (determined empirically for a tree growing under optimum conditions), with dependence on canopy-level solar radiation and the following indexed coefficients: the temperature index, which simulates the effects of monthly mean temperature on photosynthesis rates, and a soil quality index that simulates the effect of gross soil structure on tree growth. In more recent versions, soil nitrogen, depth of the water table, depth of the soil, and other soil characteristics also influence growth. The growth function is then negatively modified by a coefficient accounting for competition between trees within the same grid plot as a function of tree density and for the effects of the existing environment on each species. Species-specific recruitment depends on the species density and a coefficient defining seed production for each species.

In JABOWA, mortality affects competition for light indirectly by altering the number of tree stems on a plot. Annual mortality of trees is determined for each age class and is modeled stochastically with two algorithms. The first one is for healthy trees, in which it is assumed that 2% of the individuals of a species will, on average, reach the maximum known age for that species. The second one is for trees that are growing poorly, with a user-determined minimum growth. For these trees, a second stochastic function assumes that such a tree will, on average, survive only 10 years unless growth rises above the user-set minimum.

Realism — HIGH — JABOWA is a combination of mechanistic functions as well as site specific empirical relationships that accurately describe ecological processes in forests.

Relevance — HIGH — The model can simulate a variety of ecologically relevant endpoints, including tree biomass, forest productivity, and species richness. The model has no state variables or functions to describe physical disturbance or effects of toxic chemicals. However, the model is designed for easy addition of such algorithms. For example, it has been used to study the effects of acid rain on forests of Long Island, New York, and to examine the effects of global warming in many locations around the world (Botkin 2000, pers. comm.).

Flexibility — HIGH — JABOWA was developed as a general model of forest dynamics. The parameters of each species are based on empirical information for its entire range. The only site specific factors are the environmental factors. The model has been applied to many types of forests over a wide range of environmental conditions in North America, Siberia, Eastern Europe, and Costa Rica.

Treatment of Uncertainty — HIGH — The model includes stochastic functions for mortality and reproduction. As typically used, multiple runs starting with different random numbers of seeds are conducted as a set. The model includes a statistical analysis module that reports the mean, variance, and 95% confidence interval for each set of runs for each year the user requests output. Botkin (1993a) reported on the extensive sensitivity analyses conducted on JABOWA.

Degree of Development and Consistency — HIGH — Validation results indicate a relatively high accuracy in model output for long time periods. Botkin (1993b) provides a user's manual and software to run JABOWA.

Ease of Estimating Parameters — MEDIUM — The data requirements for JABOWA are similar to those for the more recent forest-gap models. When applying the model to a forest in the northeastern U.S., empirical functions, especially those for growth, could be conserved. The mechanistic functions in this model could be parameterized easily from site specific data.

Regulatory Acceptance — LOW — The model has no regulatory status and does not appear to have been used within a regulatory context.

Credibility — HIGH — JABOWA has a long history of development and use. Its structure served as the basis for many generations of forest-gap models. Botkin (1993a) provides a history of the development of the model.

Resource Efficiency — HIGH — JABOWA would be reasonably easy to execute within the context of the types of forest habitat for which it was developed.

REGIONAL FOREST LANDSCAPE MODEL

The regional forest landscape model is a tool for evaluating the impact of ozone exposure upon forest stands and associated water bodies (Graham et al. 1991). The model consists of two separate modules. The first module determines the impact of ozone on terrestrial landscape structure. The second module determines the impact of these changes on the acidity of receiving water bodies.

Structurally, the terrestrial model is a stochastic, spatial simulation of land cover rastered into 78,605 4-ha grid cells. Each grid cell is classified into one of 18 different types of land cover, including deciduous forest, coniferous forest, or mixed forest. A uniform concentration of ozone is applied to this landscape. Coniferous tree species are assumed to be the only targets sensitive to ozone impacts. Sufficient exposure weakens conifers until they become susceptible to bark beetle infestation, which results in tree mortality. An empirical probability density function is used to determine whether the exposure would result in the mortality of all conifers within any given grid cell. If mortality occurs within a cell, then its forest type changes. That is, a coniferous forest is converted to open land, and a mixed coniferous–deciduous forest is converted to a deciduous forest.

The water quality module of the regional forest landscape model is a very simple empirical relation between percentage land covers and expected volumes and alkalinity of resulting runoff. No mechanistic functions were included in this module.

Realism — LOW — The regional forest landscape model is a highly generalized treatment of a specific impact resulting from ozone exposure. Its realism derives solely from the accuracy of exposure–response functions and the assumption of homogeneity within any given 4-ha grid.

Relevance — HIGH — The model can simulate a variety of ecologically relevant endpoints that are indicators of forest cover, forest edge, forest interior, landscape pattern, and lake water quality. It is one of the few forest landscape models that accounts for effects of a toxic chemical.

Flexibility — HIGH — The regional forest landscape model represents a highly flexible construct. Its underlying structure is very simple and could be applied in any situation because the descriptive landscape categories are very general.

Treatment of Uncertainty — HIGH — In the model, both uncertainty and variability are conserved. Variability in response of trees to ozone is represented by probability density functions applied to model parameters. Uncertainty analysis is based on multiple scenario simulations.

Degree of Development and Consistency — MEDIUM — To our knowledge, the regional forest landscape model has not been validated. The authors have developed software for the model, but its availability is unknown.

Ease of Estimating Parameters — MEDIUM — The regional forest landscape model possesses two levels of parameter estimation. First is the landscape characterization that is obtained directly from site specific observations. Second is the development of the environmental cause-and-effect relationships. These functions include a probability density function relating conifer mortality to ozone exposure as well as an empirical function relating water quality to relative land cover. The first, although highly data intensive, could be fulfilled by using general dose–response data. The second is highly site specific and probably would require field studies.

Regulatory Acceptance — MEDIUM — The model has no regulatory status. However, it was applied in the regulatory context to fulfill risk assessment needs associated with the Oak Ridge National Laboratory's remedial investigation. Moreover, its simple structure provides for transparency that is usually necessary to attain regulatory acceptance.

Credibility — LOW — No history of use or plans for developing the regional forest landscape model were identified.

Resource Efficiency — HIGH — Application of this model would be reasonably simple within the context intended by the developers. The model could be implemented by using third-order Monte Carlo simulation.

SPATIAL DYNAMICS OF SPECIES RICHNESS MODEL

The spatial dynamics of species richness model is designed to evaluate the effects of habitat fragmentation on species richness (Wu and Vankat 1991). This evaluation focuses on the temporal dynamics of the biological community within a defined "forest island" surrounded by agricultural or urban lands or both.

The spatial dynamics of species richness model simulates changes in species richness over time by deconstructing the landscape of the forest island and its surrounding habitats. The forest island is divided into two habitat types: interior habitat and edge habitat. Edge habitat is described as being at a constant relative distance from the interface between the forest island and the surrounding habitat. Therefore, the amount of edge habitat depends on the perimeter of the forest island, whereas the interior habitat is a direct function of its volume. Resident species are defined as edge species, interior species, or generalists. Instead of tracking individual species, the model uses species richness (either edge or interior) as the principal state variable. Differential equations describing changes in species richness as a function of time are empirically based and parameterized to be representative of the eastern U.S. A time-delay function is included in the model to account for differences between the time necessary for species richness to attain steady-state and the relative rate of habitat change in the modeled environment.

Realism — LOW — The spatial dynamics of species richness model is a series of differential equations linking time-dependent changes in forest island structure with changes in species richness. Quantification of these rate equations in this nonmechanistic model is entirely empirical. Therefore, the realism of this model depends solely on the accuracy of the site specific data (or assumptions).

Relevance — MEDIUM — Although species richness is a very important endpoint in aquatic ecological risk assessment, it has not received much recognition in ecological risk assessments of chemicals in terrestrial environments. If this circumstance were to change, models such as the spatial dynamics of species richness model may become important for determining baseline conditions for species diversity evaluations. Because it is a nonmechanistic model, modifying its structure to account for effects of toxic chemicals is not practical.

Flexibility — HIGH — The model has no mechanistic functions, nor does it assume any ecological structure other than relationships between species richness, forest island area, and time. It may therefore be applied under any circumstances in any similar situation involving spatial limitations of habitat area.

Treatment of Uncertainty — LOW — The spatial dynamics of species richness model does not track uncertainty or variability.

Degree of Development and Consistency — MEDIUM — The model has not been validated. The spatial dynamics of species richness model was specifically designed to run on the STELLA platform. The coding necessary to run the model is provided in Wu and Vankat (1991).

Ease of Estimating Parameters — LOW — Changes in species number relative to changes in areas and types of habitat (both edge and interior) with time are difficult to estimate.

Regulatory Acceptance — LOW — The model has no regulatory status and does not appear to have been used in a regulatory context.

Credibility — LOW — The spatial dynamics of species richness model has no identifiable history of development.

Resource Efficiency — HIGH — Assuming the conservation of parameters associated with species change and habitat area change, the model would be extremely simple to execute.

STEPPE

STEPPE is a spatially explicit gap-dynamics model that predicts grassland productivity on the basis of competition for available resources within a semiarid environment (Coffin and Lauenroth 1989). It has been used to describe the recovery of blue grama (*Bouteloua gracilis* [H.B.K.] lag.

Ex Griffiths), which is a dominant monocot species in north-central Colorado. Although STEPPE is presently structured as a single-species model, it could be extended to multiple species.

This gap-dynamics model is similar to models developed for forests and simulates the growth and death of individual plants on a small plot (0.12 m^2) through annual time-step iteration. Recruitment in a plot is modeled as the probability that an individual of a given species will establish itself through either seed germination or vegetative propagules. The probability of mortality for each individual is determined on the basis of disturbance rate, longevity of the species, and a risk of increased mortality associated with slow growth.

Plant growth depends on the importance of belowground processes associated with the acquisition of soil water because water availability typically controls plant growth and community structure in semiarid grasslands. Belowground primary productivity contributes approximately 85% of total net primary production.

STEPPE accounts for the spatial structure of habitats by dividing the landscape into a series of plots in which processes on one plot may affect processes on others. Processes important to the recruitment of individuals of the target species are essential elements of the model. Three general processes are included, and at least one probability is associated with each process. The first is the probability that environmental conditions favorable for germination will occur within the year. This probability is simply an empirical probability constant (0.125). The second is the probability of blue grama seed being produced, which is determined on the basis of the biomass accumulation module and is principally a function of the amount of precipitation in the previous year. If annual biomass production does not reach the threshold of 49 g/m^2, the probability of seed production is assumed to be zero. If the biomass accumulation exceeds 49 g/m^2, the probability of seed production is assumed to be one. The third essential variable in the model is the probability of seeds dispersing to any given plot. This probability was determined as a function of the distance to the seed source, the height of the influorescence, and the aerodynamics of the seed.

> **Realism** — MEDIUM — STEPPE is basically a cellular automata model of the dynamics of monocot productivity in an area after physical disturbance. The principal ecological processes modeled are species-specific growth, interspecies competition, and recruitment. Disturbances are modeled as a user-defined parameter in the model, not as a mechanistic function.
>
> **Relevance** — HIGH — STEPPE mainly simulates grassland productivity, which is an ecologically relevant endpoint. STEPPE does not mechanistically model potential impacts of physical or chemical stressors. However, the seed production and recruitment functions could be modified to account for toxic effects.
>
> **Flexibility** — LOW — STEPPE is specific to a climate where availability of water (as precipitation) is the rate-limiting factor on plant growth. Furthermore, it is currently parameterized to model a single specific grass species.
>
> **Treatment of Uncertainty** — LOW — The model does not track variability or uncertainty.
>
> **Degree of Development and Consistency** — MEDIUM — STEPPE is not readily available as software. However, Coffin and Lauenroth (1989) provide sufficient details for programming and application of the model.
>
> **Ease of Estimating Parameters** — MEDIUM — Although STEPPE also requires empirically based growth functions, only three major parameters need to be estimated.
>
> **Regulatory Acceptance** — LOW — STEPPE has no regulatory status and does not appear to have been used in a regulatory context.
>
> **Credibility** — MEDIUM — STEPPE uses a commonly referenced approach and algorithms to evaluate plant growth. It has a limited history of development.
>
> **Resource Efficiency** — MEDIUM — Although STEPPE requires little effort for parameterization, the model is highly dependent on empirical site specific observations.

WILDLIFE-URBAN INTERFACE MODEL

The wildlife-urban interface model was designed to predict the effects of development on vegetation cover and wildlife habitat utilization (Boren et al. 1997). The objective of the model is to determine the probability of occurrence of selected avian species as a function of changes in landscape cover types, especially increases in agricultural and urban development. The model is structured in three components. The first estimates the probability of species occurrence in each category of land use on the basis of extensive bird and vegetation surveys performed in various types of edge habitat (i.e., on the margins between different types of land use). The second describes current landscape cover types and future changes, which depend on past trends. The third predicts the occurrence of species in relation to modeled land-use changes on the basis of relationships between avian distributions and land-use types defined in the first module.

The avian community structure module depends on the detrended correspondence analysis performed on bird survey data collected between 1966 and 1990 in Washington County, Oklahoma. Species scores are assigned according to the relation between a species' residence status and the relative proportions of habitat types present. The scores generated through this analysis are used to determine the avian species responsible for temporal shifts in community structure. In essence, this multivariate approach identifies the species that are declining or increasing within each landscape type. As part of the third module (see below), changes in these species are then used to predict shifts in avian community structure with predicted changes in landscape cover. Canonical analysis is used to determine the specific aspects of landscape structure that influence the breeding bird communities.

The landscape module uses regression analysis to estimate the probability of occurrence of landscape cover type within each location. Demographic–economic regression models provide the basis for predicting the area of all major landscape cover types. Variables include rural population density, 7-year payment of oil price, Riverside price, 5-year payment of cattle price margin, average farm size, and number of farms per county. The projected area of landscape cover types is determined by multiplying the area of the individual grid sections (50.2 ha) by the probability of occurrence of each type. The model assumes that temporal changes in landscape cover types between 1966 and 1990 would continue at the same rate until 2014.

In the third module of the wildlife-urban interface model, avian community structure is predicted on the basis of changes in landscape structure. The occurrence of each avian species responsible for shifts in community structure is related to the areas of landscape cover types by using a logistic regression model. Presence or absence of a bird species is used as the dependent variable. Based on regression methods, the independent variables — The areas of landscape cover types — are tested for linear, cubic, and quadratic effects.

Realism — MEDIUM — The wildlife-urban interface model relies on extrapolation from relationships between the past development of landscapes and associated impacts on avian communities. Results of a validation exercise indicate a reasonable level of predictive accuracy during the last 20 years.

Relevance — MEDIUM — The model has relevant ecological endpoints such as probability of occurrence of species, which can be used with landscape cover data to predict the distribution of species. The model was specifically designed to evaluate human physical disturbance of habitat and its effects on both the vegetative landscape and specific avian species. None of the grassland models, including this one, considers the effects of toxic chemicals. Because the probability of species occurrence is based on empirical relationships to land cover types and population dynamics are not addressed, accounting for the effects of toxic chemicals could be difficult.

Flexibility — MEDIUM — The wildlife-urban interface model could be applied to urban/ agricultural areas within most temperate regions.

Treatment of Uncertainty — MEDIUM — The model was developed as a probabilistic model. It examines probabilities of species occurrences on the basis of habitat types to predict community structure. Probability parameters used in the model are represented as single values and not as probability density functions.

Degree of Development and Consistency — MEDIUM — The model was partially validated in Washington County, Oklahoma. The model is not commercially available as software. However, Boren et al. (1997) provide sufficient details for programming and application of the model.

Ease of Estimating Parameters — MEDIUM — The model requires detailed estimation of habitat requirements for a large number of avian and plant species. However, when the model is applied in the context for which it was constructed, many of the parameter values may be conserved.

Regulatory Acceptance — LOW — The model has no regulatory status and does not appear to have been used in a regulatory context.

Credibility — LOW — The wildlife-urban interface model utilizes an approach that is not commonly applied in ecology. Although it uses a modeling method common in the social sciences, no citations or references specific to this model could be found other than Boren et al. (1997).

Resource Efficiency — HIGH — Many of the relationships within the model are empirical. Furthermore, the classes used for analysis of landscape cover are standard.

SLOSS

SLOSS applies biogeographic principles to evaluate steady-state distributions of species on the basis of a set of community indices derived from species–area relationships (Boecklen 1997). The author investigates the relationship between nestedness and the SLOSS indices. Nestedness is a measure of community structure used to describe the organization of subsets of species (see also Chapter 10, Ecosystem Models — Terrestrial, Nestedness Analysis Model). The SLOSS indices and the measures of nestedness depend on a database of 148 species distributions representing five major taxonomic categories: plants, invertebrates, reptiles, birds, and mammals. Species distributions are described for a series of habitat patches (or islands).

Species–area relationships are modeled from the original distribution matrix by plotting number of species against the log-transformed area. Habitat patches are then paired in all probable combinations, excluding pairs whose combined areas are larger than the largest single patch in the database. On the basis of these combinations, a series of SLOSS indices is developed. SLOSS indices are used to evaluate the degree to which a single large area of habitat contains all the species that occur within a pair of smaller habitat patches of equal total size. Each SLOSS index gives the percentage of the species pool that is either gained or lost when two small patches are compared with a single large patch of equal area.

Nestedness is calculated from the species distribution data by assuming that species occurrences are equitable for all habitat sizes but that the probability of species occurrence varies among habitat types on the basis of the observed frequencies. The major advantage of this approach is that it is both easy to calculate and independent of the size of the species distribution matrix, thereby permitting direct comparisons among different landscapes or island archipelagoes.

The nestedness indices are then compared with the SLOSS indices for various taxonomic groups and habitat patch types. As expected, several small patches of a similar type typically contain more species than a single habitat patch of equal total area. The relative advantage of pairs or trios of small habitat patches over the single large patch of equal total area varies with taxonomic group. The results indicate that nestedness indices yield poor estimates of actual species distributions.

Realism — LOW — SLOSS uses indices of species distribution derived from empirical species–area relationships to evaluate nestedness in a community. No mechanistic functions are incorporated into this approach.

Relevance — LOW — The principal function of SLOSS is to evaluate the predictive nature of nestedness as a measure of wildlife diversity. This measure would only be of peripheral interest in ecological risk assessment with regard to sensitivity of wildlife populations to potential effects. SLOSS does not explicitly address effects of toxic chemicals.

Flexibility — HIGH — SLOSS is a landscape assessment approach that is specifically designed to be independent of regional considerations.

Treatment of Uncertainty — MEDIUM — SLOSS is a probabilistic model. However, the probability is limited to the uncertainty associated with the relative predictions in the model and does not retain variability specific to the landscapes under investigation.

Degree of Development and Consistency — LOW — SLOSS is not readily available as software.

Ease of Estimating Parameters — LOW — SLOSS requires parameterization with regard to species distributions on both regional and site specific bases. Even if the results for which the nestedness index was determined were available, application of the model would require a great deal of site specific data of a type not usually available or readily obtainable during an ecological risk assessment.

Regulatory Acceptance — LOW — SLOSS has no regulatory status and does not appear to have been used in a regulatory context.

Credibility — MEDIUM — SLOSS was developed on the basis of biogeographical algorithms that have a long history of application in experimental quantitative ecology. However, this model is too new to have any lineage.

Resource Efficiency — LOW — The model has extensive requirements for parameterization and would yield predictions of only peripheral significance in ecological risk assessment.

ISLAND DISTURBANCE BIOGEOGRAPHIC MODEL

The island disturbance biogeographic model is used to evaluate the effects of perturbations on the distribution of species within a series of linked island habitats (Villa et al. 1992). Individuals are modeled in parallel, and statistics are derived to express spatial patterns of species distribution. The model was constructed on the basis of a landscape map comprising one or more habitat islands that are themselves subdivided into grid cells. The size of the grid cells is a deterministic parameter defined such that one individual occupies each cell. The distribution of each species within a habitat island is described by the spatial arrangement of inhabited or empty grid cells.

During a simulation, three main processes take place: colonization, involving the launch of individuals from the colonization front, mortality and reproduction of indigenous individuals within the habitat islands, and the perturbation that causes the deaths of settled individuals (Figure 11.8). Species distributions resulting from these processes are determined as follows. To immigrate successfully, an individual must find an empty grid cell to inhabit. The probability of this occurring depends on the distribution of empty cells and an exponential model describing the resistance to travel for a given species. Individuals who land on empty cells reproduce and die according to species-specific life-history functions. The mean frequency of reproductive events for each species is used to determine the probability of reproduction for each individual. If reproduction takes place, the number of new individuals produced is drawn from a distribution bounded by the minimum and maximum offspring number. Every new individual produced is subject to the same fate as a new immigrant (i.e., an individual progeny only survives if it can inhabit a grid cell that is not currently occupied). The probability of mortality of a given individual at each time-step is equal to the inverse of the mean lifespan for the species.

A perturbation acts directly on the proportion of grid cells occupied and is modeled as a fixed probability of death for the individuals in an occupied cell. If a cell is selected for perturbation, each individual settled within the cell is killed, thereby opening it for new settlement. All species are assumed to be equally vulnerable to perturbation. The model provides some flexibility in that such a perturbation can be parameterized with regard to periodicity and strength.

Realism — MEDIUM — The inherent simplicity of the reproduction and mortality functions in the island disturbance biogeographic model limits the realism of the model. Moreover, the assumption that all species are equally sensitive to perturbation is unrealistic. However, the model could be easily modified to enhance its potential for use in ecological risk assessment.

Relevance — HIGH — The model predicts the response of defined subpopulations to disturbances. This capability would be directly applicable in ecological risk assessment where subpopulations of affected individuals interact with a greater unaffected population. The model does not currently include toxic chemical effects, but such effects could be modeled implicitly by varying reproductive and mortality parameters.

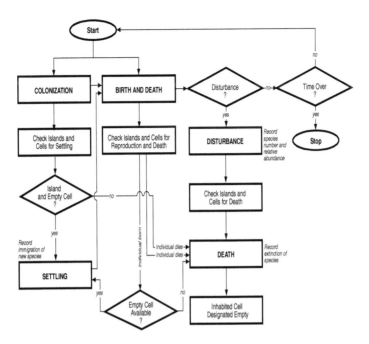

Figure 11.8 Conceptual diagram of the island distrubance biogeographic model. (Adapted from Villa et al. (1992). Understanding the role of chronic environmental disturbance in the context of island biogeographic theory. *Environ. Manage.* 16(5): 653–666. With permission by Springer-Verlag.)

Flexibility — HIGH — The model is flexible and may be applied in any situation in which mobile wildlife subpopulations interact with a greater population.

Treatment of Uncertainty — HIGH — The model is by nature a population model applied to more than one species. Although Villa et al. (1992) only reported deterministic results, uncertainty is tracked in the execution of the model.

Degree of Development and Consistency — MEDIUM — Villa et al. (1992) did not present any validation data for the model, but Whittaker (1995) independently presented data supporting the model's results. The model is available as a software package from the author.

Ease of Estimating Parameters — MEDIUM — The model requires parameterization with regard to population and subpopulation interactions as well as individual probabilities of death as the result of disturbances. Because the model deals with these interactions on a very simple level, deterministic estimates would be sufficient.

Regulatory Acceptance — LOW — The model has no regulatory status and does not appear to have been used within a regulatory context.

Credibility — HIGH — The model uses biogeographical algorithms that have a long history of development in quantitative ecology.

Resource Efficiency — HIGH — The model is easy both to construct and to parameterize (even without available software).

MULTISCALE LANDSCAPE MODEL

The multiscale landscape model is a stochastic model designed to predict landscape structure on the basis of the probability of occurrence for cover types (Johnson et al. 1999). It is principally a Markovian transition matrix incorporating probability rules that describe landscape fragmentation. It is an extension of a previous binomial model that permits the simultaneous consideration of a number of land-cover categories within each individual grid cell. The transition matrix represents

Table 11.2 An Example of a Markov Transition Matrix

		Habitat at Time t			
		A	B	C	D
Habitat at Time $t+1$	A	0.5	0.1	0.6	0.1
	B	0.2	0.1	0	0
	C	0.3	0	0.3	0
	D	0	0.8	0.1	0.9

Note: In the multiscale landscape model, each entry in the matrix represents the probability of an area changing from one habitat type to another within one time-step of the model.

the probabilities of any specific habitat (cover) type transforming into another habitat type during the next time-step. Thus, the transition matrix summarizes the probabilities of transforming from one condition to various other conditions or of staying the same (i.e., the diagonal of the matrix) as shown in Table 11.2.

As a Markov chain describes habitat transitions, the scale of change may be set to multiple spatial resolutions as determined by a user-defined parameter. Within any resolution level, a grid cell may be subdivided into a newly defined landscape (comprised of "children pixels") during a given time-step.

Parameterization of the transition matrices depends directly on site specific empirical data. The diagonal elements of the transition matrix can be considered "cell-preserving probabilities." Large values for diagonal elements represent higher patch cohesion. As the model approaches a more even distribution of probability between the rows and columns, simulated landscapes become increasingly fragmented and individual patch types are increasingly dispersed. In adapting the matrix to an actual landscape, an iterative fitting process is used.

Realism — HIGH — Assuming appropriate parameterization on the basis of empirical observations, the multiscale landscape model could yield highly realistic results.

Relevance — LOW — The requirements for predicting temporal changes in cover types and the hierarchical structure of landscapes are only peripheral to most chemical risk assessments at present. Furthermore, little information is currently available on how toxic chemicals affect landscape structure. As the field of landscape ecotoxicology develops further, this model may be useful for evaluating how land-use changes combined with chemical contamination may affect landscape pattern and fragmentation.

Flexibility — HIGH — Markov-chain models like the multiscale landscape model are inherently empirical approaches. Relationships in the model and predictions of landscape changes depend on data inputs. Because of this high dependence on initial parameterization, the model is highly flexible.

Treatment of Uncertainty — HIGH — Markov-chain models are stochastic approaches and therefore account for uncertainty. Johnson et al. (1999) also used Monte Carlo analysis to simulate various realizations of each landscape type and to derive empirical distributions of landscape indicators (e.g., contagion, interspersion, and juxtaposition).

Degree of Development and Consistency — HIGH — This multiscale landscape model is available as software.

Ease of Estimating Parameters — LOW — Because Markov models such as the multiscale landscape model are data intensive, parameter estimation is a major task.

Regulatory Acceptance — LOW — This model is not known to have any regulatory status and does not appear to have been used within a regulatory context.

Credibility — MEDIUM — The multiscale landscape model is too novel to provide a strong basis for evaluating scientific credibility. However, it incorporates modeling techniques and probabilistic analysis functions commonly in use.

Resource Efficiency — LOW — The principal structure of the model is a probability matrix. This matrix is highly data intensive and requires a very large evaluation database.

DISCUSSION AND RECOMMENDATIONS

The application of landscape modeling in ecological risk assessment accounts for large-scale spatial heterogeneity in nature and in effects of human activities. Spatially explicit approaches are especially important when addressing chemical risks because the distributions of toxic chemicals at a contaminated site or around a discharge point are heterogeneous. In aquatic systems, landscape models are most relevant for large-scale systems (large rivers, lakes, wetlands, and estuaries), where spatial homogeneity cannot be reasonably assumed. Although the same is true for terrestrial systems, substantial heterogeneity tends to occur at much smaller scales there than in aquatic ecosystems.

The four landscape models that were selected for further evaluation were the wetland model ATLSS, the landscape forest-gap models LANDIS and JABOWA, and Villa's island disturbance biogeographic model (Table 11.3). FORMOSAIC may represent a useful modeling approach as well, but because it has only been applied to Malaysian rainforests and has some serious constraints (i.e., it assumes a constant climate and an absence of droughts), it would require considerable development to apply the approach to other geographic regions and forest types. Recognizing that AQUATOX, CASM, and other aquatic ecosystem models can be applied to landscapes, we also recommend them for use in this mode. The selection of landscape models for further evaluation was made primarily according to the realism, flexibility, and relevance of the models' approaches (Table 11.3). Applying these landscape models to chemical risk assessments would require extensive site specific parameterization of the current models, addition of functional relationships to account for toxic effects, or both. Most applications of landscape models (Table 11.4) have focused on basic research issues or harvest management, not chemical risk assessment.

The availability of high-powered computers greatly increases the computational possibilities for landscape modeling and permits simulations describing changes not only in space and in time but also relative to uncertain event probabilities. It permits the expansion of both scalar and vector-nesting model architecture. Many of the experimental landscape models currently being developed are able to evaluate ecological interactions on many scales simultaneously. Unfortunately, such models tend to have very high parameter requirements; therefore, their utilization within reasonable bounds of uncertainty demands extensive databases.

The current developments in GIS technology will greatly aid in applying landscape models to ecological risk assessments. The use of GIS permits evaluation of a large amount of data and flexible parameterization on a spatial as well as a temporal basis. For example, GIS allows for variation in food-web structure or foraging parameter values with location. In a sense, the specificity in spatial structure of the model is an inherent part of the software implementation; so ecosystem models no longer have to rely upon generalized spatial functions. Another important aspect in the integration of GIS with landscape modeling is its ability to use spatial constituents as parameters themselves. Old raster-based approaches are being replaced with user-defined polygon methods that allow the definition of the spatial cellular structure as a quantifiable parameter in the model inputs. This approach has greatly increased the precision of landscape models.

Table 11.3   Evaluation of Landscape Models

Model	References	Realism	Relevance	Flexibility	Uncertainty Analysis	Degree of Development	Ease of Estimating Parameters	Regulatory Acceptance	Credibility	Resource Efficiency
Aquatic										
Marine and Estuarine										
ERSEM (benthic submodel)	Ebenhoh et al. (1995); Baretta et al. (1995)	♦♦♦	♦♦♦	♦♦	♦	♦♦	♦♦	♦	♦♦	♦
Barataria Bay Ecological	Hopkinson and Day (1977)	♦	♦	♦	♦	♦♦	♦	♦	♦♦	♦♦♦
Freshwater and Riparian										
CEL HYBRID	Nestler and Goodwin (2000)	♦♦♦	♦♦♦	♦♦♦	♦	♦	♦	♦♦	♦	♦
Delaware River Basin	Kelly and Spofford (1977)	♦	♦♦♦	♦♦♦	♦	♦♦	♦	♦♦	♦♦	♦♦♦
Patuxent watershed	Voinov et al. (1999a,b); Institute for Ecological Economics (2000)	♦♦♦	♦♦♦	♦♦	♦♦	♦♦♦	♦♦	♦	♦♦	♦
Wetland										
ATLSS	DeAngelis (1996); DeAngelis et al. (1998b)	♦♦♦	♦♦	♦♦♦	♦	♦♦♦	♦	♦♦	♦♦♦	♦
Disturbances to wetland vascular plants	Ellison and Bedford (1995)	♦	♦♦♦	♦♦	♦	♦♦	♦♦	♦	♦	♦♦
Terrestrial										
Forest										
LANDIS	Mladenoff et al. (1996); Mladenoff and He (1999)	♦♦♦	♦♦♦	♦♦♦	♦♦♦	♦♦♦	♦♦	♦	♦♦♦	♦♦
FORMOSAIC	Liu and Ashton (1998)	♦♦♦	♦♦♦	♦♦	♦	♦♦	♦♦	♦	♦♦♦	♦♦♦

Note: ♦♦♦ - high   ♦♦ - medium   ♦ - low

Table 11.3 (cont.)

Model	Reference
FORMIX	Bossel and Krieger (1991)
ZELIG	Burton and Urban (1990)
JABOWA	Botkin et al. (1972); Botkin (1993a,b)
Regional landscape	Graham et al. (1991)
Spatial dynamics of species richness	Wu and Vankat (1991)
Grassland	
STEPPE	Coffin and Lauenroth (1989); Humphries et al. (1996)
Wildlife–Urban Interface	Boren et al. (1997)
Island	
SLOSS	Boecklen (1997)
Island disturbance biogeographic	Villa et al. (1992)
Multiscale	
Multiscale landscape	Johnson et al. (1999)

Note: ♦♦♦ - high   ♦♦ - medium   ♦ - low

Table 11.4 Applications of Landscape Models

Model	Species/Ecosystem	Location/Population	Reference
Marine and Estuarine			
ERSEM	Ocean pelagic and benthic ecosystem	North Sea	Ebenhoh et al. (1995); Baretta et al. (1995)
Barataria Bay ecological	Saltmarsh ecosystems	Barataria Bay Saltmarsh, Louisiana	Hopkinson and Day (1977)
Aquatic and Riparian			
CEL HYBRID	Aquatic ecosystems	No known application	Nestler and Goodwin (2000)
Delaware River Basin	Aquatic primary producers / Aquatic primary grazers / User-defined fish	New Jersey/Delaware, Delaware River (excluding estuary)	Kelly and Spofford (1977)
Patuxent River watershed	River watershed	Patuxent River, Maryland	Voinov et al. (1999a,b); Institute for Ecological Economics (2000)
Wetland			
ATLSS	Freshwater wetland and estuary ecosystems	Florida Everglades	DeAngelis (1996)
Disturbance to wetland vascular plants model	Freshwater wetland	Wisconsin	Ellison and Bedford (1995)
Forest			
LANDIS	Hickory and cherry tree succession after alternative forest management practices	Northern Lake States	FLEL (2000a)
	Plant community succession at a regional scale after forest fires	California chaparral	FLEL (2000b)
	Transitional zone between boreal and temperate forests (aspen, birch, maple, oak, hemlock, pine, mixed-deciduous, and mixed conifers)	Northern Wisconsin	He and Mladenoff (1999)
	Mosaic of open patches, savannas, and forest stands dominated by jack pine, red pine, pin oak, and burr oak	Pine Barrens region, northwestern Wisconsin	Radeloff et al. (2000)
	Mature upland mixed forest in 60- to 90-yr age class (black oak, scarlet oak, white oak, post oak, shortleaf pine, red maple, sugar maple)	Ozark Mountains, southeastern Missouri	Gustafson et al. (2000)
FORMOSAIC	Tropical successional forests	Malaysia	Liu and Ashton (1998)
FORMIX	Tropical successional forests	Malaysia	Bossel and Krieger (1991)
ZELIG	Multiple species; user-defined	User-defined	Burton and Urban (1990)
JABOWA	User-defined (parameterized for common coniferous and deciduous tree species)	North America, Siberia, eastern Europe, and Costa Rica	Botkin et al. (1972); Botkin (1993a,b) and references therein
Regional landscape	Mixed successional forest	Adirondacks (United States)	Graham et al. (1991)
Spatial dynamics of species richness	User-defined deciduous forest island	User-defined	Wu and Vankat (1991)
Grassland			
STEPPE	Multiple grassland plant species	Colorado	Coffin and Lauenroth (1989)
Wildlife-urban interface	Multiple nonarboreal plant species / Multiple avian species	Midwestern United States	Boren et al. (1997)
Island			
SLOSS	User-defined plants, invertebrates, reptiles, aves, and mammals	User-defined	Boecklen (1997)
Island disturbance biogeographic	User-defined ecological island	User-defined	Villa et al. (1992)
Multiscale			
Multiscale landscape	Vegetation cover type progression (forest, agricultural, urban)	Pennsylvania	Johnson et al. (1999)

CHAPTER 12

Toxicity-Extrapolation Models

Jenée A. Colton

Toxicity-extrapolation models either estimate the toxicity of a chemical to one species on the basis of its toxicity to other, typically taxonomically related species, or estimate one endpoint from other test endpoints. Extrapolation models are generally nonmechanistic approaches that use empirical data to attempt to obtain a precise result for the specific question of interest. Various extrapolation models representing either species-sensitivity distributions or interspecies/interendpoint extrapolation methods are reviewed below.

Species-sensitivity distribution models estimate toxicity thresholds for biological communities from laboratory toxicity data. The basis for a species-sensitivity distribution is essentially a frequency distribution of toxicity thresholds (e.g., the median lethal concentrations [LC50s] or NOEC) for individual test species (Figure 12.1). These models have been the basis for the development of environmental quality criteria (e.g., Stephan et al. 1985; OECD 1992; Premises for Risk Management 1987, as cited in van Leeuwen 1990). Essentially, a "level of protection" for the community is selected that corresponds to a percentile of the species-sensitivity distribution. For example, the EPA ambient water quality criteria are assumed to protect 95% of the aquatic community because EPA selected the 5th percentile of the species-sensitivity distribution for a given chemical as the value of the criterion (Stephan et al. 1985).

Species interactions are not taken into account in these models. The accuracy of the extrapolation models is inherently limited because of their use of data from laboratory tests on a limited number of species that are easily maintained in the laboratory. The underlying assumption for all extrapolation models is that the tolerance of laboratory species tested under laboratory conditions is similar to that of all other related species. Also, laboratory exposure to chemicals and the sensitivity of organisms are assumed to mimic conditions in the field. Furthermore, species interactions are typically ignored when single-species toxicity tests are used, although some argue that such tests are adequately protective of natural systems (Emans et al. 1993).

In this evaluation, we assume that all of the species-sensitivity distribution models are subject to these limitations; therefore, the models are not ranked on this basis. We recognize that the selection of a 95% protection level is arbitrary, but this is the only widely accepted level. All of the reviewed species-sensitivity models choose one of three probability-density functions: log-normal, log-logistic, or log-triangular. The representativeness of each of these functions is a point of debate, but tests that examine methods that use these different models do not find statistically

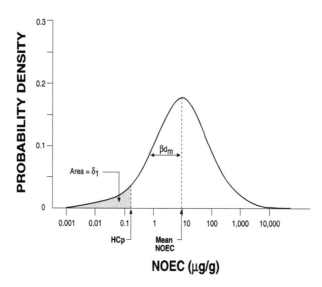

Figure 12.1 Example species-sensitivity distribution. Note: NOEC values are for cadmium exposure to soil organisms. (From van Straalen and Denneman (1989). Ecotoxological evaluation of soil quality criteria. *Ecotoxicol. Environ. Saf.* 18:241–251. With permission.)

significant differences in their results (Smith and Cairns 1993). Thus, distribution type is not used to rate the models.

Interspecies and interendpoint extrapolation models use uncertainty factors, regression techniques, or allometric scaling to extrapolate effects between species or between endpoints or to estimate chronic from acute exposure data. Uncertainty factors can be derived from data analysis or, more commonly, by professional judgment. These factors are applied to account for uncertainty in the nature of the relationships between species or between endpoints. The common practice is to apply factors as multiples of ten. The subjective derivation of uncertainty factors introduces a new source of uncertainty and makes their application vulnerable to criticism (Chapman et al. 1998).

Regression techniques (e.g., Suter et al. 1983; Linder et al. 1986) are used to extrapolate toxicity data between closely related species, although more complex regression models allow extrapolation between test species and distantly related species, which is often required in risk assessment because of toxicity data gaps. Regression techniques are also used to extrapolate between endpoints (e.g., lethal to sublethal; acute to chronic) (Mayer et al. 1994).

Allometric scaling is a very common technique that assumes a relationship between species-sensitivity to chemicals and some power of body weight (as a surrogate for rate of metabolism) (e.g., Davidson et al. 1986; Mineau et al. 1996). Allometric scaling is standard practice in the calculation of drug dosages. A weakness of the allometric scaling technique is that the relationship between sensitivity and body weight is not the same at the extremes of the range of body weights for even a single species, let alone across species. Also, scaling does not account for species differences in sensitivity.

The extrapolation models reviewed in Table 12.1 include:

- Species sensitivity distributions
  - FCV (final chronic value) model (Stephan et al. 1985)
  - HCS (hazardous concentration for sensitive species) model (Kooijman 1987)
  - HC_p (hazardous concentration for a population) model (van Straalen and Denneman 1989; Aldenberg and Jaworska 2000; Wagner and Løkke 1991; Aldenberg and Slob 1993)
- Interendpoint extrapolation methods
  - ACR (acute-to-chronic ratio) model (Kenaga 1982)

Table 12.1 Evaluation of Toxicity-Extrapolation Models

Model	Reference	Realism	Relevance	Flexibility	Treatment of Uncertainty	Degree of Development	Ease of Estimating Parameters	Regulatory Acceptance	Credibility	Resource Efficiency
Species-Sensitivity Extrapolation Models										
Final chronic value	Stephan et al. (1985)	♦♦	♦♦♦	♦♦	♦♦	♦♦♦	♦♦	♦♦♦	♦♦♦	N/A
HCS	Kooijman (1987)	♦♦	♦♦	♦♦	♦♦	♦♦	♦♦	♦♦	♦♦♦	N/A
HCp	van Straalen and Denneman (1989); Wagner and Lokke (1991); Aldenberg and Slob (1993); Aldenberg and Jaworska (2000);	♦♦	♦♦	♦♦	♦♦♦	♦♦♦	♦♦	♦♦♦	♦♦♦	N/A
Interendpoint Extrapolation Models										
ACR	Kenaga (1982)	♦	♦♦♦	♦♦	♦	♦♦	♦♦♦	♦♦	♦♦	N/A
Acute-to-chronic UF	Sloof et al. (1986)	♦♦	♦♦	♦♦♦	♦♦	♦	♦♦♦	♦	♦♦	N/A
NOEC for survival to other endpoints	Mayer et al. (1986)	♦♦	♦♦	♦♦	♦♦	♦	♦♦	♦	♦	N/A
Acute lethality to NOEC	Mayer et al. (1994)	♦♦♦	♦♦	♦♦	♦	♦	♦♦	♦	♦	N/A
Interspecies Extrapolation Models										
Allometric scaling	Davidson et al. (1986)	♦	♦♦♦	♦♦	♦	♦♦	♦♦	♦♦	♦♦	N/A
Scaling between bird species	Mineau et al. (1996)	♦	♦♦	♦♦	♦	♦♦	♦♦	♦	♦♦	N/A
Interspecies toxicity	LeBlanc (1984)	♦	♦♦	♦♦	♦♦	♦	♦♦	♦	♦♦	N/A
Species-sensitivity ratios	Hoekstra et al. (1994)	♦♦	♦♦	♦♦	♦♦	♦	♦♦	♦	♦	N/A
Interendpoint and Interspecies Extrapolation Models										
AEE	Suter et al. (1983); Linder et al. (1986)	♦♦♦	♦♦♦	♦♦	♦♦♦	♦	♦♦	♦♦	♦♦♦	N/A
Errors-in-variables regression	Suter and Rosen (1988)	♦♦	♦♦♦	♦♦	♦♦	♦	♦♦	♦	♦	N/A

Note:
♦♦♦ - high
♦♦ - medium
♦ - low
N/A - not applicable

- • Acute to chronic UF (uncertainty factor) model (Sloof et al. 1986)
- • NOEC (no-observed-effect concentration for survival to other endpoints) model (Mayer et al. 1986)
- • LC50 (acute lethality) to NOEC model (Mayer et al. 1994)
- • Interspecies extrapolation methods
  - • Scaling between bird species model (Mineau et al. 1996)
  - • Interspecies toxicity model (LeBlanc 1984)
  - • Allometric scaling model (Davidson et al. 1986)
  - • Species sensitivity ratios model (Hoekstra et al. 1994)
- • Interendpoint and interspecies methods
  - • AEE (analysis of extrapolation errors) model (Suter et al. 1983; Linder et al. 1986)
  - • Errors-in-variables regression model (Suter and Rosen 1988)

ESTIMATION OF FINAL CHRONIC VALUE MODEL

This method was developed to establish national water quality criteria applicable to most water bodies in the U.S. (Stephan et al. 1985). The final chronic value (FCV) is one of three values used to generate a criterion continuous concentration (CCC). The FCV is examined here as an example of the application of a species-sensitivity distribution method. The FCV is an estimate of the concentration corresponding to the 5th percentile of the chronic toxicity values for the genera for which acceptable chronic test results were available. The objective is to protect 95% of the genera in the aquatic community by assuming that the probability density of species-sensitivity follows a log-triangular distribution. This method also assumes that effects on species seen in standard laboratory tests will be seen in the same species in "comparable field situations" (Stephan et al. 1985). The FCV is estimated as follows:

$$FCV = e^A$$

$$A = S \times \sqrt{0.05} + L$$

$$L = \left(\sum (\ln GMCV) - S \times \left(\sum \left(\sqrt{P} \right) \right) \right) / 4$$

$$S = \sqrt{ \frac{ \left(\left(\sum (\ln GMCV)^2 \right) - \left(\left(\sum (\ln GMCV) \right)^2 \right) / 4 \right) }{ \sum (P) - \left(\left(\sum \left(\sqrt{P} \right) \right)^2 / 4 \right) } }$$

where

GMCV = genus mean chronic value
P = cumulative probability for each GMCV
S = sample standard deviation of ln (GMCV) values for the number of genera

The GMCV is the mean of the chronic toxicity values for all species within a given genus.

Realism — MEDIUM — Acceptable data are from water toxicity tests in which the organisms are not fed. Therefore, the model does not account for dietary exposure. The model can account for relationships between toxicity and physical water quality factors. The model is designed to protect almost all species (95%) in most aquatic habitats.

Relevance — HIGH — The FCV is calculated on the basis of mortality data and is therefore relevant to community and ecosystem endpoints. The model can be applied to other (nonlethal) endpoints as well.

Flexibility — MEDIUM — The FCV model can be applied to diverse aquatic habitats. However, the model requires large data sets and, thus, may be difficult to apply to terrestrial habitats and species for which such data are very limited at present. By focusing on genera, this method does not account for species differences in sensitivity.

Treatment of Uncertainty — MEDIUM — Variation between genera in effects concentrations is considered in the FCV model. However, experimental error in the original toxicity endpoints and error in parameter estimation are not considered.

Degree of Development and Consistency — HIGH — The FCV model has been tested or reviewed multiple times (e.g., Okkerman et al. 1991).

Ease of Estimating Parameters — HIGH — The only parameter required for the FCV model is the protection level, which is arbitrarily selected by the user. The toxicity data are not typically difficult to acquire, at least for aquatic species.

Regulatory Acceptance — HIGH — The method is currently encoded in U.S. environmental regulations and is in use by EPA. Ambient water quality criteria derived by using this method are the basis for state water quality standards.

Credibility — HIGH — The FCV model is widely known and used in the U.S. in setting National Ambient Water Quality Criteria. This method was reviewed favorably by OECD (1992) and has been critically reviewed (OECD 1992; Smith and Cairns 1993). In a review paper comparing this method with the HC_p model of van Straalen and Denneman (1989), this model was not preferred (Okkerman et al. 1991).

Resource Efficiency — Not applicable (N/A).

HCS

This extrapolation method defines a term called the *hazardous concentration for the most sensitive species* (HCS) that equals the geometric mean of a sample of LC50s divided by a statistically derived application factor. Kooijman (1987) developed an algorithm to derive HCS from LC50 (or median effect concentration [EC50]) data and, in doing so, intended to protect 50% of the population of the most sensitive species. This model assumes that all species sensitivities follow the same log-logistic distribution and have the same mean and variance. The HCS value depends on the number of species in a community and generally decreases with an increasing number of species. This model was designed for application to aquatic habitats.

Realism — MEDIUM — The HCS method assumes that all species sensitivities follow the same logistic distribution and that the value for one species is a randomly selected sample from the probability distribution of sensitivities of all species in a community. The algorithm developed for HCS includes a term called an *application factor* (T) that is not arbitrarily chosen but is statistically derived. T is dependent on the number of species in the community. The model does not account for relationships between toxicity and physical factors. Kooijman (1987) does not specify how to select toxicity data for use in the model.

Relevance — MEDIUM — The HCS model is calculated on the basis of mortality data and is therefore relevant to ecological risk assessment. This model is designed to protect the most sensitive species in a community and therefore produces an extremely conservative, perhaps overprotective, HCS.

Flexibility — HIGH — The number of test species used and the number of species in the community can be varied in the HCS method. The HCS approach was designed with aquatic species in mind but can be applied to terrestrial biota. Far fewer LC50 data exist for terrestrial biota, however, so application may necessarily be limited to aquatic biota at present. The level of protection, expressed as the probability that the LC50 value of the most sensitive species will be less than the HCS, is specified by the user. The model allows small toxicity data sets to be used and can be applied to a number of aquatic habitats.

Treatment of Uncertainty — MEDIUM — The HCS method ignores experimental error by assuming the LC50 from a given toxicity test is a fixed number. It does account for uncertainty in the mean and variance of a group of LC50 values from multiple experiments for the test species.

Degree of Development and Consistency — MEDIUM — The HCS model has been tested or validated at least once (van Straalen and van Leeuwen in progress).

Ease of Estimating Parameters — MEDIUM — The parameters necessary for the HCS model are the location (α) and dispersion (β) values, which require minimal calculation to estimate, and the application factor (T). Calculation of the application factor requires estimation of the number of species in the community. This information is rarely available and needs to be estimated, but Kooijman (1987) only suggests an algorithm developed on the basis of the surface area of habitat. The toxicity data are not typically difficult to acquire, at least for aquatic species.

Regulatory Acceptance — MEDIUM — The Health Council of the Netherlands has reviewed and used parts of the method for its proposal on risk assessment methods.

Credibility — HIGH — The HCS model has been used as the basis for other extrapolation models (van Straalen and van Leeuwen in progress; van Leeuwen 1990).

Resource Efficiency — N/A.

HC_p

Three models have been developed as modifications of the Kooijman (1987) method. These models define hazardous concentrations for a certain percentage of species (HC_p). The van Straalen and Denneman (1989) model applies NOEC data instead of LC50 (or EC50) data in the calculation of an HC_p (Figure 12.1). Instead of developing a hazardous concentration on the basis of the most sensitive species, van Straalen and Denneman define an HC_p that protects a certain percentage (p) of the species. The model algorithm is as follows:

$$HC_p = \frac{\exp(X_m)}{T}$$

where

$$T = \exp\left[\frac{3d_m S_m}{\pi^2} \ln \frac{1-\delta_1}{\delta_1}\right]$$

and

X_m = sample mean of ln chronic NOEC values for m test species
S_m = sample standard deviation of ln (NOEC values) for m test species
δ_1 = fraction of ecosystem that is not protected
d_m = value such that the probability of $(S_m > d_m) = 0.05$
T = an application factor between HC_p and $\exp(X_m)$

This method assumes that sensitivity distributions for the sample community and for all species in the world are the same. Therefore, estimating community size is not necessary, as with the Kooijman (1987) method.

The HC_p 1 model of Wagner and Løkke (1991) applies NOEC values in the calculation of HC_p but applies tolerance limits to the HC_p estimate. The probability density of species sensitivity is assumed to follow a log-normal distribution.

The HC_p 2 model of Aldenberg and Slob (1993) also estimates a toxicity criterion from species-sensitivity distribution but considers confidence limits on HC_p. The authors recommend using the lower 95% and 50% confidence limits as a range of uncertainty. The authors choose to apply NOEC data and define the probability distribution of species sensitivity as a log-logistic distribution.

Aldenberg and Jaworska (2000) derived Bayesian and classical uncertainty estimates for the HC_p and the fraction of species affected at a given concentration according to a normal distribution of species sensitivity. These authors found that the results from the Bayesian and classical approaches were numerically identical for the case of the normal distribution of species sensitivity. The confidence limits for HC_p and the fraction of species affected depend largely on the number of species tested. Aldenberg and Jaworska (2000) advocate using the species-sensitivity distribution to estimate the fraction of species affected at a given exposure concentration for a risk assessment instead of deriving an environmental quality objective (e.g., from an HC_p model) and using that in a hazard quotient approach.

Realism — HIGH — The HC_p model assumes that all species sensitivities follow the same log-logistic distribution and that the sensitivity of a single species is a randomly selected sample from the probability distribution of sensitivities of all species in a community. The algorithm developed for HC_p includes an application or safety factor (T) that is not arbitrarily chosen but is statistically derived. A NOEC value is selected rather than a percentage-mortality endpoint such as the LC50. van Straalen and Denneman (1989) specify criteria to use in selecting NOEC data and how to normalize for differences in test conditions. The model protects a percentage of the species in a community by assuming that communities can persist despite some small adverse effects. The protection level and the probability of overprotection are flexible and assigned by the user.

Relevance — HIGH — The toxicity endpoints (growth, survival, and reproduction) are directly applicable to risk assessment. The model targets protection of the community and allows some level of adverse effects on species.

Flexibility — MEDIUM — The HC_p method by van Straalen and Denneman (1989) was originally modified from Kooijman (1987) for application to soil invertebrates and does not include microflora. The Wagner and Løkke (1991) and Aldenberg and Slob (1993) methods were developed for aquatic habitats. All of these models could be modified for other biota with moderate effort if all the assumptions were held and applicable relationships between biota and physical environment were developed. The models allow small toxicity data sets to be used.

Treatment of Uncertainty — HIGH — The HC_p methods ignore experimental error by assuming the NOEC from a given toxicity test is a fixed number. They do account for uncertainty in the mean and variance of the NOEC values from multiple experiments for the test species. The methods of Wagner and Løkke (1991), Aldenberg and Slob (1993), and Aldenberg and Jaworska (2000) allow a full estimate of uncertainty in the HC_p and the fraction of species affected at a given concentration.

Degree of Development and Consistency — HIGH — The HC_p models have been tested or validated multiple times (van Straalen and van Leeuwen in progress, van Leeuwen 1990; Okkerman et al. 1993; van de Plassche et al. 1999).

Ease of Estimating Parameters — HIGH — Parameter estimation of the HC_p models requires minimal calculation. The toxicity data are not typically difficult to acquire, at least for aquatic species.

Regulatory Acceptance — HIGH — The van Straalen and Denneman (1989) method was used in the Netherlands to derive environmental quality criteria. The Wagner and Løkke (1991) method has been reviewed alongside the models of van Straalen and Denneman, Aldenberg and Slob, and Kooijman for application in the derivation of national environmental quality criteria (OECD 1992). The Aldenberg and Slob (1993) method is currently used in the Netherlands to derive environmental quality criteria (DGEPMH 1989).

Credibility — HIGH — The HC_p methods have been critically reviewed and supported in the literature (van Straalen and van Leeuwen in progress; Smith and Cairns 1993; OECD 1992; Okkerman et al. 1991; van Leeuwen 1990).

Resource Efficiency — N/A.

ACR

The ACR model of Kenaga (1982) defines an acute-to-chronic ratio (ACR) as the inverse of the application factor (AF) defined by Mount and Stephan (1967). The ACR value is the acute LC50 value divided by the maximum acceptable toxicant concentration (MATC). It was developed as a basis for predicting chronic toxicity from acute toxicity. The model uses toxicity data from both static and flow-through systems.

Realism — LOW — The ACR model develops an extrapolation factor that is simply a ratio of the LC50 to the MATC, which is not derived statistically. The author does not attempt to define the nature of the relationship between chronic and acute toxicity. Point estimates (without variances) are used for both the LC50 and the MATC. These aspects of this model can lead to unrealistic results.

Relevance — MEDIUM — The endpoints (e.g., mortality) are pertinent to risk assessment. Estimates of chronic toxicity often are needed because of gaps in toxicity databases.

Flexibility — MEDIUM — Although the ACR model was originally developed for aquatic species, this model can be used for species in other habitats. This model does not incorporate individual or species differences.

Treatment of Uncertainty — LOW — The ACR model does not account for extrapolation or experimental error.

Degree of Development and Consistency — MEDIUM — The ACR model has been validated (Roex et al. 2000).

Ease of Estimating Parameters — HIGH — There are no parameters to estimate for the ACR model. The variables of the MATC and the LC50 either exist or do not exist; thus, substantial calculation is not required.

Regulatory Acceptance — HIGH — The ACR model is used to derive water quality guidelines in Canada (Canadian Council of Ministers of the Environment 1991, as cited in Chapman et al. 1998) and in calculation of water quality criteria in the U.S. (60 Fed. Reg. 56 Part 132).

Credibility — MEDIUM — The ACR model has been frequently applied and critically reviewed in the literature (e.g., Länge et al. 1998; Chapman et al. 1998; Roex et al. 2000). As an empirical method, the accuracy of the results depends entirely on the toxicity database used to develop the ACR value and the appropriateness of the application.

Resource Efficiency — N/A.

ACUTE-TO-CHRONIC UF MODEL

The acute-to-chronic UF model calculates UFs for extrapolation of an acute LC50 or EC50 to a chronic NOEC by using a regression of log-transformed toxicity data (Sloof et al. 1986). The UF is calculated on the basis of the 95% prediction limits around the regression and is defined as the minimum ratio of the estimated toxicity value and its 95% upper or lower prediction limit after back transformation from log space.

Realism — MEDIUM — The acute-to-chronic UF model uses the lowest reported toxicity values to develop a regression relationship between acute and chronic data and develops the uncertainty factor from the 95% prediction limits. The test results used in the model are not screened and may overpredict or underpredict the actual toxicity value.

Relevance — HIGH — The endpoints are pertinent to risk assessment as long as the endpoints used are related to survival, growth, and reproduction.

Flexibility — HIGH — The model was originally applied to aquatic habitats but may be adapted to terrestrial habitats if toxicity data are available.

Treatment of Uncertainty — MEDIUM — The model develops 95% uncertainty factors on the basis of the estimation of HC_p. It ignores experimental error but incorporates natural variation in sensitivity.

Degree of Development and Consistency — LOW — No validation studies were found.

Ease of Estimating Parameters — HIGH — Parameter estimation for the acute-to-chronic UF model requires minimal calculation. The toxicity data are not typically difficult to acquire, at least for aquatic species.

Regulatory Acceptance — LOW — This model has not been adopted by a regulatory agency.

Credibility — MEDIUM — This model has been critically reviewed in the literature (van Leeuwen 1990).

Resource Efficiency — N/A.

NOEC FOR SURVIVAL TO OTHER ENDPOINTS MODEL

The extrapolation method of Mayer et al. (1986) predicts NOECs for a specified endpoint from the NOEC for survival. Ratios of NOEC for survival to NOECs for other endpoints were calculated and subjected to frequency analysis. In addition, least-squares regression analyses were used to determine relationships between NOECs for growth and for survival (r = 0.949 to 0.974). The authors propose using the endpoint correlations to predict the NOEC for growth from the NOEC for survival. A factor of 0.2 is applied to the NOEC for survival on the basis of frequency analysis results that indicate that the NOECs for other endpoints are usually five times less than the NOEC for survival. Univariate analysis indicated a range for this factor of 0.13 to 0.22, and the authors recommend further evaluation to more accurately estimate the appropriate factor.

Realism — HIGH — Mayer and colleagues (1986) calculated a NOEC for each endpoint as the geometric mean of the highest concentration that caused no effect and the lowest statistically significant effect concentration. The authors attempted to control for the effects of different test methods by screening out studies with unusual or questionable test conditions. Errors-in-variables regression is presented by the authors as a more appropriate method for extrapolation with toxicity data than least-squares regression. Most of the design features of the model are considered realistic.

Relevance — MEDIUM — The endpoints used include growth, survival, and reproduction but also include endpoints less relevant for risk assessment (e.g., histopathology, tissue lesions, biochemical symptoms).

Flexibility — MEDIUM — The model was designed and tested on freshwater fish species. Considerable testing would be required to apply the model to other biotic groups to verify the relationships between endpoints.

Treatment of Uncertainty — MEDIUM — Selected data are obtained from similar test conditions. Uncertainty is included in the estimation of the NOEC for growth in the form of 95% confidence intervals, but an uncertainty factor is applied to the final value.

Degree of Development and Consistency — LOW — No validation studies were found.

Ease of Estimating Parameters — HIGH — The model allows for variably sized data sets. A minimum of three data points are required. All parameters are easily estimated.

Regulatory Acceptance — LOW — This model has not been adopted by a regulatory agency.

Credibility — LOW — The model has not been critically reviewed or commonly cited.

Resource Efficiency — N/A.

ACUTE LETHALITY TO NOEC MODEL

Mayer and colleagues (1994) developed a two-step linear regression approach to predict a chronic NOEC for selected endpoints (e.g., growth and other sublethal endpoints) from acute lethality data (LC50). It considers the chemical concentration, degree of response, and time course of effects. In the first regression, a concentration–response curve is derived by using probit analysis of the results for each time period of exposure. In the second step, the LCO1 (0.01% per response) values estimated for different time periods using the results of the first step are regressed against the reciprocal of time. The y-intercept of the second regression is the predicted NOEL for chronic lethality. In comparison to MATC values, the predicted chronic NOEC values are highly accurate according to the authors (within a factor of 3.6 in 97% of cases). Accuracy may depend on the use of large data sets from a common laboratory.

Realism — HIGH — The LC50 data are selected on the basis of their origination from one of two federal laboratories so that data collected by highly standardized test methods could be used. This model acknowledges the possible change in species response over the course of exposure. The model can use all the information from an acute toxicity test instead of only one point estimate for a single time period of exposure. All of the design features in this model are realistic.

Relevance — MEDIUM — According to the authors, this model should not be used to estimate chronic reproductive effects because acute lethality effects have not been successfully related to chronic reproductive effects.

Flexibility — MEDIUM — The model was designed for and tested on freshwater and marine fish species. Considerable testing would be required to apply the model to other biotic groups to verify the relationships between acute and chronic responses.

Treatment of Uncertainty — LOW — The model does not incorporate uncertainty in experimental testing or NOEC estimation.

Degree of Development and Consistency — LOW — No validation studies were found.

Ease of Estimating Parameters — HIGH — The model does allow for variably sized data sets. A minimum of three data points are required. All parameters are easily estimated.

Regulatory Acceptance — LOW — This model has not been adopted by a regulatory agency.

Credibility — LOW — The model has not been critically reviewed or commonly cited.

Resource Efficiency — N/A.

ALLOMETRIC SCALING MODEL

Davidson et al. (1986) defined a general allometric equation for extrapolation of biological data between species as a power function of body weight. This model has been applied to relate various physiological, morphological, and toxicological characteristics to body weight for extrapolating toxicity data between species.

Realism — LOW — The model assumes that toxicity is directly related to metabolic rate and that metabolic rate is allometrically related to body weight. This model is very simple and does not account for many influencing factors such as diet. Moreover, the basic assumption may be incorrect when the toxicity of bioaccumulative chemicals is examined.

Relevance — HIGH — The model is commonly used in risk assessment to estimate the exposure of one species on the basis of that of another or to extrapolate toxicity thresholds for specified endpoints between species.

Flexibility — HIGH — This model can be applied to any mammalian species for many different data types (e.g., toxicity values, ingestion rates).

Treatment of Uncertainty — LOW — This model does not account for uncertainty.

Degree of Development and Consistency — MEDIUM — This model was validated by Sample and Arenal (1999) and Peters-Volleberg and colleagues (1994).

Ease of Estimating Parameters — HIGH — Allometric relationships for toxicity are established for many mammal and bird species.

Regulatory Acceptance — MEDIUM — The model is sometimes applied by EPA.

Credibility — HIGH — This model has been critically reviewed in the literature (Paxton 1995; Sample and Arenal 1999).

Resource Efficiency — N/A.

SCALING BETWEEN BIRD SPECIES MODEL

The model of Mineau et al. (1996) empirically derives scaling factors for birds for extrapolation of acute toxicity data between species. It derives an exponential scaling coefficient of 1.0 to 1.55 for birds on the basis of the regression of median lethal doses (LD50s) for a series of pesticides. Log-transformed LD50 data are fitted with linear regression to a power curve:

$$\log(LD50) = \log(a) + b*\log(weight)$$

The resulting slopes in the avian model for various pesticides average closer to 1.0 than to the 0.6 to 0.7 that is seen for mammals.

Realism — LOW — LD50 data are screened to decrease experimental variation, but the method of dosing is not one of the screening criteria. Thus, toxicity data from tests using artificial dosing methods, such as gavage, are included. Test animal weights are not available so mean species weights are applied from the literature.

Relevance — MEDIUM — Scaling done on the basis of LD50 data is not as useful in risk assessment as that done on the basis of NOEL data. If enough data were available, this exercise could be repeated by using NOEL data instead.

Flexibility — HIGH — The scaling factor in the model of Mineau et al. (1996) applies to birds. The method (but not the scaling factor) could be applied to other biotic groups.

Treatment of Uncertainty — LOW — The model chooses single values of LD50 for each combination of bird species and pesticide; error in estimation of LD50 and experimental error are ignored. No confidence limits on the scaling factor are presented or recommended. The standard deviation of the average slope is calculated by the authors but is not used.

Degree of Development and Consistency — MEDIUM — The model is validated in one study by Sample and Arenal (1999).

Ease of Estimating Parameters — HIGH — The LD50 data are usually easily obtained.

Regulatory Acceptance — MEDIUM — This method has been used by EPA but not officially adopted by a regulatory agency.

Credibility — MEDIUM — Because the results of this study indicate that the scaling factor should be near 1.0, risk assessors sometimes choose not to apply a scaling factor for birds as they do for mammals.

Resource Efficiency — N/A.

INTERSPECIES TOXICITY MODEL

LeBlanc (1984) used polynomial regression to develop interspecies extrapolation relationships for several aquatic species. These relationships are based on acute LC50s for a wide range of chemicals (pesticides, metals, nonpesticide organics).

Realism — LOW — LeBlanc's (1984) method did not specify screening criteria for evaluating the toxicity data sets. Many of the design features for the other models do not apply to this model, and no emphasis is placed on careful selection of appropriate toxicity data.

Relevance — MEDIUM — This model will not be directly applicable to most risk assessments because of its reliance on acute LC50 data. Nevertheless, the approach could be applied to data sets for other endpoints.

Flexibility — HIGH — This method could be applied to chronic or acute data and to any species.

Treatment of Uncertainty — LOW — The model does not account for error in estimation of interspecies relationships between acute LC50s. Experimental error is ignored.

Degree of Development and Consistency — LOW — No validation studies were found.

Ease of Estimating Parameters — HIGH — The LC50 data are readily available.

Regulatory Acceptance — LOW — This model has not been adopted by a regulatory agency.

Credibility — MEDIUM — This model has been critically reviewed in the literature (Calabrese and Baldwin 1993).

Resource Efficiency — N/A.

SPECIES-SENSITIVITY RATIOS MODEL

Hoekstra and colleagues (1994) developed sensitivity ratios ($SR_{95:5}$ and $SR_{50:5}$) that indicate the width of the LC50 distribution and then determined the sensitivity of organisms to chemicals on the basis of their taxonomic position. The sensitivity ratio is simply the ratio of the 95th or 50th percentile to the 5th percentile of the distribution of LC50s. These ratios are independent of sample size. Thus, the species-sensitivity ratios model is an empirical method that derives a simple measure of variability in toxicity thresholds among species. The model is derived from basic statistical concepts and can be applied to a variety of species.

Realism — HIGH — Toxicity data are screened for test conditions (e.g., pH, duration of test, test chemical purity, hardness) to exclude data obtained from tests with exceptional conditions. Data from different life stages are treated as independent observations because their variation is similar to interspecies variation. The protection level of 5% may be considered overprotective. Most of the design features in this model are realistic.

Relevance — MEDIUM — The species-sensitivity ratios describe the spread of the distribution of acute LC50 data, which is not as useful in most risk assessments as chronic NOEC (or NOEL) data. The model could be applied to other endpoints when sufficient data are available.

Flexibility — HIGH — The model can be applied to any species in any habitat.

Treatment of Uncertainty — MEDIUM — The model incorporates interspecies variation in sensitivity. It ignores experimental error.

Degree of Development and Consistency — LOW — No validation studies could be identified.

Ease of Estimating Parameters — HIGH — The model only requires LC50 data.

Regulatory Acceptance — LOW — The model has not been formally approved or used by a regulatory body.

Credibility — LOW — The model has not been critically reviewed or commonly cited.

Resource Efficiency — N/A.

AEE

The AEE model (Suter et al. 1983; Linder et al. 1986) was developed at the Oak Ridge National Laboratory. This model calculates a toxicity endpoint such as an LC50 for one species by extrapolating from acute LC50 data for a test species that is closely related to the species of interest. This model combines two extrapolations: extrapolation from acute LC50 to chronic MATC effects and from one species to another. Errors-in-variables estimation is used to find the linear relationships for these extrapolations. The variance from the regression analyses is combined with the best estimate of a chronic MATC to derive a probability distribution for the MATC.

Realism — HIGH — Extrapolation is done between taxa having the next higher taxonomic level in common. Toxicity data are selected as LC50 or MATC pairs from the same study of several chemicals using two species. From the information presented about this model, the design choices seem to be realistic.

Relevance — HIGH — The model endpoints directly relate to risk assessment endpoints.

Flexibility — HIGH — The model was designed originally for freshwater organisms, but Linder et al. (1986) have also found it appropriate for marine fish and crustaceans. Suter et al. (1983) claim the model is applicable to any species or chemical.

Treatment of Uncertainty — HIGH — The model accounts for experimental and extrapolation error. It uses a major-axis regression technique because the independent variable (LC50 values) is subject to error.

Degree of Development and Consistency — LOW — No validation studies were found.

Ease of Estimating Parameters — HIGH — The model allows for variably sized data sets. All parameters are easily estimated.

Regulatory Acceptance — MEDIUM — This model has been applied by the EPA.

Credibility — HIGH — This model has been critically reviewed in the literature (Calabrese and Baldwin 1993).

Resource Efficiency — N/A.

ERRORS-IN-VARIABLES REGRESSION MODEL

Suter and Rosen (1988) use an errors-in-variables regression model to extrapolate between taxa from different environments (freshwater to marine taxa) and between test endpoints (LC50 to MATC).

Realism — MEDIUM — Interspecies extrapolations are only done for species that share the next-highest taxonomic level, although the variation at the order level and higher is almost double. Extrapolations are presented in terms of probabilities or likelihood and not as point estimates. Species differences in sensitivity are not represented in the parameters. Sensitivity is assumed to be the same for all closely related species.

Relevance — HIGH — The model endpoints (survival, growth, and fecundity) are directly related to risk assessment endpoints.

Flexibility — MEDIUM — The model can be applied to fish and crustacean species in freshwater and marine environments.

Treatment of Uncertainty — MEDIUM — The model incorporates variance in estimating the predicted value as 95% confidence intervals. It attempts to minimize variation in experimental design. In this model, a replicate test result is randomly selected when more than one existed to prevent artificial reduction of variance.

Degree of Development and Consistency — LOW — No validation studies were found.

Ease of Estimating Parameters — HIGH — The model allows for variably sized data sets. All parameters are easily estimated.

Regulatory Acceptance — LOW — This method has not been adopted by a regulatory agency.

Credibility — LOW — The model has not been critically reviewed or commonly cited.

Resource Efficiency — N/A.

DISCUSSION AND RECOMMENDATIONS

Toxicity-extrapolation models were originally developed to allow estimation of toxicity thresholds for chemicals with limited data. Extrapolation models are commonly used in screening risk assessments and in developing environmental quality criteria. They may also be useful in higher tier risk assessments to support a toxicity evaluation within the context of a population, ecosystem, or landscape model. Nevertheless, for site specific assessments, extrapolation methods should be avoided if possible, especially when additional modeling at the population, ecosystem, or landscape level is done. Efforts expended on higher level modeling may be wasted if the toxicity assessment is made on the basis of a poorly founded extrapolation. Resources may be better spent in obtaining relevant toxicity data before complex ecological-effects models are applied.

The various extrapolation methods are intended for different purposes, such as extrapolating from acute to chronic endpoints (Kenaga 1982; Sloof et al. 1986), from an LC50 to a NOEC (Mayer et al. 1994); between species (Suter et al. 1983; Linder et al. 1986; Suter and Rosen 1988); or across a community spectrum (i.e., species-sensitivity distributions) (Stephan et al. 1985; van Straalen and Denneman 1989; Aldenberg and Slob 1993). In most cases, the choice of a method depends only on the objective, and the most recently developed approach is likely to be applicable. Notable exceptions include the relatively higher rating of the UF for acute to chronic extrapolation (Sloof et al. 1986), of the AEE for interendpoint and interspecies extrapolation (Suter 1983; Linder et al. 1986), and of the HC_p models for species-sensitivity distributions (van Straalen and Denneman 1989; Wagner and Løkke 1991; Aldenberg and Slob 1993).

The use of arbitrary uncertainty factors of 10, 100, or 1000 for extrapolating toxicity data between species and between endpoints is generally not recommended (Chapman et al. 1998). When statistical methods are not possible for interendpoint and interspecies extrapolations, the uncertainty factor methods reviewed in this document can serve as improvements to generic uncertainty factors (i.e., 10, 100, 1000), which are not chemical or species-specific. However, the user should keep in mind that error is not tracked in the use of any uncertainty factor approach. The more extrapolation steps required (e.g., between acute and chronic and between species), the more undefined error and inaccuracy will be present in the result.

CHAPTER 13

Profiles of Selected Models

Robert A. Pastorok

Table 13.1 summarizes the models selected for further development and use in ecological risk assessment in the near future. These models were rated relatively high in the evaluation within model type. Profiles of models selected for further development and use are shown in Tables 13.2 through 13.7.

Toxicity extrapolation models are intended mainly for use in screening-level risk assessments and for developing generic environmental criteria (e.g., EPA's ambient water quality criteria). They are not recommended for use alone in a detailed (e.g., baseline) risk assessment. Nevertheless, they may be useful in supporting population, ecosystem, or landscape models as part of a detailed assessment.

Of the population models, stochastic scalar abundance models (either discrete or continuous-time) and deterministic life-history matrix models are most appropriate for screening-level ecological risk assessments (Table 13.1). Some simple food-web models, founded on RAMAS Ecosystem or Populus, for example, may be appropriate for screening-level assessments at larger, complex sites, especially where disruption of food-web structure may be an issue. More complex ecosystem models and landscape models are not recommended for screening-level assessments because of the relatively high level of effort and expense involved in developing and running these models.

For detailed assessments, stochastic life-history matrix models and metapopulation models (e.g., RAMAS GIS and VORTEX) are recommended. These models, as well as aquatic ecosystem models like AQUATOX, CASM, and IFEM, aquatic landscape models like ATLSS, AQUATOX, and CASM, and terrestrial landscape models like LANDIS, JABOWA, and the disturbance bioge-ography model, are suitable for application in detailed ecological risk assessments. Several model categories lack specific examples of available models for detailed assessments. Further development of such models could include integration of metapopulation models with food-web models and other ecosystem models and with landscape models (Table 13.1).

The selection of specific models for addressing an ecological risk issue depends on the balance between model complexity and the availability of data, the degree of site-specificity of available models, and the issue, ecosystem, endpoints, and chemicals of interest. The models must be appropriate for the context, whether for the evaluation of risks associated with new chemicals and their uses, of ecological impacts and risks associated with past uses, or of clean-up and restoration issues. Because the selection of models is specific to the issue, chemical, and site of interest, we have provided only general guidance for selecting ecological models (see Chapter 1, Introduction,

Table 13.1 Ecological-Effects Models Selected for Further Development and Use in Chemical Risk Assessment

| Level of Effort | Population | Ecosystem | | | Landscape | | | Toxicity-Extrapolation |
		General Food-Web	Aquatic	Terrestrial	General	Aquatic	Terrestrial	
Screening	Stochastic scalar abundance Life-history matrix (deterministic)	Populus RAMAS Ecosystem	NA	NA	NA	NA	NA	Interendpoint interspecies Species-sensitivity distribution N/A[a]
Detailed	Life-history matrix (stochastic) Metapopulation RAMAS GIS VORTEX	Integration of RAMAS Ecosystem with spatially explicit metapopulation models	AQUATOX CASM IFEM	Integration of spatially explicit metapopulation and food-web models	Integration of spatially explicit metapopulation and landscape models	ATLSS AQUATOX CASM IFEM	LANDIS JABOWA Island disturbance biogeographic	

Note: N/A - not applicable.

[a]Relationships for use with ecological models.

Table 13.2 Profiles of Selected Population Models — Scalar Abundance Models

Model: Stochastic Continuous-Time Models	
References:	Poole (1974), Goel and Richter-Dyn (1974), May (1974), Ginzburg et al. (1982), Iwasa (1998), Tanaka (1998), Hakoyama and Iwasa (1999)
Stressor:	Any stressor whose impacts can be summarized as a change in the mean or variance of the population growth rate or density-dependence parameters
Type of ecosystem:	Various (terrestrial, freshwater, estuarine, marine)
Temporal resolution:	Determined by resolution of available census data and generation time of the species; time step is usually between 1 week and 1 decade
Time scale:	Determined by the modeler according to the question addressed, the quality and amount of available population data (e.g., period during which the data were collected), and generation length of the species; time period is usually between several weeks and a century
Spatial resolution:	Not spatially resolved
Geographic scale:	Implicitly identified with the geographical scale of modeled population
Biological scale:	Population
Model type:	Simple scalar abundance model
Level of detail:	Screening
Endpoints:	Expected future population abundance; statistical distribution of abundance; risk of a population decline of a specified size; time to a population decline of a specified size
Comments:	Has minimum data requirements for any population-level, risk-analytic model; can include effects of density dependence or omit them in a conservative assessment; can integrate effects of chemical and non-chemical stressors through time; methodology can be used now but would benefit from further development of allometric relationships to inform data parameterizations when empirical information is sparse

Model: Stochastic Discrete-Time Models	
References:	Poole (1974), Capocelli and Ricciardi (1974), Goel and Richter-Dyn (1974), May (1974), Ginzburg et al. (1982), Ferson (1999)
Stressor:	Any stressor whose impacts can be summarized as a change in the mean or variance of the population growth rate or density-dependence parameters
Type of ecosystem:	Various (terrestrial, freshwater, estuarine, marine)
Temporal resolution:	Determined by resolution of available census data and generation time of the species; timestep is usually between 1 week and 1 decade
Time scale:	Determined by the modeler according to the question addressed, the quality and amount of available population data (e.g., period during which the data were collected), and generation length of the species; time period is usually between several weeks and a century
Spatial resolution:	Not spatially resolved
Geographic scale:	Implicitly identified with the geographical scale of the modeled population
Biological scale:	Population
Model type:	Simple scalar abundance model
Level of detail:	Screening
Endpoints:	Expected future population abundance; statistical distribution of abundance; risk of a population decline of a specified size; time to a population decline of a specified size
Comments:	Has minimum data requirements for any population-level, risk-analytic model; can include effects of density dependence or omit them in a conservative assessment; can integrate effects of chemical and non-chemical stressors through time; methodology can be used now but would benefit from further development of allometric relationships to inform data parameterizations when empirical information is sparse; easier both to implement in software and to parameterize than continuous stochastic models

The Process of Ecological Modeling for Chemical Risk Assessment). Moreover, the models that we selected for further development and use are not necessarily the only models that would be useful for ecological risk assessments in the near future.

Many of the models reviewed here are applicable for answering ecological risk questions. We urge the reader to consider his/her needs, select a general category of models according to the endpoint

Table 13.3 Profiles of Selected Population Models — Life-History Models

Model: General Matrix Models	
Reference:	Caswell (2001)
Stressor:	Any (including chemical and thermal pollution, harvest, entrainment, impingement, habitat loss, habitat fragmentation); depends on implementation
Type of ecosystem:	Various (terrestrial, freshwater, estuarine, marine)
Temporal resolution:	Theoretically any temporal resolution is allowed; practically, temporal resolution of the model depends on temporal resolution of the data; generally ranges from 1 day to 10 years
Spatial resolution:	Not spatially explicit, but can be embedded within a metapopulation model
Geographic scale:	Not spatially explicit
Biological scale:	Population
Time scale:	Theoretically any; practically, depends on temporal resolution and question being asked (endpoint); generally ranges from several weeks to several decades
Model type:	Structured population
Level of detail:	Tier 2
Endpoints:	Depends on implementation (e.g., which software package is being used); possibilities include expected future abundance, risk of decline in abundance, risk of extinction, time to extinction, time to decline, harvested biomass, as well as measures of uncertainty about these endpoints
Comments:	Highly recommended as a general class of models; assumptions are realistic and endpoints are comprehensive; implementation generally requires the use of a software package or programming; matrix models have been embedded in metapopulation models in which each subpopulation is modeled with a matrix; new theory that will add to the utility of matrix models is the development of risk-based sensitivity analysis

Model: RAMAS Age, Stage, Metapop, or Ecotoxicology	
Reference:	Applied Biomathematics (2000)
Stressor:	Any stressor, including chemical and thermal pollution, harvest, entrainment, impingement, habitat loss, habitat fragmentation
Type of ecosystem:	Various (terrestrial, freshwater, estuarine, marine)
Temporal resolution:	Theoretically any temporal resolution is allowed; practically, temporal resolution of the model depends on temporal resolution of the data; generally ranges from 1 day to 10 years
Spatial resolution:	Not spatially explicit (but see comments)
Geographic scale:	Not spatially explicit (but see comments)
Biological scale:	Population
Time scale:	Theoretically any; practically, depends on temporal resolution and question being asked (endpoint); generally ranges from several weeks to several decades
Model type:	Structured population
Level of detail:	Tier 2
Endpoints:	Expected future abundance, risk of decline in abundance, risk of extinction, time to extinction, time to decline, harvested biomass, and measures of uncertainty around each endpoint
Comments:	These programs comprehensively apply matrix models; Age and Stage are only available in DOS, whereas Metapop and Ecotoxicology are Windows®-based; only Ecotoxicology explicitly models the effects of toxic chemicals; the other programs can be manipulated to model toxicant effects by running several simulations (e.g., with and without toxic chemicals)

desired, and review the ratings of individual models. For specific needs, the ratings may be differentially weighted, and the best models may differ from the ones selected by us for further development.

Table 13.4 Profiles of Selected Population Models — Metapopulation Models

Model: RAMAS GIS (and Metapop)

Reference:	Akçakaya (1998a,b); Applied Biomathematics (2000)
Stressor:	Any stressor, including chemical and thermal pollution, harvest, entrainment, impingement, habitat loss, habitat fragmentation
Type of ecosystem:	Various (terrestrial, freshwater, estuarine, marine)
Temporal resolution:	Variable; determined by resolution of available demographic data (e.g., census frequency) and generation length of the species; typically between 1 week and 1 decade
Spatial resolution:	Variable; determined by resolution of available spatial data (e.g., size of raster map cells) and spatial behavior of the species (e.g., home range size, daily movement distance, etc.); typically between one meter and several kilometers
Geographic scale:	Variable; determined by the range of the species, the question addressed, and the availability of geographic data; typically between one hectare and thousands of square kilometers
Biological scale:	Population or metapopulation
Time scale:	Variable; determined by the question addressed, the quality and amount of available demographic data (e.g., period during which the data were collected), and generation length of the species; typically between several weeks and several centuries
Model type:	Metapopulation
Level of detail:	Tier 2
Endpoints:	Expected future abundance, risk of decline in abundance, risk of extinction, time to extinction, time to decline, harvested biomass
Comments:	Has high realism, relevance, and flexibility and can be used for risk assessment without additional development; however, further development that will likely make it more suitable includes linking it to a GIS-based hydrodynamic model, adding sex structure, genetics, and different types of density-dependence; GIS/Metapop enables one to model several structured subpopulations that together comprise a metapopulation

Model: VORTEX

Reference:	Lacy (1993, 1999)
Stressor:	Any stressor, including chemical and thermal pollution, harvest, entrainment, impingement, habitat loss
Type of ecosystem:	Various, but mostly terrestrial
Temporal resolution:	Variable; determined by resolution of available demographic data (e.g., census frequency) and generation length of the species; typically between 1 week and 1 decade
Spatial resolution:	Not spatially explicit; spatial structure based on predefined populations
Geographic scale:	Variable; determined by the range of the populations modeled and the question addressed
Biological scale:	Population or metapopulation
Time scale:	Variable; determined by the question addressed, the quality and amount of available demographic data (e.g., period during which the data were collected), and generation length of the species; typically between several weeks and several centuries
Model type:	Individual-based metapopulation
Level of detail:	Tier 2
Endpoints:	Expected future abundance, risk of extinction, time to extinction
Comments:	Has medium level of realism (in comparison with other individual-based models reviewed), high relevance, and high resource efficiency; parameters in the model can be estimated with relative ease; may not be suitable for plants, highly fecund species, and very abundant species; does not incorporate habitat relationships

Table 13.5 Profiles of Selected Ecosystem Models — Food-Web Models

Model: RAMAS Ecosystem	
Reference:	Spencer and Ferson (1997a); Applied Biomathematics (2000)
Stressor:	Chemical, physical
Type of ecosystem:	Various (terrestrial, freshwater, estuarine, marine)
Temporal resolution:	Any temporal resolution is allowed by the program, but generally ranges from weeks to years
Spatial resolution:	Not spatially explicit
Geographic scale:	Not spatially explicit
Biological scale:	Food web, food chain, or simple predator–prey interaction
Time scale:	Any time frame is allowed, but the general range is between several months and several decades
Model type:	Food web/food chain
Level of detail:	Tier 2
Endpoints:	Expected future abundance, risk of decline in abundance, risk of extinction, time to extinction, time to decline, toxicant concentration in the environment, toxicant concentration in the organisms
Comments:	This program is quite flexible and easy to use; it explicitly models the effects of toxic chemicals, and does not assume that populations are at equilibrium (which is in its favor); several relevant endpoints are available; a generalization to allow an intermediate between prey-dependent and ratio-dependent functional responses would increase its utility

Model: Populus	
Reference:	Alstad et al. (1994a,b); Alstad (2001)
Stressor:	Chemical, physical
Type of ecosystem:	Various (terrestrial, freshwater, estuarine, marine)
Temporal resolution:	Any temporal resolution is allowed by the program, but generally ranges from weeks to years
Spatial resolution:	Not spatially explicit
Geographic scale:	Not spatially explicit
Biological scale:	Food web, food chain, or simple predator–prey interaction
Time scale:	Any time frame is allowed, but the general range is between several months and several decades
Model type:	Food-web/food-chain
Level of detail:	Tier 2
Endpoints:	Expected future abundance
Comments:	Endpoints are limited, but the program is flexible in that any kind of predator–prey model can be implemented, as equations can be defined by the user; also offers several predefined predator–prey models; however, Populus does not explicitly model the effects of toxic chemicals and has no treatment of uncertainty

Table 13.6 Profiles of Selected Ecosystem Models — Aquatic Ecosystem Models

Model: AQUATOX

Reference:	Park et al. (1974); Park (1998); U.S. EPA (2000a,b,c)
Stressor:	User-defined toxic chemicals
Type of ecosystem:	Freshwater, estuarine
Temporal resolution:	Days
Spatial resolution:	Not applicable[a]
Geographic scale:	Not applicable[a]
Biological scale:	Guild
Time scale:	User-defined
Model type:	Multicompartment; linear additive integral
Level of detail:	Tier 2
Endpoints:	Guild biomass, body burdens of contaminants; specific fish type abundance
Comments:	AQUATOX was specifically designed to model the ecological impact of chemical contaminants on aquatic ecosystems; it not only models changes in productivity and standing stock abundance but also the transfer of contaminants between media and biota as well as through various trophic levels; AQUATOX is supported by EPA and is available on a software platform that is still undergoing revision and development

Model: CASM (Modified SWACOM)

Reference:	DeAngelis et al. (1989); Bartell et al. (1992, 1999)
Stressor:	Nutrients and user-defined toxic chemicals
Type of ecosystem:	Freshwater, estuarine
Temporal resolution:	User-defined periodicity
Spatial resolution:	Not applicable[a]
Geographic scale:	User-defined; area treated as a homogeneous unit[a]
Biological scale:	Guild
Time scale:	User-defined
Model type:	Multicompartment; linear additive integral
Level of detail:	Tier 2
Endpoints:	Nutrient concentrations, productivity, abundance, and diversity of fish
Comments:	This model is very similar to AQUATOX; unlike AQUATOX, SWACOM was initially developed to model nutrient impacts

Model: IFEM

Reference:	Bartell et al. (1988)
Stressor:	User-defined; established for naphthalene
Type of ecosystem:	Freshwater, estuarine
Temporal resolution:	Days
Spatial resolution:	Not spatially explicit[a]
Geographic scale:	Not applicable[a]
Biological scale:	Guild
Time scale:	User-defined
Model type:	Compartment dynamic; integrated linear additive
Level of detail:	Tier 2
Endpoints:	Guild productivity, toxic impact, guild-specific body burdens of contaminants
Comments:	This model is an iterative simulation that describes food-web transfer for both carbon and contaminants and estimates both primary and secondary impacts of toxicity

[a] Spatially aggregated model. May also be applied as a landscape model.

Table 13.7 Profiles of Selected Landscape Models

	Model: ATLSS
Reference:	DeAngelis (1996)
Stressor:	Developed for mercury, but may be modified for other contaminants
Type of ecosystem:	Large-scale wetlands (Florida Everglades)
Temporal resolution:	Variable; executed on user-defined cycles
Spatial resolution:	Currently based on 100-m² grids
Geographic scale:	Variable; based on multiple grids
Biological scale:	Variable; guild at lower trophic levels and species at upper trophic levels
Time scale:	Open
Model type:	Spatially explicit; low trophic levels are process models, middle trophic levels are structured cohort models, and high trophic levels are individual-based models
Level of detail:	Tier 2
Endpoints:	Abundance of guild or species; intertrophic carbon flow and populations (abundance and diversity) of keystone species; mercury bioaccumulation
Comments:	ATLSS is one of the most comprehensive landscape models available; its ambition is balanced by its flexibility in that the creators recognized that different trophic levels require different types of model formats; this allows the simulations to be tailored not only based on the life histories of a particular species but also on the available data and understanding of that receptor's behavior; although this model was originally developed for application to the Florida Everglades, it could be used in other biomes where there is a terrestrial food web reliant on the productivity of a large-scale aquatic habitat

	Model: LANDIS
Reference:	Mladenoff et al. (1996); Mladenoff and He (1999)
Stressor:	Wind and fire disturbances; harvesting; any physical disturbance
Type of ecosystem:	Northern temperate zone forests
Temporal resolution:	Hundreds of years
Spatial resolution:	10 × 10 m
Geographic scale:	Thousands of hectares
Biological scale:	Tree species as the presence or absence of 10-year age cohorts (not as individual stems)
Time scale:	Hundreds of years
Model type:	Stochastic, spatially explicit model of forest landscape processes; raster-based
Level of detail:	Tier 2
Endpoints:	Tree species presence/absence, stand age structure and species richness, potential impact of toxic chemicals on species-specific life-history traits (e.g., mortality, seed dispersal, etc.)
Comments:	LANDIS is an elaboration of the VAFS/LANDSIM model; however, LANDIS operates in raster mode, and the spatial interactions are based on distances (instead of polygon neighborhoods); operation can be at different scales of resolution and is programmed in C++ by using hierarchical, object-oriented data structures; LANDIS uses a free-standing spatial analysis package, reads and writes ERDAS raster files, and incorporates wind-throw and fire disturbance, which are absent from other models

Table 13.7 (cont.)

Model: JABOWA	
Reference:	Botkin et al. (1972); West et al. (1981); Bodkin (1993a,b)
Stressor:	Physical disturbance, primarily harvesting
Type of ecosystem:	Multispecies forests throughout the world
Temporal resolution:	Annual time step
Spatial resolution:	10-m × 10-m grid sections (default value, which is user-adjustable in the early versions of the model)
Geographic scale:	User-defined landscape
Biological scale:	Individual trees of various species
Time scale:	User-defined
Model type:	Spatially explicit model of forest landscape with stochastic functions for mortality and reproduction
Level of detail:	Tier 2
Endpoints:	Tree biomass, stand biomass, stand age structure, forest productivity, and species richness based on growth, reproduction, and mortality of individual trees; species and biomass distribution in space
Comments:	JABOWA was developed as a general model of forest dynamics; it was among the first successful multispecies computer simulations of terrestrial ecosystems and has gone through extensive modification and use during the 30 years since it originated; one of the best-documented and most flexible models of its type

Model: Island Disturbance Biogeographic Model	
Reference:	Villa et al. (1992)
Stressor:	Variable; manifest as impacts to populations
Type of ecosystem:	Various; model must be tailored by the user
Temporal resolution:	Periodicity is user defined
Spatial resolution:	Grid size and resolution are user defined
Geographic scale:	Grid size and resolution are user defined
Biological scale:	Communities across a landscape
Time scale:	Periodicity is user defined
Model type:	Spatially explicit; quantitative rule-based model based on iterative Lotka-Volterra kinetics
Level of detail:	Tier 2
Endpoints:	Biodiversity and temporal species sensitivity
Comments:	The Villa model uses the same principles as most island biogeographic models; however, this implementation (1) is spatially explicit, (2) has implicit integration of the disturbance function, (3) has multiple population interactions, and (4) calculates immigration relative to island carrying capacities; this model, as presented, is highly theoretical; this increases its flexibility but would require explicit modifications to tailor it to a specific situation; it does not currently have functions to explicitly model toxicity

CHAPTER 14

Enhancing the Use of Ecological Models in Environmental Decision-Making

Lev R. Ginzburg and H. Resit Akçakaya

Although ecological models have been used to better understand natural systems to support environmental management (e.g., Holling 1978; Hilborne and Mangel 1997; Walters 1986; Breininger et al. in press), relatively few applications to toxic chemical issues exist. One of the most important factors preventing widespread use of models in decision-making is a lack of modeling training, which prevents many managers from determining the best kind of model and scale of resolution for solving a given problem (Ginzburg and Akçakaya in press; Breininger et al. in press). Another important factor is the lack of appropriate and relevant data (Ginzburg and Akçakaya in press). Moreover, untrained managers often do not know what types of data to collect to optimize model use.

As shown in our review of models above, many ecological models already exist that can be applied to practical problems in environmental management of toxic chemicals. Ginzburg and Akçakaya (in press) identified four different areas that require additional investment to enhance the use of ecological modeling in decision-making: training, applying existing models, integrating existing models, and developing new, case-specific models.

TRAINING AND EDUCATION

Teaching environmental managers how to use existing models is the most efficient way to enhance model use in the short term. Jackson et al. (2000) provide an excellent introduction to the practice of ecological modeling that is suitable for any audience. However, the specific content and format of training courses should be based on the audience. Audiences for such training courses may be divided into two general categories: environmental managers and technical support personnel.

The first potential audience for training courses includes environmental managers — management and research decision-makers in government agencies and industry. Managers often are unfamiliar with the potential of modeling. Topics for educating managers in the use of ecological models include:

1-56670-574-6/01/$0.00+$1.50
© 2002 by CRC Press LLC

- The types of questions that can be addressed by using ecological models
- Selecting the appropriate model
- The types of data needed for different types of models
- Interpreting the results of models
- Modeling, interpreting, and communicating risk and uncertainty
- Identifying inappropriate use of models
- Examples of successful applications of models

Technical personnel are another audience who may be developing or reviewing models. To judge the technical merits of models, they need to have a basic understanding of the fundamentals of modeling. Topics for technical personnel should include all of the topics for managers, as well as the following topics for more advanced researchers and technical personnel:

- Components of different types of ecological models
- Trade-offs between complexity (realism) and practicality (data availability)
- Collecting the appropriate data
- Analyzing data to estimate model parameters
- Incorporating variability and uncertainty
- Presenting model results

APPLYING EXISTING ECOLOGICAL MODELS

The simplest way to develop a model is to apply an existing model to a management question. "Applying" means estimating the parameters of the model from data that are specific to the location, species, and system in question and using the model to address a specific management question. Many existing models have been successfully used to guide management decisions (Holling 1978; Walters 1986; Bartell et al. 1999).

Population and metapopulation models have been successfully applied to many environmental issues, especially questions regarding recovery of threatened and endangered species and management of harvesting. For example, in May 1995, the Oregon Department of Fish and Wildlife used an age-structured model to report on the biological status of the marbled murrelet (*Brachyramphus marmoratus*) in response to a petition to list the species under Oregon's Endangered Species Act (Oregon Department of Fish and Wildlife 1995). This species was later listed as threatened under the act. In another case, a metapopulation model was used to evaluate the effectiveness of translocation as a management tool for the endangered helmeted honeyeater (*Lichenostomus melanops cassidix*) and to support the decision regarding the timing of the release of this species into its new environment (Akçakaya et al. 1995). Data from a GIS and a metapopulation model were used to determine viable population size for the Florida scrub jay (*Aphelocoma coerulescens*) (Root 1998) in the context of four reserve designs developed in a Habitat Conservation Planning process for nonfederal scrub habitat in Brevard County, Florida (Brevard County Office of Natural Resources 1995). Another metapopulation model was applied to the redhorse (*Moxostoma* sp.) populations in the Muskingum River in Ohio (Root et al. 1997) to model the thermal impact of a proposed increase in power plant operation, which was later approved by the Ohio EPA.

Many existing ecological models can be applied to support or guide management decisions. Such applications require collection and statistical analysis of site specific data to estimate model parameters. Once the model parameters (and their uncertainties from measurement error and natural variability) have been determined, the application of an existing model requires minimal research effort. Therefore, the primary scientific issues in the application of existing models involve data analysis methods. These include survival estimation methods based on mark–recapture data, methods for estimating spatial, temporal, and error variance components, and variance due to age and sex. Applications to toxic chemical issues require use of available toxicity data, preferably

Table 14.1 Estimated Effort and Expenditure Required for Application of Ecological Models[a]

Model Category	Effort Required (in time)	Required Expenditure (in $1000s)
Toxicity-Extrapolation models	0.2–2 months	2–40
Population models — scalar abundance	0.5–2 months	10–40
Population models — life history	1–3 months	20–80
Population models — individual-based (with software)	2–5 months	40–100
Population models — individual-based (without software)	0.5–1.5 years	100–200
Population models — metapopulation	1–3 months	20–80
Ecosystem models — food web	2–4 months	40–100
Ecosystem models — aquatic	0.3–1 year	40–200
Ecosystem models — terrestrial	0.3–1 year	40–200
Landscape models — aquatic and terrestrial	0.5–2 years	60–500

[a] The estimates of effort and expenditure shown in the table depend on the assumption that the ecological model is run to assess the effects of a single chemical. They do not encompass the toxicity assessment, which varies depending on the number and types of chemicals addressed. The toxicity assessment and derivation of dose– or concentration–response curves may require effort and expenditure on the order of that shown in the table for applying the ecological models. This table does not address collection of field data or other parts of an ecological risk assessment (e.g., problem formulation, exposure assessment, and risk characterization).

dose– or concentration–response curves, or the collection of appropriate toxicity data. Most ecological models that do not already have functions for incorporating toxic chemical effects can be easily modified to account for such effects. Often, constraints on the immediate use of ecological models to address toxic chemical problems will arise from lack of toxicity data, not from lack of an appropriate model.

Table 14.1 specifies the expected effort (in time) and expenditure (in thousands of dollars) required to use models from the broad categories for a specific application. For each model category, we assume that collection of field data would not be required. Also, at least one model within the category must exist that is appropriate for the task without requiring substantial research or major modifications for model development. All effort and cost figures are approximate and include compilation and analysis of data from the literature, and parameterization, calibration, and sensitivity analysis of the model for a single toxic chemical.

INTEGRATING EXISTING MODELS

In the recent past, model development has involved integrating existing models — for example, habitat models and demographic models or life-history matrix models and spatial models (e.g., Akçakaya 1998a,b). One of several such potential developments involves linking simple (scalar) population models to models based on allometric relationships. The result can be used in screening assessments with minimal or no field data. An important related research issue is testing whether the level of conservatism (precaution) of this approach is comparable with the level required in a screening test.

Another type of integration links fate-and-transport models to ecological models. Although this approach has been used in specific cases, a general modeling platform is needed to link physical/chemical models (e.g., hydrological models), dose–response models, and population or metapopulation models (Reinert et al. 1998; Ginzburg and Akçakaya in press).

General models are also needed to integrate food-web and metapopulation models. Such integration would allow the modeling of trophic interactions in a spatially structured habitat with different metapopulation structures for different species. With this approach, coordinating temporal and spatial scales and resolutions is an important research issue.

Temporal and spatial scales are also important in linking metapopulation and landscape models. Incorporating landscape dynamics in the spatial structure of metapopulation models allows the effects of landscape management options on the viability of key species to be evaluated.

Integration may also involve several models of the same type. For example, linking population models for a set of several target species allows decisions to be made in a multispecies context without constructing a complex food-web or ecosystem model.

DEVELOPING NEW, CASE-SPECIFIC MODELS

In most cases, an existing model can be used to address management questions. However, many cases require developing a new model. Such a requirement can often be met by integrating existing models but, in rare cases, the management question may require a completely new modeling approach.

INVESTMENT TRADE-OFFS

Investments in enhancing the use of ecological models in environmental decision-making involve three trade-offs represented by the diamond-shaped boxes in Figure 14.1 (Ginzburg and Akçakaya in press). The first trade-off is between research and training. Of these two options, investment in training yields the greatest short-term return. For managers, a one-day workshop is the most suitable approach. For technical staff and researchers, the most cost-efficient approach is Internet-based teaching supplemented by telephone.

The second trade-off is between model development and application (Figure 14.1). The best option depends on the time constraints for management decisions. Both options require collecting and analyzing data, after which minimal additional investment of time and research effort is required for application. Therefore, for short to medium time-horizons, the most efficient way to develop models is to use existing models with the data needed to answer the specific question at hand. In most cases, the availability of data, and not the model, is the limiting factor.

The third trade-off is between integrating existing models and creating new models (Figure 14.1). Integrating existing models results in models with enhanced capabilities in the midterm, whereas creating new models results in new models in the longterm. The resources required to develop new models depend on the type of model and the population, ecosystem, or landscape being modeled.

As investment resources increase for enhancing the use of ecological modeling in environmental decision-making, the type of activity emphasized should shift from training to model application and development (Figure 14.2). Shifts in activity as a result of changes in funding targets also correspond to a sequence in these activities over time. With small investments, funding should focus on training of managers and technical staff. With improved knowledge as a result of training, increasing investments can be focused on application of existing models. Over time, the increased knowledge from applying models allows integration of existing models and development of new models.

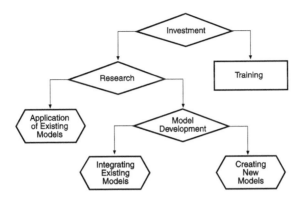

Figure 14.1   Series of three decisions for making research investments to enhance the use of ecological models in environmental decision-making. Note: Diamond-shaped box indicates a decision point. (From Ginzburg, L.R. and H.R. Akçakaya (in press). Science and management investments needed to enhance the use of ecological modeling in decision-making. In V. Dale, Ed., *Ecological Modeling for Environmental Management*, Springer-Verlag, New York. Redrawn with permission.)

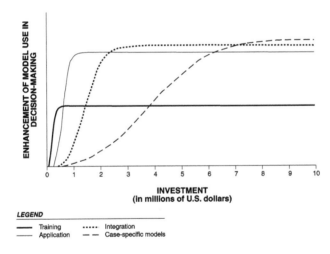

Figure 14.2   Efficient investment approach for enhancing the use of ecological models in environmental decision-making. (From Ginzburg, L.R. and H.R. Akçakaya (in press). Science and management investments needed to enhance the use of ecological modeling in decision-making. In V. Dale, Ed., *Ecological Modeling for Environmental Management*, Springer-Verlag, New York. Redrawn with permission.)

CHAPTER 15

Conclusions and Recommendations

Robert A. Pastorok and Lev. R. Ginzburg

Population, ecosystem, and landscape models are generally mechanistic models that predict ecological state variables by using mathematical relationships to represent components and processes in environmental systems. In many cases, the state variables estimated by these models are relevant endpoints for ecological risk assessment (e.g., organism abundance or biomass, recruitment, or population growth rate). Thus, these models are directly applicable in a risk characterization for assessing the significance of estimated risks initially expressed in terms of individual-level endpoints. In this mode, ecological models aid in translating risks for individual-level endpoints to more relevant endpoints at higher levels of biological organization.

In contrast with population and higher-level models, toxicity-extrapolation models are generally nonmechanistic, statistical models that attempt to extrapolate as precisely as possible the toxicity of a chemical from one endpoint to another, from one species to another, or across a whole community of species (e.g., by formulating species-sensitivity distributions). Toxicity-extrapolation models may be applied in an effects assessment to support the use of a population, ecosystem, or landscape model.

Currently, many ecological risk assessments are limited by a failure to consider population-, ecosystem-, or landscape-level endpoints. The typical hazard quotient approach compares an exposure estimate to a toxicity threshold for some individual-level endpoints such as organism survival, growth, or reproduction. Some authors interpret the hazard quotient for individual-level endpoints to infer population-level risks. This approach often leads to an overestimation of risk at the population level but in some cases may lead to an underestimation of population risk. For example, in their summary of toxicity test data with both individual-level endpoints and population parameters, Forbes and Calow (1999) found that the population growth parameter was more sensitive than individual-level endpoints in 5 of 99 cases when the lowest-observed-effect concentrations were compared among endpoints. When the percentage change in an endpoint was used as the measure of effect, 66 of 81 cases were found in which the relative change in the population growth parameter was less than the change in the most sensitive individual-level endpoint, two cases were found in which the percentage changes were equal; and 13 cases were found in which the relative change in the population growth parameter was greater than the change in the most sensitive individual-level endpoint.

Our review of ecological models shows that population and ecosystem modeling have been applied successfully in past ecological risk assessments, especially for toxic chemical issues

(Barnthouse et al. 1986; Bartell et al. 1992; Spromberg et al. 1998; Spencer et al. 1999) and for population viability analysis in conservation biology (Boyce 1992; Burgman et al. 1993; Norton 1995; Brook et al. 2000). Although their use is not presently widespread in the assessment of toxic chemical contamination, ecological models can contribute new perspectives and enhance the value of risk assessments in support of environmental management. Increasingly, ecotoxicologists are calling for an evaluation of population-level effects (Barnthouse et al. 1986; Emlen 1989; Sibley 1996; Caswell 1996; Barnthouse 1998; Forbes and Calow 1999; Snell and Serra 2000; Kuhn et al. 2000). Forbes and Calow concluded that the basic population growth parameter, r, is "a better measure of response to toxicants than individual-level effects because it integrates potentially complex interactions among life-history traits and provides a more relevant measure of ecological impact." Population models are currently a cost-effective approach for addressing many risk assessment issues (Ferson et al. 1996; Barnthouse 1998). Indeed, incorporation of simple, scalar population modeling into screening-level ecological risk assessments could greatly improve the results of such assessments; in some cases, they could help to avoid unnecessary expenditures on a higher-tier risk assessment. Examined over a hierarchical spectrum of endpoints, modeling of populations and metapopulations gives the highest combination of ecological relevance and tractability (Figure 15.1). We therefore recommend that population modeling be included in most ecological risk assessments either at the screening level or in detailed assessments.

We recommend scalar population models, life-history models, and metapopulation models for widespread use in ecological risk assessment. Scalar population models are still very crude, but they address the essential elements of risk to the basic ecological unit (i.e., a group of individuals of the same species constituting an interacting population). Although they ignore biological complexity, they are tractable and widely accepted, and their use in screening-level ecological risk assessments is clearly valuable. These and other population models can be implemented in a full probabilistic mode. In particular, we recommend the development of stochastic differential equation models and stochastic discrete-time models. Life-history models are more realistic than scalar population models, and they are essential in cases in which the receptors of concern form an age or stage class within the population. Matrix models in particular have a long history of use, and their behavior is well understood. Macroecology databases (e.g., allometric relationships and other summary statistical approaches) should be developed further to enable simple scalar population models and life-history models to be used efficiently in ecological risk assessments.

Metapopulation models are particularly relevant for addressing physical disruption of habitats because they account for the effects of habitat fragmentation. Because physical habitat features, including fragmentation patterns, can greatly affect the response of a population to toxic chemicals (Spromberg et al. 1998), we recommend the development of such models for use in ecological risk assessments of toxic chemicals. In particular, metapopulation models should be linked to GIS platforms to integrate fate-transport analysis with an evaluation of ecological effects.

Individual-based population models have not generally been applied to chemical risk issues, so most available models do not incorporate functions or rules to account for the effects of toxic chemicals. In some cases, these models are being developed further to address toxic chemical issues (e.g., as part of the ATLSS approach to evaluate mercury contamination in the Everglades), but their transference to other systems is questionable. One can implicitly model chemical effects by running different scenarios with individual-based models and changing organism dispersal, fecundity, and growth parameters to account for toxicity. However, the use of individual-based models is presently limited because of the high level of effort necessary to develop such models for specific species and sites.

Although ecosystem models provide valuable conceptual tools for analyzing ecological systems subjected to stress, they are expensive to develop and apply to particular risk issues (Figure 15.1; Table 14.1). Ecosystem models have been developed to fulfill two basic roles. First, ecologists have used them as descriptive constructs to evaluate the sensitivity of ecosystems to specific environmental parameters. Second, they have been developed as predictive tools to evaluate environmental

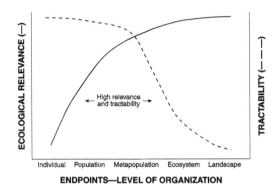

Figure 15.1 Relevance and tractability of ecological models in relation to endpoints.

management alternatives. Within either of these contexts, ecosystem models require substantial development in specific cases to provide predictions with the level of precision often desired for decision-making. Koelmans et al. (2001) discussed several general problems with complex models, including the lack of a general theory concerning the structural detail needed for accurate description of ecosystem dynamics, the difficulty of coupling representations of different trophic levels that function on a wide range of space and time scales, and the risk of obtaining a good model fit with the wrong parameter setting. Moreover, uncertainties increase in integrated models such as IFEM and AQUATOX, which superimpose toxicological effects models for food webs on representations of contaminant transport and fate. Koelmans et al. (2001) recommended that use of integrated fate and effects models of ecosystems in a predictive mode be restricted to forecasting the short-term effects of acutely toxic chemicals until further progress can be made in understanding and simulating the dominant processes in complex systems.

Ecosystem and landscape models are best utilized as heuristic tools for understanding basic ecological processes and for identifying sources of uncertainty in predictive outcomes. Some models, such as IFEM, incorporate Monte Carlo analysis to yield probabilistic expressions of risk as their final output. Other computational methods such as neuronets, Bayesian analysis, and maximum likelihood predictors have been introduced recently into ecosystem models to allow the results to be expressed in terms of probability. This step has allowed much more transparent treatment of uncertainty. Unfortunately, most of the models utilizing these new techniques are highly experimental and would not be directly applicable in current ecological risk assessment. Aquatic ecosystem models are generally better developed than terrestrial ecosystem models (aside from the forest-gap models, which have been as well researched as the aquatic models). In the near future, ecosystem and landscape models are likely to be applied to only the most complex sites and issues.

We recommend the following activities to further the development of ecological models for use in chemical risk assessment:

- Educating environmental managers and risk assessors on the application of ecological models through workshops and the development of ecological modeling courses offered via the Internet
- Developing several generic models of populations or ecosystems for routine use in registering new chemicals
- Modifying available, spatial models of populations (i.e., RAMAS GIS and VORTEX) so the effects of toxicants are explicitly represented
- Developing software and guidance for use of stochastic scalar abundance models in ecological risk assessments of toxic chemicals
- Integrating metapopulation models with ecosystem models and with landscape models
- Linking available fate and transport models with selected ecological models
- Applying selected models in site specific test cases

Table 15.1 Selected Ecological Effects Software for Risk Assessment

Model Type	Web Site
Populations and Metapopulations	
RAMAS® GIS	www.ramas.com
VORTEX	pw1.Netcom.com/~rlacy/vortex.html
Food Web	
Populus	www.cbs.umn.edu/populus
RAMAS® Ecosystem	www.ramas.com
Ecosystem	
AQUATOX	www.epa.gov/ost/models/aquatox
CASM	www.cadmusgroup.com
IFEM	www.cadmusgroup.com
Landscape	
ATLSS	http://atlss.org/
LANDIS	http://landscape.forest.wisc.edu/Projects/LANDIS_overview/landis_overview.html
JABOWA	www.naturestudy.org/
	http://eco.wiz.uni-kassel.de/model_db/mdb/jabowa.html

Further information about the recommended ecological models can be obtained from the Internet web sites listed in Table 15.1.

Although our review emphasizes user-friendly software that is specifically designed with built-in optional equations to support population and ecosystem modeling, all of the model types discussed in this report can also be implemented using generalized modeling software such as STELLA (High Performance Systems 2001), ModelMaker (ModelKinetix 2001), and MATLAB (MathWorks 2001). In all applications, selection of the most appropriate model is the primary concern; the specific software used to implement the model is a secondary concern. Moreover, development of generic models for use in applications such as registering new chemicals should not be seen as an excuse for relying on a fixed set of general models. If such models are developed, they should be regularly evaluated and updated on the basis of current reviews of available approaches. In most cases, a single model is insufficient for all applications, even when a generic approach is appropriate. Several complementary models (e.g., within population and ecosystem or landscape categories) should be developed for each purpose so potential users have options depending on their specific assessment needs.

CHAPTER **16**

Summary

Robert A. Pastorok and H. Resit Akçakaya

In current practice, ecological risk assessments for toxic chemicals are typically made on the basis of individual-organism endpoints such as survival, growth, or reproductive measures. Recognizing the limitations of this approach, many ecologists advocate the use of population and ecosystem modeling for assessing risks of toxic chemicals. Such models are used to translate the results of risk characterization for individual-organism endpoints into estimates of effects on population, ecosystem, and landscape endpoints. These ecological endpoints include species richness, population abundance or biomass, population growth rate or reproductive output, population age structure, and productivity.

We report here the results of a critical evaluation of ecological-effects models that are potentially useful for chemical risk assessment. After candidate models were compiled, they were classified as toxicity-extrapolation, population, ecosystem, or landscape models. Toxicity-extrapolation models are simple empirical, sometimes statistical, means of extrapolating toxicity thresholds or of ordering species sensitivity to toxic chemicals. Population models typically deal with the dynamics of the abundance or distribution of single species and sometimes with explicit descriptions of endpoints in time and space. Ecosystem models are mathematical expressions that are intended to describe ecological systems composed of interacting species (food webs) with or without abiotic environmental factors. Spatially explicit, multispecies models, which generally include abiotic factors, were defined as landscape models, whereas spatially explicit models of single-species populations were defined as metapopulation models. Other model types and formulations that might be defined by others as ecological models were excluded from the initial compilation because they did not predict relevant ecological endpoints or because they addressed spatial or temporal scales beyond the scope of interest. Among these excluded model types are some that are nevertheless very important for exposure assessment, including chemical fate and transport models.

Evaluation criteria included model realism and complexity, prediction of relevant ecological endpoints, treatment of uncertainty, ease of estimating parameters, degree of model development, regulatory acceptance, credibility, and resource efficiency. Models selected for further development were identified on the basis of their relatively high ratings with respect to the evaluation criteria. Detailed evaluations were conducted within each model category. The ratings should be interpreted with caution and are only comparable within each evaluation table. Profiles of the recommended models were prepared by describing the following model attributes: stressor, habitat type, temporal resolution, spatial resolution, geographic scale, biological scale (e.g., individual, population, community), time scale, model type, level of detail (i.e., tier in an ecological risk assessment), and endpoints.

1-56670-574-6/01/$0.00+$1.50
© 2002 by CRC Press LLC

SELECTING AND USING ECOLOGICAL MODELS
IN ECOLOGICAL RISK ASSESSMENT

The selection of specific models for addressing an ecological risk issue depends on the habitat, endpoints, and chemicals of interest, the balance between model complexity and the availability of data, the degree of site specificity of available models, and the risk issue. The model must be appropriate for the context, whether for the evaluation of risks associated with new chemicals and their uses, of ecological impacts and risks associated with past uses, or of clean up and restoration issues. Many of the models discussed in this report could be applied in any of these contexts. The selection of the best model to apply for a specific problem depends on the risk hypotheses and the management objectives, which ultimately drive the receptors, endpoints, and levels of protection. The final selection of ecological models then depends on the desired level of detail in the analysis and the ease with which such models can interface with the chemical fate and transport model being used. Moreover, the complexity of the model selected to address a particular issue depends on the level of realism and precision desired as well as the quality and quantity of data.

RESULTS OF THE EVALUATION OF ECOLOGICAL MODELS

Table 13.1 summarizes the models selected for further development and use in ecological risk assessments in the near future. The selected models were identified on the basis of their relatively high ratings with respect to the evaluation criteria, and further development of models was recommended at two levels: screening and detailed assessment.

Population and ecosystem modeling have been applied successfully in past ecological risk assessments, especially for addressing toxic chemical issues and for conducting population viability analysis in conservation biology. Although their use is not presently widespread for chemical risk assessments, ecological models can provide a fresh perspective and enhance the value of assessment results in supporting environmental management decisions. Population-level assessments provide a better measure of response to toxicants than assessments of individual-level effects because they integrate potentially complex interactions among life-history traits and provide a more relevant measure of ecological impact. Moreover, population models are currently a cost-effective approach for addressing most risk assessment issues, including screening-level evaluations.

For screening-level ecological risk assessments, stochastic scalar abundance models and deterministic life-history matrix models are most appropriate. Some simple food-web models, developed on the basis of RAMAS Ecosystem or Populus, for example, may be appropriate for screening-level assessments at large, complex sites, especially where disruption of the food-web structure may be an issue.

For detailed assessments, stochastic life-history matrix models and metapopulation models (e.g., RAMAS GIS and VORTEX) are recommended. These models, as well as aquatic ecosystem models like AQUATOX, CASM, and IFEM, aquatic landscape models like ATLSS, and terrestrial landscape models like LANDIS, JABOWA, and the disturbance biogeography model, are suitable for detailed ecological risk assessments.

Currently, we view applying population models to ecological risk assessments for toxic chemicals as more cost-effective than using ecosystem and landscape models. Although ecosystem models provide valuable conceptual tools for analyzing ecological systems subjected to stress, they are expensive to develop and apply to particular risk issues. Ecosystem models have been developed to fulfill two basic roles. First, ecologists have used them as descriptive constructs to evaluate the sensitivity of ecosystems to specific environmental parameters. Second, they have been developed as predictive tools for evaluating environmental management alternatives. Ecosystem models are best utilized as heuristic tools for understanding basic ecological processes and for identifying sources of uncertainty in predictive outcomes. Aquatic ecosystem models are generally better developed than terrestrial ecosystem models (aside from the forest gap models, which have been as well researched as the aquatic models). In the near future, ecosystem and landscape models are likely to be applied to only the most complex sites and chemical risk issues.

Despite their limitations, toxicity-extrapolation methods are commonly used in screening risk assessments and in developing environmental quality criteria. The various extrapolation methods are intended for different purposes, such as extrapolating from acute-to-chronic endpoints, from an LC50 to a no-observed-effects concentration (NOEC), between species, or across a community spectrum (i.e., for developing species-sensitivity distributions). We recommend that the extrapolation methods be used mainly to support toxicity evaluations within the context of a population, ecosystem, or landscape model. The use of arbitrary uncertainty factors of 10, 100, or 1000 is generally not recommended.

We estimated the expected effort (in time) and expenditure required to apply specific categories of ecological models in chemical risk assessment. Estimates may be as little as 0.2 to 2 months and $2000 to $40,000 for applying toxicity-extrapolation models or scalar-abundance population models. In contrast, it may require as many as 0.3 to 2 years and as much as $40,000 to $2,000,000 for applying complex ecosystem and landscape models. Applications of other population models and food-web models require about 1 to 6 months and $40,000 to $500,000. The estimates of effort and expenditure are approximate and are made on the assumption that the ecological model is run to assess the effects of a single chemical. They do not encompass the basic toxicity assessment, which varies depending on the number and types of chemicals addressed. The estimates also do not include collection of field data or other parts of an ecological risk assessment (e.g., problem formulation, exposure assessment, and risk characterization).

Several categories of models lack specific examples of available models for detailed assessments. Further development of such models could include the integration of metapopulation models with food-web and other ecosystem models and with landscape models.

Several activities are recommended to further the development of ecological models for use in chemical risk assessment. First, workshops and Internet-based courses on ecological modeling should be developed to educate environmental managers and risk assessors. Second, selected ecological models should be enhanced to facilitate their application to risk assessment. For example, one or a few generic models of populations or ecosystems should be developed for routine use in registering new chemicals. Available spatially explicit models of populations should be modified so that the effects of toxicants are explicitly represented. Software and guidance for use of stochastic scalar abundance models in ecological risk assessments of toxic chemicals should be developed. Available fate and transport models should be linked with the selected ecological models. The models considered most useful for chemical risk assessments should be applied in specific cases and their performance and value documented.

Constraints on the immediate use of ecological models to address chemical risk issues will most likely arise from lack of toxicity data, not from lack of an appropriate model. Although we have emphasized the user-friendly software that is now available for ecological modeling, all of the models reviewed could be implemented using generalized modeling software. In all applications, selecting the most appropriate model is more important that the choice of specific software to implement it.

References

Abbott, C.A., M.W. Berry, J.C. Dempsey, E.J. Comiskey, L.J. Gross, and H.-K. Luh. 1995. Computational Models of White-tailed Deer in the Florida Everglades. UTK-CS Technical Report No. CS-95-296. University of Tennessee, Knoxville.

Aberg, P. 1992. Size-based demography of the seaweed *Ascophyllum nodosum* in stochastic environments. *Ecology* 73:1488–1501.

Akçakaya, H.R. 1998a. RAMAS GIS: Linking Landscape Data with Population Viability Analysis (ver. 3.0). Applied Biomathematics, Setauket, NY.

Akçakaya, H.R. 1998b. RAMAS Metapop: Viability Analysis for Stage-structured Metapopulations (ver. 3.0). Applied Biomathematics, Setauket, NY.

Akçakaya, H.R. 2000. Population viability analyses with demographically and spatially structured models. *Ecol. Bull.* 48:23–38.

Akçakaya, H.R. and B. Baur. 1996. Effects of population subdivision and catastrophes on the persistence of a land snail metapopulation. *Oecologia.* 105:475–483.

Akçakaya, H.R. and J.L. Atwood. 1997. A habitat-based metapopulation model of the California gnatcatcher. *Conserv. Biol.* 11:422–434.

Akçakaya, H.R. and M.G. Raphael. 1998. Assessing human impact despite uncertainty: viability of the northern spotted owl metapopulation in the northwestern USA. *Biodiv. Conserv.* 7:875–894.

Akçakaya, H.R. and P. Baker. 1998. Zebra Mussel Demography and Modeling: Preliminary Analysis of Population Data from Upper Midwest Rivers. Contract Report EL-98-1. U.S. Army Corps of Engineers, Waterways Experiment Station, Vicksburg, MS.

Akçakaya, H.R. and S. Ferson. 1999. RAMAS Red List: Threatened Species Classifications under Uncertainty. Applied Biomathematics, Setauket, NY.

Akçakaya, H.R. and P. Sjögren-Gulve. 2000. Population viability analysis in conservation planning: an overview. *Ecol. Bull.* 48:9–21.

Akçakaya, H.R., M.A. McCarthy, and J. Pearce. 1995. Linking landscape data with population viability analysis: management options for the helmeted honeyeater. *Biol. Conserv.* 73:169–176.

Akçakaya, H.R., M.A. Burgman, and L.R. Ginzburg. 1999. *Applied Population Ecology, Principles and Computer Exercises Using RAMAS Ecolab.* 2nd ed. Sinauer Associates, Sunderland, MA.

Albers, P.H., G.H. Heinz, and H.M. Ohlendorf (Eds.). 2001. Environmental Contaminants and Terrestrial Vertebrates: Effects on Populations, Communities, and Ecosystems. SETAC Special Publication Series. Society of Environmental Toxicology and Chemistry. Pensacola, FL.

Aldenberg, T. and W. Slob. 1993. Confidence limits for hazardous concentrations based on logistically distributed NOEC toxicity data. *Ecotoxicol. Environ. Saf.* 25:48–63.

Aldenberg, T. and J.S. Jaworska. 2000. Uncertainty of the hazardous concentration and fraction affected for normal species sensitivity distributions. *Ecotoxicol. Environ. Saf.* 46:1–18.

Allee, W.C. 1938. *Cooperation among Animals.* Henry Schuman, New York.

Allee, W.C. and E. Bowen. 1932. Studies in animal aggregations: mass protection against colloidal silver among goldfishes. *J. Exper. Zool.* 61:185–207.

Allee, W.C., A.E. Emerson, O. Park, T. Park, and K.P. Schmidt. 1949. *Principles of Animal Ecology.* W.B. Saunders, Philadelphia.

Alstad, D.N., C. Bratteli, and L. Goehring. 1994a. Populus 3.4: Simulations of Population Biology. Department of Ecology, Evolution and Behavior, University of Minnesota, St. Paul. [Software available from http://www.bioweb.uncc.edu/popgen/ pop34p.exe; see http://www.bioweb.uncc.edu/faculty/leamy/popgen/populus.htm and http://dino.wiz.uni-kassel.de/model_db/mdb/populus.html.]

Alstad, D.N., J. Curtsinger, P. Abrams, and D. Tilman. 1994b. Populus. http://biosci.cbs.umn.edu/software/populus.html.

Alstad, D.N. 2001. *Basic Populus Models of Ecology*. Prentice-Hall, Upper Saddle River, NJ. http://www.cbs.umn.edu/populus/.

Amthor, J.S. and E.A. Davidson. 1997. Generalizing the Core Experiment Work at Harvard Forest to Address Global Change Issues through Mechanistic Modeling of Forest Metabolism. http://nigec.ucdavis.edu/publications/annual94/northeast/project03.html, updated May 29, 1997. Northeast Regional Center Director's Report, University of California at Davis.

Anastacio, P.M., S.N. Nielsen, J.C. Marquez, and S.E. Jorgensen. 1993. Rice and Crayfish Production in the Lower Mondego River Valley (Portugal): A Management Model for Farmers. *Procedures International Congress on Modelling and Simulation* 4:1687–1692.

Andrewartha, H.G. and L.C. Birch. 1954. *The Distribution and Abundance of Animals*. The University of Chicago Press, Chicago.

ANL. 1998. Argonne National Laboratory's Web Site for Natural Resource Systems and Assessments. http://www.ead.anl.gov/~web/newead/prgprj/nrs/nrs.html.

Applied Biomathematics. 2000. RAMAS Ecological Software. http://www.ramas.com, last updated March 3, 2000. Applied Biomathematics, Setauket, NY.

Arditi, R. and L.R. Ginzburg. 1989. Coupling in predator-prey dynamics: ratio-dependence. *J. Theor. Biol.* 139:311–326.

Armbruster, P. and R. Lande. 1993. A population viability analysis for African elephant (*Loxodonta africana*): how big should reserves be? *Conserv. Biol.* 7:602–610.

Ashley, J. 1998. Habitat Use and Trophic Status as Determinants of Hydrophobic Organic Contaminant Bioaccumulation. Dissertation. University of Maryland at College Park.

ASTM. 1990. Standard terminology relating to biological effects and environmental fate. E943-90. In ASTM. 1990. *Annual Book of ASTM Standards*, Section 11, Water and Environmental Technology. American Society for Testing and Materials, Philadelphia.

Ault, J.S., J.G. Luo, S.G. Smith, J.E. Serafy, J.D. Wang, R. Humston, and G.A. Diaz. 1999. A spatial dynamic multistock production model. *Can. J. Fish. Aquat. Sci.* 56(Suppl. 1):4–25.

Bailey, N.T.J. 1968. Stochastic birth, death and migration processes for spatially distributed populations. *Biometrika* 55:189–198.

Barber, M.C., L.A. Suarez, and R.R. Lassiter. 1991. Modelling bioaccumulation of organic pollutants in fish with an application to PCBs in Lake Ontario salmonids. *Can. J. Fish. Aquat. Sci.* 48:318–337.

Baretta, J.W., W. Ebenhöh, and P. Ruardij. 1995. The European regional seas ecosystem model, a complex marine ecosystem model. *Neth. J. Sea Res.* 33:233–246.

Barnthouse, L.W. 1993. Population level effects. pp. 447–274. In *Ecological Risk Assessment*. G.W. Suter II (Ed.). Lewis Publishers, Chelsea, MI.

Barnthouse, L.W. 1998. Modeling ecological risks of pesticides: a review of available approaches. pp. 769–798. In *Ecotoxicology*. G. Schürmann and B. Markert (Eds.). John Wiley & Sons, Inc., New York.

Barnthouse, L.W., G.W. Suter II, S.M. Bartell, J.J. Beauchamp, R.H. Gardner, E. Linder, R.V. O'Neill, and A.E. Rosen. 1986. User's Manual for Ecological Risk Assessment. Environmental Sciences Division Publication No. 2679. U.S. Environmental Protection Agency, Office of Research and Development, Washington, D.C., and Oak Ridge National Laboratory, Oak Ridge, TN.

Bartell, S.M. 1986. Description, application, and analysis of the fates of aromatics model (FOAM). Pp. 173–193. In Environmental Modeling for Priority Setting among Existing Chemicals. Workshop Proceedings, November 11–13, 1985, Munich. EcoMed, Neuherberg, Germany.

Bartell, S. 2000. Personal Communication (Conversation with R. Pastorok, Exponent, Bellevue, WA, in July 2000, regarding Modification of *Daphnia* Model for Application to *Acartia*). Cadmus Group, Oak Ridge, TN.

Bartell, S.M., P.F. Landrum, and J.P. Giesy. 1981. Simulated transport of polycyclic aromatic hydrocarbons in artificial streams. pp.133 –144. In *Energy and Ecological Modelling*. W.J. Mitsch, R.W. Bosserman, and J.M. Klopatek (Eds.). Elsevier, New York.

Bartell, S.M., R.H. Gardner, and R.V. O'Neill. 1988. An integrated fates and effects model for estimation of risk in aquatic systems. pp. 261–274. In *Aquatic Toxicology and Hazard Assessment:* Vol. 10, ASTM STP 971. American Society for Testing and Materials, Philadelphia.

Bartell, S.M., R.H. Gardner, and R.V. O'Neill. 1992. *Ecological Risk Estimation*. Lewis Publishers, Chelsea, MI.

Bartell, S.M., J.S. LaKind, J.A. Moore, and P. Anderson. 1998. Bioaccumulation of hydrophobic organic chemicals by aquatic organisms: a workshop summary. *Int. J. Environ. Pollut.* 9(1):3–25.

Bartell, S.M., G. Lefebvre, G. Kaminski, M. Carreau, and K.R. Campbell. 1999. An ecosystem model for assessing ecological risks in Québec rivers, lakes, and reservoirs. *Ecol. Model.* 124:43–67.

Bartell, S.M., K.R. Campbell, C.M. Lovelock, S.K. Nair, and J.L. Shaw. 2000. Characterizing aquatic ecological risks from pesticides using a diquat dibromide case study. III. Ecological process models. *Environ. Toxicol. Chem.* 19(5):1441–1453.

Barwick, D.H., T.C. Folsom, L.E. Miller, and S.S. Howie. 1994. Assessment of Fish Entrainment at the Bad Creek Pumped Storage Station. Duke Power Company, Huntersville, NC.

Baveco, J.M. and A.M. de Roos. 1996. Assessing the impact of pesticides on lumbricid populations: an individual-based modelling approach. *J. Appl. Ecol.* 33:1451–1468.

Beach, R.B., P.M. Craig, R. DiNitto, A.S. Donigian, G. Lawrence, R.A. McGrath, R.A. Park, A. Stoddard, S.C. Svirsky, W.D. Tate, and C.M. Wallen. 2000. Modeling Framework Design — Modeling Study of PCB Contamination in the Housatonic River. Prepared for U.S. Environmental Protection Agency, Region I, Boston. EcoModeling, Diamondhead, MI.

Bearlin, A.R., M.A. Burgman, and H.M. Regan. 1999. A stochastic model for seagrass *Zostera mulleri* in Port Phillip Bay, Victoria, Australia. *Ecol. Model.* 178:131–148.

Beddington, J.R., M.P. Hassell, and J.H. Lawton. 1976. The components of arthropod predation. II. The predator rate of increase. *J. Anim. Ecol.* 45:165–186.

Begon, M. and M. Mortimer. 1981. *Population Ecology*. Sinauer, Sunderland, MA.

Beissinger, S.R. 1995. Modeling extinction in periodic environments: Everglades water levels and snail kite population viability. *Ecol. Appl.* 5:618–631.

Benndorf, J. and F. Recknagel. 1982. Problems of application of the ecological model SALMO to lakes and reservoirs having various trophic states. *Ecol. Model.* 17:139–145.

Benndorf, J., R. Koschel, and F. Recknagel. 1985. The pelagic zone of Lake Stechlin: An approach to a theoretical model. pp. 433–453. In *Lake Stechlin: A Temperate Oligotrophic Lake*. J. Casper (Ed.). Dr. W. Junk Publishers, Dordrecht.

Benndorf, J. 2000. Personal Communication (E-mail to R. Pastorok, Exponent, Bellevue, WA, on October 19, 2000 regarding SALMO and its applications). Technische Universität Dresden, Institut für Hydrobiologie, D-01062 Dresden.

Benton, T.G., A. Grant, and T.H. Clutton-Brock. 1995. Does environmental stochasticity matter? Analysis of red deer life-histories on rum. *Evol. Ecol.* 9:559–574.

Berryman, A.A. and J.A. Millstein. 1992. Population Analysis System, PAS-P1a Single Species Analysis. Version 4.0. Ecological System Analysis, Pullman, WA.

Beudels, R.C., S.M. Durant, and J. Harwood. 1992. Assessing the risk of extinction for local populations of roan antelope *Hippotragus equinus. Biol. Conserv.* 61:107–116.

Beverton, R.J.H. and S.J. Holt. 1959. A review of the lifespans and mortality rates of fish in nature, and their relation to growth and other physiological characteristics. *Ciba Found. Colloq. Aging* 5:142–177.

Bierman, V.J., Jr., J.V. DePinto, T.C. Young, P.W. Rodgers, S.C. Martin, and R. Raghunathan. 1992. Development and Validation of an Integrated Exposure Model for Toxic Chemicals in Green Bay, Lake Michigan. Final Report for EPA Cooperative Agreement CR-8148 85. ERL-Duluth, Large Lakes and Rivers Research Branch, Grosse Ile, MI.

Bledsoe, S. 2000. Personal Communication (E-mail and Conversation with S. Ferson, Applied Biomathematics, Setauket, NY, during July 2000, regarding Ecotox Software). Department of Civil and Environmental Engineering, University of California, Davis.

Bledsoe, L.J. and B.A. Megrey. 1989. Chaos and pseudoperiodicity in the dynamics of a bioenergetic food web model. *Am. Fish. Soc. Sym.* 6:121–137.

Bledsoe, L.J. and J. Yamamoto. 1996. Ecotox. Report # 00-1, Center for Environmental and Water Resources Engineering, University of California, Davis. [see http://ecology.ucdavis.edu/ecolan/Ecotox.html]

Bledsoe, L.J. and J. Yamamoto. (In prep.) Application of Ecotox Model to Wetland Food Web in Kesterson Wildlife Refuge, San Joaquin Valley, California. Project funded by California Environmental Protection Agency. Department of Environmental Engineering, University of California, Davis.

Boecklen, W.J. 1997. Nestedness, biogeographic theory, and the design of nature reserves. *Oecologia* 112(1):123–142.

Boersma, M., C.P. van Schaik, and P. Hogeweg. 1991. Nutrient gradients and spatial structure in tropical forests: a model study. *Ecol. Model.* 55:219–240.

Bolger, D.T., A.C. Alberts, R.M. Sauvajot, P. Potenza, C. McCalvin, D. Tran, S. Mazzoni, and M.E. Soule. 1997. Response of rodents to habitat fragmentation in coastal southern California. *Ecol. Appl.* 7(2):552–563.

Boren, J.C., D.M. Engle, and R.E. Masters. 1997. Vegetation cover type and avian species changes on landscapes within a wildland-urban interface. *Ecol. Model.* 103(2–3):251–266.

Bosch, M., D. Oro, F.J. Cantos, and M. Zabala. 2000. Short-term effects of culling on the ecology and population dynamics of the yellow-legged gull. *J. Appl. Ecol.* 37(2):369–385.

Bossel, H. and H. Krieger. 1991. Simulation model of natural tropical forest dynamics. *Ecol. Model.* 59:37–71.

Bossel, H. and H. Schafer. 1989. Generic simulation model of a forest growth, carbon and nitrogen dynamics, and application to tropical acacia and European spruce. *Ecol. Model.* 48:221–265.

Botkin, D.B. 1993a. *Forest Dynamics: An Ecological Model.* Oxford University Press, New York.

Botkin, D.B. 1993b. JABOWA-II: A Computer Model of Forest Growth (software and manual). Oxford University Press, New York.

Botkin, D.B. 2000. Personal Communication (E-mail to R. Pastorok, Exponent, Bellevue, WA, concerning the Forest Model JABOWA). Department of Ecology, Evolution and Marine Biology, University of California, Santa Barbara.

Botkin, D.B., J.F. Janak, and J.R. Wallis. 1972. Some ecological consequences of a computer model of forest growth. *J. Ecol.* 60(3):849–872.

Bowler, P.A., C.A. Watson, J.R. Yearsley, and P.A. Cirone. 1992. Assessment of Ecosystem Quality and Its Impact on Resource Allocation in the Middle Snake River Subbasin. In Proc. 1992 Annual Symposium, Volume XXIV, ISSN 1068-0381, Desert Fishes Council, Bishop, CA.

Boyce, M.S. 1992. Population viability analysis. *Annu. Rev. Ecol. Syst.* 23:481–506.

Boyce, M. 1996. Review of RAMAS/GIS. *Q. Rev. Biol.* 71:167–168.

Brault, S. and H. Caswell. 1993. Pod-specific demography of killer whales (*Orcinus orca*). *Ecology* 74:1444–1454.

Breininger, D.R., M.A. Burgman, H.R. Akçakaya, and M.A. O'Connell. (in press). Use of metapopulation models in conservation planning. In *Concepts and Applications of Landscape Ecology in Biological Conservation.* K. Gutzwiller (Ed.). Springer-Verlag, New York.

Brevard County Office of Natural Resources. 1995. Scrub Conservation and Development Plan. Melbourne, FL.

Bro, E., F. Sarrazin, J. Clobert, and F. Reitz. 2000. Demography and the decline of the grey partridge *Perdix perdix* in France. *J. Appl. Ecol.* 37(3):432–448.

Brook, B.W., L. Lim, R. Harden, and R. Frankham. 1997. How secure is the Lord Howe Island Woodhen? A population viability analysis using VORTEX. *Pac. Conserv. Biol.* 3:125–33

Brook, B.W., J.J. O'Grady, A.P. Chapman, M.A. Burgman, H.R. Akçakaya, and R. Frankham. 2000. Predictive accuracy of population viability analysis in conservation biology. *Nature* 404:385–387.

Brown, L.C. and T.O. Barnwell. 1987. The Enhanced Stream Water Quality Model QUAL2E and QUAL2E-UNCAS. Documentation and User Model. EPA-600-3-87-007.

Bugmann, H. 1997. An efficient method for estimating the steady-state species composition of forest gap models. *Can. J. Forest Res.* 27(4):551–556.

Bugmann, H. and W. Cramer. 1998. Improving the behaviour of forest gap models along drought gradients. *Forest Ecol. Manage.* 103(2–3):247–263.

Bugmann, H., A. Fischlin, and F. Kienast. 1996. Model convergence and state variable update in forest-gap models. *Ecol. Model.* 89(1–3):197–208.

Bulak, J.S., D.S. Wethey, and M.G. White III. 1995. Evaluation of management options for a striped bass population in the Santee-Cooper system. *N. Am. J. Fish. Manage.* 15:84–94.

Burgman, M., S. Ferson, and H.R. Akçakaya. 1993. *Risk Assessment in Conservation Biology.* Population and Community Biology Series. Chapman & Hall, London.

Burgman, M.A. and V.A. Gerard. 1989. A stage-structured, stochastic population model for the giant kelp, *Macrocytis pyrifera. Marine Biol.* 105:15–23.

Burmaster, D.E. and K.E. von Stackelberg. 1989. Quantitative Uncertainty Analysis in Exposure and Dose–Response Assessments in Public Health Risk Assessments using Monte Carlo Techniques. pp. 82–85. In Proc. 10th National Conference, Superfund '89. November 27–29, 1989. Sponsored by The Hazardous Materials Control Research Institute.

Burnham, K.P. and W.S. Overton. 1979. Robust estimation of population size when capture probabilities vary among animals. *Ecology* 60(5):927–936.

Burton, P.J. and D.L. Urban. 1990. An Overview of ZELIG, a Family of Individual-Based Gap Models Simulating Forest Succession. pp. 92–96. In FRDA Report. Canadian Forestry Service.

Butterworth, D.S. 2000. Possible interpretation problems for the current CITES listing criteria in the context of marine fish species under commercial harvest. *Popul. Ecol.* 42:29–35.

Calabrese, E.J. and L.A. Baldwin. 1993. *Performing Ecological Risk Assessments.* Lewis Publishers, Boca Raton.

Canadian Council of Ministers of the Environment. 1991. A Protocol for the Derivation of Water Quality Guidelines for the Protection of Aquatic Life. Appendix 9. Canadian water quality guidelines. Winnipeg, MB, Canada.

Canales, J., M.C. Trevisan, J.F. Silva, and H. Caswell. 1994. A demographic study of an annual grass (*Andropogon brevifolius* Schwarz). *Acta Oecologia* 15:261–273.

Capocelli, R.M. and L.N. Ricciardi. 1974. A diffusion model for population growth in random environments. *Theor. Popul. Biol.* 5:28–41.

Carpenter, S.R., J.F. Kitchell, and J.R. Hodgson. 1985. Cascading trophic interactions and lake productivity. *BioScience* 35:634–639.

Carpenter, S.R. and J.F. Kitchell. 1988. Consumer control of lake productivity. *BioScience* 38:764–769.

Caswell, H. 2001. *Matrix Population Models: Construction, Analysis and Interpretation.* 2nd ed. Sinauer Associates, Sunderland, MA.

Caswell, H. 1996. Demography meets ecotoxicology: untangling the population level effects of toxic substances. pp. 255–292. In *Ecotoxicology: A Hierarchical Treatment.* M.C. Newman and C.H. Jagoe (Eds.). Lewis Publishers, Boca Raton.

Caswell, H. and A.M. John. 1992. From the individual to the population in demographic models. In *Individual-Based Models and Approaches in Ecology: Populations, Communities and Ecosystems.* D.L. DeAngelis and L.J. Gross (Eds.). Chapman & Hall, London.

Cazelles, B., D. Fontvieille, and N.P. Chau. 1991. Self-purification in a lotic ecosystem: A model of dissolved organic carbon and benthic microorganisms dynamic. *Ecol. Model.* 58:91–117.

CFBMG. 1977. Conifer: A Model of Carbon and Water Flow through a Coniferous Forest. Bulletin No. 8. Coniferous Forest Biome Modeling Group, University of Washington, Seattle, and University of Oregon, Corvallis.

Chambers, R.C., K.A. Rose, and J.A. Tyler. 1995. Recruitment and recruitment processes of winter flounder, *Pleuronectes americanus*, at different lattitudes: implications of an individual-based simulation model. *Neth. J. Sea Res.* 34:19–43.

Chapman, D.G. 1981. Evaluation of marine mammal population models. pp. 277–296. In *Dynamics of Large Mammal Populations.* C.W. Fowler and T.D. Smith (Eds.). John Wiley & Sons, New York.

Chapman, P.M., A. Fairbrother, and D. Brown. 1998. A critical evaluation of safety (uncertainty) factors for ecological risk assessment. *Environ. Toxicol. Chem.* 17:99–108.

Chen, C.W. and G.T. Orlob. 1975. Ecologic simulation for aquatic environments. pp. 475–588. In *Systems Analysis and Simulation in Ecology,* Volume III. B.C. Patten (Ed.). Academic Press, New York.

Chesser, R.K., K.B. Willis, and N.E. Mathews. 1994. Impacts of toxicants on population dynamics and gene diversity in avian species. pp. 171–188. In *Wildlife Toxicology and Population Modeling, Integrated Studies of Agroecosystems.* R.J. Kendall and T.E. Lacher, Jr. (Eds.), Proceedings of the Ninth Pellston Workshop, July 22–27, 1990. SETAC Special Publication Series. Society of Environmental Toxicology and Chemistry. Lewis Publishers, Boca Raton.

Chrostowski, P.C., S.A. Foster, and D. Dolan. 1991. Monte Carlo Analysis of the Reasonable Maximum Exposure (RME) Concept. pp. 577–584. In Proc. of the 12th National Conference, Hazardous Materials Control/Superfund '91. December 3–5, 1991. Sponsored by The Hazardous Materials Control Research Institute.

Clifford, P.A., Barchers, D.E., Ludwig, D.F., Sielken, R.L., Klingensmith, J.S., Graham, R.V., and Banton, M.I. 1995. An approach to quantifying spatial components of exposure for ecological risk assessment. *Environ. Toxicol. Chem.* 14(5): 895–906.

Coffin, D.P. and W.K. Lauenroth. 1989. Disturbances and gap dynamics in a semiarid grassland: a landscape-level approach. *Landscape Ecol.* 3(1):19–27.

Cohen, J.E. and C.M. Newman. 1985. A stochastic theory of community food webs. I. Models and aggregated data. *Proc. R. Soc. London B* 224:421–448.

Cohen, J.E., S.W. Christensen, and C.P. Goodyear. 1983. A stochastic age-structured model of striped bass (*Morone saxatilis*) in the Potomac River. *Can. J. Fish. Aquat. Sci.* 40:2170–2183.

Comiskey, E.J., L.J. Gross, D.M. Fleming, M.A. Huston, O.L. Bass, H.-K. Luh, and Y. Wu. 1997. A Spatially-Explicit Individual-Based Simulation Model for Florida Panther and White-Tailed Deer in the Everglades and Big Cypress landscapes. pp. 494–503. In Proc. Florida Panther Conference, Fort Myers FL, Nov. 1–3, 1994. D. Jordan (Ed.). U.S. Fish and Wildlife Service.

Congleton, W.R., B.R. Pearce, and B.F. Beal. 1997. A C++ implementation of an individual/landscape model. *Ecol. Model.* 103(1):1–17.

Coniglio, L. and R. Baudo. 1989. Life-tables of *Daphnia obtuse* (Kurz) surviving exposure to toxic concentrations of chromium. *Hydrobiologia* 188/189:407–410.

Cook, R.R. and J.F. Quinn. 1998. An evaluation of randomization models for nested species subsets analysis. *Oecologia* 113(4):584–592.

Copeland, T.L., D.J. Paustenbach, M.A. Harris, and J. Otani. 1993. Comparing the results of a Monte Carlo analysis with EPA's reasonable maximum exposed individual (RMEI): a case study of a former wood treatment site. *Reg. Toxicol. Pharmacol.* 18:275–312.

Costanza, R., F.H. Sklar, and M.L. White. 1990. Modeling coastal landscape dynamics. *BioScience* 40:91–107.

Crookston, N.L. 1990. User's Guide to the Event Monitor: Part of Prognosis Model Version 6. Gen. Tech. Rep. INT-275. U.S. Department of Agriculture, Forest Service, Intermountain Research Station, Ogden, UT.

Crookston, N.L. 1997. SUPPOSE: An Interface to the Forest Vegetation Simulator. In Proc. Forest Vegetation Simulator Conference. February 3–7, 1997, Fort Collins, CO. R. Teck, R., Moeur, and J. Adams (Eds.). Gen. Tech. Rep. INT-GTR-373. U.S. Department of Agriculture, Forest Service, Intermountain Research Station, Ogden, UT.

Crookston, N.L. and A.R. Stage. 1991. User's Guide to the Parallel Processing Extension of the Prognosis Model. Gen. Tech. Rep. INT-281. U.S. Department of Agriculture, Forest Service, Intermountain Research Station, Ogden, UT.

Crowder, L.B., D.T. Crouse, S.S. Heppell, and T.H. Martin. 1994. Predicting the impact of turtle excluder devices on loggerhead sea turtle populations. *Ecol. Appl.* 4:437–445.

Crutchfield, J. and S. Ferson. 2000. Predicting recovery of a fish population after heavy metal impacts. *Environ. Sci. Policy* 3:S183–S189.

Cullen, A.C. 1994. Measures of compounding conservatism in probabilistic risk assessment. *Risk Anal.* 14:389–393.

Dale, V.H. and R.H. Gardner. 1987. Assessing regional impacts of growth declines using a forest succession model. *J. Environ. Manage.* 24:83–93.

Daniel, T.C. and J. Vining. 1983. Methodological issues in the assessment of landscape quality. In *Behaviour and the Natural Environment*. I. Altman (Ed.).

Davidson, I.W.F., J.C. Parker, and R.P. Beliles. 1986. Biological basis for extrapolation across mammalian species. *Regul. Toxicol. Pharmacol.* 6:211–237.

Davidson, J. 1938. On the growth of the sheep population in Tasmania. *Trans. Royal Soc. South Aust.* 62:342–346.

DeAngelis, D. 1996. Across Trophic Level Simulation System (ATLSS). http://atlss.org. University of Tennessee, Institute for Environmental Modeling.

DeAngelis, D.L. and L.J. Gross (Eds.). 1992. *Individual-Based Models and Approaches in Ecology: Populations, Communities, and Ecosystems*. Chapman & Hall, London.

DeAngelis, D.L. and K.A. Rose. 1992. Which individual-based approach is most appropriate for a given problem? In *Individual-Based Models and Approaches in Ecology: Populations, Communities and Ecosystems*. D.L. DeAngelis and L.J. Gross (Eds.). Chapman & Hall, London.

DeAngelis, D.L., R.A. Goldstein, and R.V. O'Neill. 1975. A model for trophic interaction. *Ecology* 56:881–892.

DeAngelis, D.L., S.M. Bartell, and A.L. Brenkert. 1989. Effects of nutrient recycling and food-chain length on resilience. *Am. Nat.* 134(5):778–805.

DeAngelis, D.L., L. Godbout, and B.J. Shuter. 1991. An individual-based approach to predicting density-dependent dynamics in smallmouth bass populations. *Ecol. Model.* 57:91–115.

DeAngelis, D.L., H. Gaff, L. Gross, and M. Shorrosh. 1998a. ATLSS Landscape Fish Model (ALFISH) Brief Description — November 1997. Report from The Institute for Environmental Modeling, University of Tennessee, Knoxville. http://atlss.org/ bluebook.197.draft.txt.

DeAngelis, D.L, L.J. Gross, M.A. Huston, W.F. Wolff, D.M. Fleming, E.J. Comiskey, and S.M. Sylvester. 1998b. Landscape modelling for Everglades ecosystem restoration. *Ecosystem* 1:64–75.

Dejak, C., D. Franco, R. Pasters, and G. Pecenik. 1989. A steady state achieving 3D eutrophication-diffusion submodel. *Environ. Software* 4(2):94–101.

Dennis, R.L.H. and T.G. Shreeve. 1997. Diversity of butterflies on British islands: Ecological influences underlying the roles of area, isolation and the size of the faunal source. *Biol. J. Linnean Soc.* 60(2):257–275.

DGEPMH. 1989. Premises for Risk Management. Risk Limits in the Context of Environmental Policy. Annex to the Dutch National Environmental Policy Plan. Directorate General for Environmental Protection at the Ministry of Housing, Hague, The Netherlands.

Diamond, J.M. 1975. Assembly of species communities. pp. 342–344. In *Ecology and Evolution of Communities*. M.L. Cody and J.M. Diamond (Eds.). Harvard University Press, Cambridge, MA.

Diffendorfer, J.E. 1998. Testing models of source-sink dynamics and balanced dispersal. *Oikos* 81(3):417–433.

Dixon, A.M., G.M. Mace, J.E. Newby, and P.J.S. Olney. 1991. Planning for the re-introduction of scimitar-horned oryx (*Oryx dammah*) and addax (*Addax nasomaculatus*) into Niger. *Symp. Zool. Soc. London* 62:201–216.

Dixon, K.R., T-Y. Huang, K.T. Rummel, L.L. Sheeler-Gordon, J.C. Roberts, J.F. Fagan, D.B. Hogan, and S.R. Anderson. 1999. An individual-based model for predicting population effects from exposure to agro-chemicals. In *Aspects of Applied Biology 53: Challenges in Applied Population Biology*. M.B. Thomas and T. K. Edwards (Eds.). The Association of Applied Biologists, Warwick, U.K.

Doak, D., P. Kareiva, and B. Klepetka. 1994. Modeling population viability for the desert tortoise in the western Mojave Desert. *Ecol. Appl.* 4:446–460.

Dobson, A.P., G.M. Mace, J. Poole, and R.A. Brett. 1992. Conservation biology: the ecology and genetics of endangered species. pp. 405–430. In *Genes in Ecology*. R.J. Berry, T.J. Crawford, and G.M. Hewitt (Eds.). Blackwell, Oxford, U.K.

Doherty, F.G. 1983. Interspecies correlations of acute aquatic median lethal concentration for four standard testing species. *Environ. Sci. Technol.* 17:661–665.

Drechsler, M., B.B. Lamont, M.A. Burgman, H.R. Akçakaya, and E.T.F. Witowksi. 1999. Modeling the persistence of an apparently immortal *Banksia* species after fire and land clearing. *Biol. Conserv.* 88:249–259.

Duarte, P. and J.G. Ferreira. 1997. Dynamic modeling of photosynthesis in marine and estuarine ecosystems. *Environ. Model. Assess.* 2:83–93.

Duffy, K.J., B.R. Page, J.H. Swart, and V.B. Bajic. 1999. Realistic parameter assessment for a well-known elephant-tree ecosystem model reveals that limit cycles are unlikely. *Ecol. Model.* 121:(2–3)115–125.

Durant, S.M. and J. Harwood. 1992. Assessment of monitoring and management strategies for local populations of the Mediterranean monk seal *Monachus monachus*. *Biol. Conserv.* 61:81–91.

Ebenhoh, W., C. Kohlmeier, and P.J. Radford. 1995. The benthic biological submodel in the European regional seas ecosystem model. *Neth. J. Sea Res.* 33(3/4):423–452.

ECOLECON. 1993. An Ecological-Economic model for species conservation in complex forest landscapes. *Ecol. Model.* 70:63–87.

Einarsen, A.S. 1945. Some factors affecting ring-necked pheasant population density. *Murrelet* 26:39–44.

Ellison, A.M. 2000. Personal Communication (E-mail to R. Pastorok, Exponent, Bellevue, WA, on October 10, 2000 regarding the Disturbance to Wetland Vascular Plants Model). Department of Biological Sciences, Mount Holyoke College, South Hadley, MA.

Ellison, A.M. and B.L. Bedford. 1995. Response of a wetland vascular plant community to disturbance: a simulation study. *Ecol. Appl.* 5(1):109–123.

Ellner, S. and P. Turchin. 1995. Chaos in a noisy world: new methods and evidence from time series analysis. *Am. Nat.* 145:343–375.

Emans, H.J.B., E.J.V.D. Plassche, and J.H. Canton. 1993. Validation of some extrapolation methods used for effect assessment. *Environ. Toxicol. Chem.* 12:2139–2154.

Emlen, J.M. 1989. Terrestrial population models for ecological risk assessment: a state-of-the-art review. *Environ. Toxicol. Chem.* 8(9):831–842.

Emlen, J.M. Unpublished. Information on INTASS. U.S. Geological Survey, Biological Resources Division, Seattle, WA.

Emlen, J.M., D.C. Freeman, and F. Wagstaff. 1989. Interaction assessment: rationale and a test using desert plants. *Evol. Ecol.* 3:115–149.

Emlen, J.M., D.C. Freeman, M.B. Bain, and J. Li. 1992. Interaction assessment II: a tool for population and community management. *J. Wildl. Manage.* 56:708–717.

Enderle, M.J. 1982. Impact of a Pumped Storage Station on Temperature and Water Quality of the Two Reservoirs. Third International Conference on State of the Art in Ecological Modeling, May 24–28, 1982, Fort Collins, CO. *ISEM J.* 4(3–4):87–95.

EPRI. 1982. Proceedings: Workshop on Compensatory Mechanisms in Fish Populations. Electric Power Research Institute, Inc., Palo Alto, CA.

EPRI. 1996. EPRI's Compmech Suite of Modeling Tools for Population-Level Impact Assessment. Technical Brief. http://www.esd.ornl.gov/programs/COMPMECH/ brief.html. Electric Power Research Institute, Inc., Palo Alto, CA.

EPRI. 2000. Tools for Individual-Based Stream Fish Models: Improving the Cost Effectiveness and Credibility of Individual-Based Approaches for Instream Flow Assessment. TR-114006. Electric Power Research Institute, Inc., Palo Alto, CA.

ESA. 1994. Population Analysis System (Version 4.0) web page. http://esa.palouse. net/PasWeb/. Ecological Systems Analysis, Pullman, WA.

Etter, D. and T. Van Deelen. 1999. Suburban Deer Model. Illinois Natural History Survey, IL.

Fagan, W.F., E. Meir, and J.L. Moore. 1999. Variation thresholds for extinction and their implications for conservation strategies. *Am. Nat.* 154:510–520.

Fenchel, T. 1974. Intrinsic rate of natural increase: the relationship with body size. *Oecologia* 14:317–326.

Ferreira, J.G. 1995. EcoWin — an object-oriented ecological model for aquatic ecosystems. *Ecol. Model.* 79:21–34.

Ferriere, R., F. Sarrazin, S. Legendre, and J.P. Baron. 1996. Matrix population models applied to viability analysis and conservation: theory and practice using the ULM software. *Acta Oecologia* 17(6):629–656.

Ferson, S. 1990. *RAMAS Stage: Generalized Stage-Based Modeling for Population Dynamics.* Applied Biomathematics, Setauket, NY.

Ferson, S. 1993. *RAMAS Stage: Generalized Stage-Based Modeling for Population Dynamics.* Applied Biomathematics, Setauket, NY.

Ferson, S. 1999. Ecological Risk Assessment Based on Extinction Distributions. In Proc. Second International Workshop of Risk Evaluation and Management of Chemicals. Japan Science and Technology Corporation, Yokohama, Japan.

Ferson, S. and H.R. Akçakaya. 1988. RAMAS Age: Modeling Fluctuations in Age-Structured Populations. Applied Biomathematics, Setauket, NY.

Ferson, S., L.R. Ginzburg, and R.A. Goldstein. 1996. Inferring ecological risk from toxicity bioassays. *Water Air Soil Pollut.* 90:71–82.

Fitz, H. C., E. B. DeBellevue, R. Costanza, R. Boumans, T. Maxwell, and L. Wainger. 1996. Development of a general ecosystem model (GEM) for a range of scales and ecosystems. *Ecol. Model.* 88:263–295.

FLEL. 2000a. LANDIS Forest Management. Modeling the Implications of Forest Management Scenarios for Landscape Structure and Composition. Web page at http://landscape.forest.wisc.edu/Projects/ LANDIS_overview/LD_Forest_Management/ ld_forest_management.html. Forest Landscape Ecology Laboratory, University of Wisconsin, Madison.

FLEL. 2000b. LANDIS in Chapparal. Fire Regimes and Landscape Scale Vegetation Patterns in Southern California. Web page at http://landscape.forest.wisc.edu/ Projects/LANDIS_overview/ LD_Chapparal/ld_chapparal.html. Forest Landscape Ecology Laboratory, University of Wisconsin, Madison.

Forbes, V.E. and P. Calow. 1999. Is the per capita rate of increase a good measure of population-level effects in ecotoxicology? *Environ. Toxicol. Chem.* 18:1544–1556.

Friend, A.D., H.H. Shugart, and S.W. Running. 1993. A physiology-based gap model of forest dynamics. *Ecology* 79:792–797.

Friend, A.D., A.K. Stevens, R.G. Knox, and M.G.R. Cannell. 1997. A process-based, biogeochemical, terrestrial biosphere model of ecosystem dynamics (Hybrid v3.0). *Ecol. Model.* 95:249–287.

Fries, C., M. Carlsson, B. Dahlin, T. Lamas, and O. Sallnas. 1998. A review of conceptual landscape planning models for multiobjective forestry in Sweden. *Can. J. For. Res.* 28(2):159–167.

Frolking, S. 1999. Modeling the Carbon Balance of a Boreal Spruce/Moss Forest. www.gac.sr.unh.edu/Steve/model.html. University of New Hampshire.

Funk, W.H., H.L. Gibbons, G.C. Bailey, S. Mawson, and M. Gibbons. 1982. Final Report for Liberty Lake Restoration Project. Washington State University, Department of Civil and Environmental Engineering, Environmental Research Laboratory, Pullman.

Gaff, H., D.L. DeAngelis, L.J. Gross, R. Salinas, and M. Shorrosh. 2000. A dynamic landscape model for fish in the Everglades and its application to restoration. *Ecol. Model.* 127:33–52.

Gallopin, G.C. 1971. A generalized model of a resource population system. *Oecologia* 7:382–413.

Gause, G.F. 1934. *The Struggle for Existence*. Williams & Wilkins, Baltimore.

Getz, W.M. and R.G. Haight. 1989. *Population Harvesting: Demographic Models of Fish, Forest, and Animal Resources*. Princeton University Press, Princeton, NJ.

Gil Friend and Associates. 1995. EcoAudit: Ecosystem Model. hhtp://www.webcom. com/manoman/gfa/EcoAudit/model.html.

Gilpin, M.E. and I. Hanski (Eds.). 1991. *Metapopulation Dynamics*. Academic Press, London.

Ginzburg, L.R. and H.R. Akçakaya. (in press). Science and management investments needed to enhance the use of ecological modeling in decision-making. In *Ecological Modeling for Environmental Management*. V. Dale (Ed.). Springer-Verlag, New York.

Ginzburg, L.R. and E.M. Golenberg. 1985. *Lectures in Theoretical Population Biology*. Prentice-Hall, Englewood Cliffs, NJ.

Ginzburg, L.R., L.B. Slobodkin, K. Johnson, and A.G. Bindman. 1982. Quasiextinction probabilities as a measure of impact on population growth. *Risk Anal.* 21:171–181.

Ginzburg, L.R., S. Ferson, and H.R. Akçakaya. 1990. Reconstructibility of density dependence and the conservative assessment of extinction risks. *Conserv. Biol.* 4:63–70.

Glaser, D. and J.P. Connelly. 2000. The use of ecotoxicology and population models in natural remediation. pp. 121–157. In *Natural Remediation of Environmental Contaminants: Its Role in Ecological Risk Assessment and Management*. M. Swindoll, R.G. Stahl, Jr. and S.J. Ells (Eds.). SETAC General Publications Series. Society of Environmental Toxicology and Chemistry, SETAC Press, Pensacola, FL.

Gobas, F.A.P.C. 1993. A model for predicting the bioaccumulation of hydrophobic organic chemicals in aquatic food webs: application to Lake Ontario. *Ecol. Model.* 69:1–17.

Goel, N.S. and N. Richter-Dyn. 1974. *Stochastic Models in Biology*. Academic Press, New York.

Goldingay, R. and H.P. Possingham. 1995. Area requirements for viable populations of the Australian gliding marsupial, *Petaurus australis*. *Biol. Conserv.* 73:161–167.

Goodman, E.D. 1982. Modeling effects of pesticides on populations of soil/litter invertebrates in an orchard ecosystem. *Environ. Toxicol. Chem.* 1:45–60.

Gosselin, F. and J.-D. Lebreton. 2000. Potential of branching processes as a modeling tool for conservation biology. pp. 199–225. In *Quantitative Methods for Conservation Biology*. S. Ferson and M. Burgman (Eds.). Springer-Verlag, New York.

Gotelli, N.J. 1991. Demographic models for *Leptogorgia virgulata*, a shallow-water gorgonian. *Ecology* 72:457–467.

Gotelli, N.J. 1998. *A Primer of Ecology*. 2nd ed. Sinauer Associates, Sunderland, MA.

Gragnani, A., M. Scheffer, and S. Rinaldi. 1999. Top-down control of cyanobacteria: a theoretical analysis. *Am. Natural.* 153(1):59–72.

Graham, R.L., C.T. Hunsaker, R.V. O'Neill, and B.L. Jackson. 1991. Ecological risk assessment at the regional scale. *Ecol. Appl.* 1(2):196–206.

Gross, K., J.R. Lockwood, III, C.C. Frost, and W.F. Morris. 1998. Modeling controlled burning and trampling reduction for conservation of *Hudsonia montana*. *Conserv. Biol.* 12:1291–1301.

Gulland, J.A. 1972. Population Dynamics of World Fisheries. Washington Sea Grant Program, University of Washington, Seattle, WA.

Gurney, W.S.C., E. McCauley, R.M. Nisbet, and W.W. Murdoch. 1990. The physiological ecology of *Daphnia*: a dynamic model of growth and reproduction. *Ecology* 71(2):716–732.

Gustafson, E.J. and G.R. Parker. 1994. Using an index of habitat patch proximity for landscape design. *Landsc. Urban Plan.* 29(2–3):117–130.

Gustafson, E.J., S.R. Shifley, D.J. Mladenoff, K.K. Nimerfro, and H.S. He. 2000. Spatial simulation of forest succession and harvesting using LANDIS. *Can. J. Forest Res.* 30:32–43.

Haefner, J.W. 1996. Modeling Biological Systems: Principles and Applications. Chapman & Hall, London.

Hairston, N.G., F.E. Smith, and L.S. Slobodkin. 1960. Community structure, population control, and competition. *Am. Nat.* 94:421–425.

Hairston, N.G., D.W. Tinkle, and H.W. Wilbur. 1970. Natural selection and the parameters of population growth. *J. Wildl. Manage.* 34:681–690.

Hakoyama, H. and Y. Iwasa. 1998. Ecological risk assessment: A New Method of Extinction Risk Assessment and Its Application to a Freshwater Fish (*Carassius auratus* subsp.). pp. 93–110. In Proc. First International Workshop of Risk Evaluation and Management of Chemicals. J. Nakanishi (Ed.). Japan Science and Technology Corporation, Yokohama.

Hakoyama, H. and Y. Iwasa. 1999. Ecological Risk Assessment: Bootstrap Estimates of the Extinction Risk Based on a Stochastic Model. pp. 117–127. In Proc. Second International Workshop of Risk Evaluation and Management of Chemicals. J. Nakanishi (Ed.). Japan Science and Technology Corporation, Yokohama.

Hakoyama, H. and Y. Iwasa. 2000. Extinction risk of a density-dependent population estimated from a time series of population size. *J. Theor. Biol.* 204:337–359.

Hakoyama, H., Y. Iwasa, and J. Nakanishi. 2000. Comparing risk factors for population extinction. *J. Theor. Biol.* 204:327–336.

Hallam, T. 2000. Personal Communication. (E-mail to S. Ferson, Applied Biomathematics, Setauket, NY, on July 3, 2000, regarding Hallam's research). Department of Ecology and Evolutionary Biology, University of Tennessee, Knoxville.

Hallam, T.G. and E.T. Funasaki. 1999. Complexity and emergence in models of chemically stressed populations. *Ann. N.Y. Acad. Sci.* 879:302–311.

Hallam, T.G. and S.A. Levin. 1986. *Mathematical Ecology: An Introduction*. Springer-Verlag, Berlin.

Hallam, T.G., C.E. Clark, and R.R. Lassiter. 1983a. Effects of toxicants on populations: a qualitative approach. I. Equilibrium environmental exposure. *Ecol. Model.* 18:291–304.

Hallam, T.G., C.E. Clark, and G.S. Jordan. 1983b. Effects of toxicants on populations: a qualitative approach. II. First order kinetics. *J. Math. Biol.* 18:25–37.

Hallam, T.G., R.R. Lassiter, J. Li, and W. McKinney. 1990. Toxicant-induced mortality in models of *Daphnia* populations. *Environ. Toxicol. Chem.* 9:597–621.

Hallam, T.G., T.L. Trawick, and W.F. Wolff. 1996. Modeling effects of chemicals on a population: application to a wading bird nesting colony. *Ecol. Model.* 92:155–178.

Hanratty, M.P. and F.S. Stay. 1994. Field evaluation of the littoral ecosystem risk assessment model's predictions of the effects of chlorpyrifos. *J. Appl. Ecol.* 31:439–453.

Hanratty, M.P. and K. Liber. 1996. Evaluation of model predictions of the persistence and ecological effects of dibenzuron in a littoral ecosystem. *Ecol. Model.* 90:79–95.

Hanski, I. 1994. A practical model of metapopulation dynamics. *J. Animal Ecol.* 63:151–162.

Hanski, I. and M. Gilpin. 1991. Metapopulation dynamics: brief history and conceptual domain. *Biol. J. Linnean Soc.* 42:3–16.

Hanski, I. and E. Korpimaki. 1995. Microtine rodent dynamics in northern Europe — parameterized models for the predator–prey interaction. *Ecology* 76(3):840–850.

Hanski, I., L. Hannson, and H. Hentonnen. 1991. Specialist predators, generalist predators, and the microtine rodent cycle. *J. Anim. Ecol.* 60(1):353–367.

Hanski, I., P. Turchin, E. Korpimaki, and H. Henttonen. 1993. Population oscillations of boreal rodents — regulation by mustelid predators leads to chaos. *Nature* 364(6434):232–235.

Hanski, I., A. Moilanen, and M. Gyllenberg. 1996. Minimum viable metapopulation size. *Am. Nat.* 147:527–541.

Hanson, J.D., J.W. Skiles, and W.J. Parton. 1988. A multispecies model for rangeland plant communities. *Ecol. Model.* 44:89–123.

Harris, J.R.W., A.J. Bale, B.L. Bayne, R.F.C. Mantoura, A.W. Morris, L.A. Nelson, P.J. Radford, R.J. Uncles, S.A. Weston, and J. Widdows. 1983. A preliminary model of the dispersal and biological effect of toxins in the Tamar Estuary, England. *Ecol. Model.* 22:253–284.

Harris, R.B., L.H. Metzgar, and C.D. Bevins. 1986. GAPPS: Generalized Animal Population Projection System. Version 3.0 User's Manual. University of Montana Missoula.

Harrison, G.W. 1995. Comparing predator–prey models to Luckinbill's experiment with *Didinium* and *Paramecium*. *Ecology* 76(2):357–374.

Harvey, B.C. 1998. Influence of large woody debris on retention, immigration and growth of coastal cutthroat trout (*Oncorhynchus clarki clarki*) in stream pools. *Can. J. Fish. Aquat. Sci.* 55(8):1902–1908.

Harvey, B.C., R.J. Nakamoto, and J.L. White. 1999. The influence of large woody debris and a bankfull flood on movement of adult resident coastal cutthroat trout (*Oncorhynchus clarki*) during fall and winter. *Can. J. Fish. Aquat. Sci.* 56(11):2161–2166.

Hassell, M.P. and G.C. Varley. 1969. New inductive population model for insect parasites and its bearing on biological control. *Nature* 223:1133–1136.

He, H.S. and D.J. Mladenoff. 1999. Spatially explicit and stochastic simulation of forest landscape fire disturbance and succession. *Ecology* 80:81–99.

Heasley, J.E., W.K. Lauenroth, and J.L. Dodd. 1981. Systems analysis of potential air pollution impacts on grassland ecosystems. pp. 347–359. In *Energy and Ecological Modelling*. W.J. Mitsch, R.W. Bosserman, and J.M. Klopatek (Eds.). Elsevier Scientific Publishing Company, New York.

Heath, R.T., R. Sturtevant, D. Shoup, and P. Enflo. 1999. Modeling the Effects of Nutrient Concentrations on Ecosystem Stability: Framework of a Great Lakes Model. http://www.ijc.org/boards/cglr/modsum/heath.html.

Heppell, S.S., L.B. Crowder, and D.T. Crouse. 1996. Models to evaluate headstarting as a management tool for long-lived turtles. *Ecol. Appl.* 6:556–565.

High Performance Systems. 2001. STELLA Software for Mathematical Modeling. Hanover, NH. http://www.hps-inc.com/edu/index.htm.

Hilborne, R. and M. Mangel. 1997. The Ecological Detective: Confronting Models with Data. Monographs in Population Biology Number 28. Princeton University Press, Princeton, NJ.

Hill, J.L., P.J. Curran, and G.M. Foody. 1994. The effect of sampling on the species-area curve. *Global Ecol. Biogeogr. Lett.* 4(4):91–106.

Hobbs, R.J. 1994. Dynamics of vegetation mosaics: can we predict responses to global change. *Ecoscience* 1(4):346–356.

Hoekstra, J.A., M.A. Vaal, J. Notenboom, and W. Sloof. 1994. Variation in the sensitivity of aquatic species to toxicants. *Bull. Environ. Toxicol.* 53:98–105.

Holling, C.S. 1959. Some characteristics of simple types of predation and parasitism. *Can. Entomol.* 91:385–398.

Holling, C.S. 1966. The functional response of invertebrate predators to prey density. *Mem. Entomol. Soc. Can.* 48:1–86.

Holling, C.S. 1978. *Adaptive Environmental Assessment and Management*. John Wiley & Sons, London.

Hopkinson, Jr., C.S. and J.W. Day, Jr. 1977. A model of the Barataria Bay salt marsh ecosystem. Chapter 10. In *Ecosystem Modeling in Theory and Practice*. C.A.S. Hall and J.W. Day, Jr. (Eds.). John Wiley & Sons, New York.

Horn, W. and J. Benndorf. 1980. Field investigations and model simulation of the dynamics of zooplankton population in fresh waters. *Int. Revue Ges. Hydrobiol.* 65:209–222.

Horwitz, R.J. 1981. The direct and indirect impacts of entrainment on estuarine communities — transfer of impacts between trophic levels. pp. 185–197. In *Energy and Ecological Modelling*. W.J. Mitsch, R.W. Bosserman, and J.M. Klopatek (Eds.). Elsevier Scientific Publishing Company, New York.

Hough, M.N. 1993. The growing and grazing season in the United Kingdom. *Grass Forage Sci.* 48:26–37.

Hudson, P. 2001. Evaluating a Population's Viability. 30WB Course Notes. 30WB (Autumn): Wildlife Conservation Biology. http://www.stir.ac.uk/Departments/ NaturalSciences/DBMS/coursenotes/30WB/pva.htm. Department of Natural Sciences, University of Stirling, Stirling, Scotland.

Humboldt State University. 1999. The California Individual-Based Fish Simulation System: A Tool for Building, Testing, and Conducting Experiments with Individual-Based Models. Technical Brief. Department of Mathematics, Humboldt State University, Arcata, CA. http://weasel.cnrs.humboldt.edu/~simsys/.

Humphries, H.C., D.P. Coffin, and W.K. Lauenroth. 1996. An individual-based model of alpine plant distributions. *Ecol. Model.* 84(1/3):99–126.

Hunt, H.W. 1992. GEM. *Ecol. Model.* 53:205–245.

Huston, M. and T. Smith. 1987. Plant succession: life history and competition. *Am. Nat.* 130(2):168–198.

IERM. 1999. Ecological Modelling and Knowledge-Based Systems Web Page. http://helios.bto.ed.ac.uk/ierm/research/ecomod.htm, updated May 9, 1999. Institute of Ecology and Resource Management, University of Edinburgh.

Imaging Systems Laboratory. 2000. SmartForest — II. hhttp://imlab9.landarch.uiuc.edu/ SF2/smfor.html. University of Illinois at Urbana-Champaign.

Innis, G.S. (Ed.). 1978. *Grassland Simulation Model.* Springer-Verlag, New York.

Institute for Ecological Economics. 2000. http://iee.umces.edu/PLM/PLM1.html. Website contact; Alexey Voinov. University of Maryland, Chesapeake Bay Laboratory.

IUCN. 1994. IUCN Red List Categories. IUCN — The World Conservation Union, Gland, Switzerland.

Ivlev, V.S. 1961. *Experimental Ecology of the Feeding of Fishes.* Yale University Press, New Haven, CT.

Iwasa, Y. 1998. Ecological Risk Assessment by the Use of the Probability of Species Extinction. pp. 42–49. In Proc. First International Workshop of Risk Evaluation and Management of Chemicals. J. Nakanishi (Ed.). Japan Science and Technology Corporation, Yokohama.

Iwasa, Y. and H. Hakoyama. 1998. Extinction rate of a population with both demographic and environmental stochasticity. *Theor. Popul. Biol.* 53:1–15.

Iwasa, Y. and M. Nakamaru. 1999. How to Combine the Extinction Risk of Different Populations. pp. 151–155. In Proc. Second International Workshop of Risk Evaluation and Management of Chemicals. J. Nakanishi (Ed.). Japan Science and Technology Corporation, Yokohama.

Iwasa, Y., H. Hakoyama, M. Nakamaru, and J. Nakanishi. 2000. Estimate of population extinction risk and its application to ecological risk management. *Popul. Ecol.* 42:73–80.

Jackson, L.J., A.S. Trebitz, and K.L. Cottingham. 2000. An introduction to the practice of ecological modeling. *BioScience* 50:694–706.

Jager, H.I., D.L. DeAngelis, M.J. Sale, W. Van Winkle, D.D. Schmoyer, M.J. Sabo, D.J. Orth, and J.A. Lukas. 1993. An individual-based model for smallmouth bass reproduction and young-of-year dynamics in streams. *Rivers* 4:91–113.

Jager, H.I., K. Lepla, J. Chandler, P. Bates, and W. Van Winkle. 1999. Population Viability Analysis for Riverine Fishes. In Proc. EPRI Conference on Hydropower Impacts on Aquatic Resources, Atlanta.

Jagoe, R.H. and M.C. Newman. 1997. Bootstrap estimation of community NOEC values. *Ecotoxicology* 6:293–306.

Janse, J.H. and L. van Liere. 1995. PCLAKE: a modelling tool for the evaluation of lake restoration scenarios. *Water Sci. Technol.* 31:371–374.

Jaworska, J.S., K.A. Rose, and A.L. Brenkert. 1997. Individual-based modeling of PCBs effects on young-of-the-year largemouth bass in southeastern U.S. reservoirs. *Ecol. Model.* 99:113–135.

Jeffries, C. 1989. *Mathematical Modeling in Ecology.* Birkhausen, Boston.

Jeltsch, F., C. Wissel, S. Eber, and R. Brandl. 1992. Oscillating dispersal patterns of tephritid fly populations. *Ecol. Model.* 60:63–75.

Jenkinson, I.R. and T. Wyatt. 1992. Selection and control of Deborah numbers in plankton ecology. *J. Plankton Res.* 14(12):1697–1721.

Jensen, A.L. 1991. Simulation of fish population responses to exploitation. *Ecol. Model.* 55(3–4):203–218.

Johnson, G.D., W.L. Myers, and G.P. Patil. 1999. Stochastic generating models for simulating hierarchically structured multi-cover landscapes. *Landscape Ecol.* 14:413–421.

Jones, M.L., J.F. Koonce, and R. O'Gorman. 1993. Sustainability of hatchery-dependent salmonine fisheries in Lake Ontario: the conflict between predator demand and prey supply. *Trans. Am. Fish. Soc.* 122:1002–1018.

Jørgensen, S.E. 1976. A eutrophication model for a lake. *Ecol. Model.* 2:147–165.

Jørgensen, S.E. 1992. Development of models able to account for changes in species composition. *Ecol. Model.* 62:195–208.

Jørgensen, S.E. 1994. *Fundamentals of Ecological Modelling*. 2nd ed. Elsevier, Amsterdam.

Jørgensen, S.E. 1997. *Integration of Ecosystem Theories: A Pattern*. Kluwer Academic Publishers, Dordrecht.

Jørgensen, S.E. 2000. State-of-the-art of ecological modeling with emphasis on development of structural dynamic models. *Ecol. Model.* 120:75–96.

Jørgensen, S.E. and R. de Bernardi. 1997. The application of a model with dynamic structure to simulate the effect of mass fish mortality on zooplankton structure in Lago de Annone. *Hydrobiologia* 356:87–96.

Jørgensen, S.E., L.A. Jørgensen, L.K. Nielsen, and H.F. Mejer. 1981. Parameter estimation in eutrophication modeling. *Ecol. Model.* 13:111–129.

Jørgensen, S.E., B. Halling-Sorensen, and S.N. Nielsen. 1996. *Handbook of Environmental and Ecological Modeling*. Lewis Publishers, New York.

Jørgensen, S.E., L. Barnthouse, D.L. DeAngelis, J. Emlen, and K. van Leeuwen. 2000. Improvements in the Application of Models in Ecological Risk Assessment: Conclusions of the Expert Review Panel. Prepared for American Chemistry Council, Washington, D.C.

Jørgensen, S.E., S.N. Nielsen, and L.A. Jørgensen. 1991. *Handbook of Ecological Parameters and Ecotoxicology*. Elsevier Press, Amsterdam.

Kammenga, J.E., C.A.M. van Gestel, and E. Hornung. 2001. Switching life-history sensitivities to stress in soil invertebrates. *Ecol. Applic.* 11:226–238.

Kareiva, P., J. Stark, and U. Wennergren. 1996. Using demographic theory, community ecology and spatial models to illuminate ecotoxicology. pp. 13–23. In *Ecotoxicology: Ecological Dimensions*. D.J. Baird, L. Maltby, P.W. Grieg-Smith, and P.E.T. Douber (Eds.). Chapman & Hall, London.

Katona, M.A., T.F. Long, and K.R. Trowbridge. 1997. Risk Assessment of Secondary Poisoning by Organophosphate Pesticides and Environmental Stressor Impacts. Poster presented at the 18th SETAC Annual Meeting, San Francisco, CA. November 16–20, 1997.

Kaye, T.N., K. Pendergrass, and K.K. Finley. 1994. Population biology of *Lomatum bradshawii*: I. Population Viability Analysis under Three Prairie Burning Treatments. Oregon Department of Agriculture.

Keane, R.E., S.F. Arno, and J.K. Brown. 1989. FIRESUM — An Ecological Process Model for Fire Succession in Western Conifer Forests. Res. Pap. INT-266. U.S. Department of Agriculture, Forest Service, Intermountain Research Station, Ogden, UT.

Kellomäki, S. and H. Väisänen. 1991. Application of a gap model for the simulation of forest ground vegetation in boreal conditions. *Forest Ecol. Manage.* 42:35–147.

Kellomäki, S., H. Väisänen, H. Hänninen, T. Kolström, R. Lauhanen, U. Mattila, and B. Pajari. 1992. SIMA: A model for forest succession based on the carbon and nitrogen cycles with application to silvicultural management of the forest ecosystem. Silva Carelica 22. University of Joensuu, Faculty of Forestry, Gummerus Kirjapaino Oy, Jyväskylä.

Kelly, R.A. and W.O. Spofford, Jr. 1977. Application of an ecosystem model to water quality management: the Delaware estuary. Chapter 18. In *Ecosystem Modeling in Theory and Practice*. C.A.S. Hall and J.W. Day, Jr. (Eds.). John Wiley & Sons, New York.

Kenaga, E.E. 1979. Aquatic test organisms and methods useful for assessment of chronic toxicity of chemicals. pp. 101–111. In *Analyzing the Hazard Evaluation Process*. K.L. Dickson, A.W. Maki, and J. Cairns, Jr. (Eds.). American Fisheries Society.

Kenaga, E.E. 1982. Predictability of chronic toxicity from acute toxicity of chemicals in fish and aquatic invertebrates. *Environ. Toxicol. Chem.* 1:347–358.

Kendall, R.J. and T.E. Lacher, Jr. 1994. *Wildlife Toxicology and Population Modeling: Integrated Studies of Agroecosystems*. Proceedings of the Ninth Pellston Workshop, July 22–27, 1990. SETAC Special Publication Series. Society of Environmental Toxicology and Chemistry. Lewis Publishers, Boca Raton.

Kienast, F. 1987. FORECE — A Forest Succession Model for Southern Central Europe. Report ORNL/TM 10575. Publication No. 2989. Oak Ridge National Laboratory, Environmental Sciences Division. Oak Ridge, TN.

Kindvall, O. 1995. Ecology of the Bush Cricket *Metrioptera bicolor* with Implications for Metapopulation Theory and Conservation. Thesis. Uppsala University, Sweden.

Kindvall, O. 1999. Dispersal in a metapopulation of the bush cricket, *Metrioptera bicolor* (Orthoptera:Tettigoniidae). *J. Anim. Ecol.* 68:172–185.

Kingston, T. 1995. Valuable modeling tool: RAMAS/GIS. *Conserv. Biol.* 9:966–968.

Kirsta, Y.B. 1992. Time-dynamic quantization of molecular-genetic, photosynthesis, and ecosystem hierarchical levels of the biosphere. *Ecol. Model.* 62:259–274.

Kleyer, M. 1998. Individual-based modelling of plant functional type successions on gradients of disturbance intensity and resource supply. [German] *Verhandlungen Gesellschaft für Ökologie* 28:175–181.

Knox, R.G., V.L. Kalb, E.R. Levine, and D.J. Kendig. 1997. A problem-solving workbench for interactive simulation of ecosystems. *IEEE Comput. Sci. Eng.* 4(3):52–60.

Koelmans, A.A., A. van der Heijde, L. Knijff, and R.H. Aalderink. 2001. Integrated modeling of eutrophication and contaminant fate and effects in aquatic ecosystems: A review. *Water Res.*

Kohler, P. and A. Huth. 1998. The effects of tree species grouping in tropical rainforest modelling: simulations with the individual-based FORMIND. *Ecol. Model.*

Kohrs, S. 1996. Verteilte Simulation eines Zeitdiskreten, Individuenorientierten Modells. Diploma Thesis, Carl von Ossietzky University, Oldenburg, Germany.

Kokko, H., J. Lindström, and E. Ranta. 1997. Risk analysis of hunting of seal populations in the Baltic. *Conserv. Biol.* 11:917–927.

Koncsos, L. and L. Somlyody. 1991. An Interactive Model System for Operating the Kis-Balaton Reservoir. XVI General Assembly of European Geographical Society. Wiesbaden.

Kooijman, S.A.L.M. 1987. A safety factor for LC50 values allowing for differences in sensitivity among species. *Water Res.* 21(3):269–276.

Kooijman, S.A.L.M. 1993. Dynamics and Energy Budgets in Biological Systems. Cambridge University Press, Cambridge, U.K..

Kooijman, S.A.L.M. and J.A.J. Metz. 1984. On the dynamics of chemically stressed populations: the deduction of population consequences from effects on individuals. *Ecotoxicol. Environ. Saf.* 8:254–274.

Kooijman, S.A.L.M., J.J.M. Bedaux, A.A.M. Gerritsen, H. Oldersama, and A.O. Hanstveit. 1998. Dynamic measures for ecotoxicity. In *Risk Assessment: Logic and Measurement*. M.C. Newman and C.L. Strojan (Eds.). Ann Arbor Press, Chelsea, MI.

Koonce, J.F. and A.B. Locci. 1995. Lake Erie Ecological Modeling Project. Final Report on Phase II Development of the Lake Erie Ecosystem Model. International Joint Commission, Windsor, Ontario.

Koponen, J. and E. Alasaarela. 1993. Use of mathematical models for solving environmental management problems. *CSC News* 5(3): 3–6. August. (Center for Scientific Computation, P.O. Box 40, 02101 Espoo, Finland).

Kuhn, A., W.R. Munns, Jr., S. Poucher, D. Champlin, and S. Lussier. 2000. Prediction of population-level response from mysid toxicity test data using population modeling techniques. *Environ. Toxicol. Chem.* 19(9):2364–2371.

Lacy, R.C. 1993. VORTEX: a computer simulation model for population viability analysis. *Wildl. Res.* 20:45–65.

Lacy, R.C. 1999. VORTEX: Simulation Model of Population Dynamics and Viability. Documentation File. Chicago Zoological Society, IL. http://pw1.netcom.com/~rlacy/ vortex.html.

Lacy, R.C. and D.B. Lindenmayer. 1995. A simulation study of the impacts of population subdivision on the mountain brushtail possum, *Trichosurus caninus* Ogilby (Phalangeridae: Marsupialia), in southeastern Australia. II. Loss of genetic variation within and between subpopulations. *Biol. Conserv.* 73:131–142.

LaHaye, W.S., R.J. Gutiérrez, and H.R. Akçakaya. 1994. Spotted owl metapopulation dynamics in southern California. *J. Anim. Ecol.* 63:775–785.

LAII. 1998. ATLAS: Arctic Transitions in the Land-Atmosphere System Web Page. http://www.laii.uaf.edu/atlas/atlas.html, updated 3/98. Arctic System Science Land-Atmosphere-Ice Interactions Investigators, University of Alaska, Fairbanks.

Lande, R. 1993. Risks of population extinction from demographic and environmental stochasticity and random catastrophes. *Am. Nat.* 142:911–927.

Lande, R. and S.H. Orzack. 1988. Extinction dynamics of age-structured populations in a fluctuating environment. *Proc. Nat. Acad. Sci. U.S.A.* 85:7413–7421.

Landis, W.G. 2000. The pressing need for population-level risk assessment. *SETAC Globe* 1(2):44–45.

Landis, W.G. and M-H. Yu. 1995. *Introduction to Environmental Toxicology: Impacts of Chemicals upon Ecological Systems*. CRC Press–Lewis Publishers, Boca Raton.

Länge, R., T.H. Hutchinson, N. Scholz, and J. Solbé. 1998. Analysis of the ECETOC aquatic toxicity (EAT) database II — comparison of acute to chronic ratios for various aquatic organisms and chemical substances. *Chemosphere* 36(1):115–127.

Langton, C., R. Burkhart, M. Daniels, and A. Lancaster. 1999. The SWARM Simulation System. Santa Fe Institute. http://www.swarm.org/.

Lasch, P., M. Lindner, B. Ebert, M. Flechsig, F.-W. Gerstengarbe, F. Suckow, and P.C. Werner. 1999. Regional impact analysis of climate change on natural and managed forests in the Federal state of Brandenburg, Germany. *Environ. Modeling Assessment* 4:273–286.

Lawton, J.H. 1999. Are there general laws in ecology? *Oikos* 84:177–192.

Layton, D.W., B.J. Mallon, D.H. Rosenblatt, and M.J. Small. 1987. Deriving allowable daily intakes for systemic toxicants lacking chronic toxicity data. *Regul. Toxicol. Pharmacol.* 7:96–112.

LeBlanc, G.A. 1984. Interspecies relationships in acute toxicity of chemicals to aquatic organisms. *Environ. Toxicol. Chem.* 3:47–60.

Lebreton, J. 1982. Applications of Discrete Time Branching Processes to Bird Population *dynamics modelling*. pp. 115–133. In ANAIS da 10 Conferencia Internacional de Biometria. EMBRAP-DID/DMQ/Sociedade Internacional de Biometria.

Lefkovitch, L.P. 1965. ULM, a software for conservation and evolutionary biologists. *J. Appl. Stat.* 22:817–834.

Legendre, S. 1999. Demographic stochasticity: a case study using the ULM software. *Bird Study* 46:140–147.

Leslie, P.H. 1945. On the use of matrices in certain population mathematics. *Biometrika* 35:213–245.

Leslie, P.H. and R.M. Ranson. 1940. The mortality, fertility, and rate of natural increase of the vole (*Microtus agrestis*) as observed in the laboratory. *J. Anim. Ecol.* 9:27–92.

Lett, P.F., R.K. Mohn, and D.F. Gray. 1981. Density-dependent processes and management strategy for the Northwest Atlantic harp seal population. pp. 135–157. In *Dynamics of Large Mammal Populations*. C.W. Fowler and T.D. Smith (Eds.). John Wiley & Sons, New York.

Levin, L., H. Caswell, T. Bridges, C. DiBacco, D. Cabrera, and G. Plaia. 1996. Demographic responses of estuarine polychaetes to pollutants: life table response experiments. *Ecol. Appl.* 6:1295–1313.

Li, C. and M.J. Apps. 1996. Effects of contagious disturbance on forest temporal dynamics. *Ecol. Model.* 87(1–3):143–151.

Li, H., J.F. Franklin, F.J. Swanson, and T.A. Spies. 1993. Developing alternative forest cutting patterns: a simulation approach. *Landsc. Ecol.* 8(1):63–75.

Lindenmayer, D.B. and R.C. Lacy. 1995a. Metapopulation viability of Leadbeater's possum, *Gymnobelideus leadbeateri*, in fragmented old-growth forests. *Ecol. Appl.* 5:164–182.

Lindenmayer, D.B. and R.C. Lacy. 1995b. A simulation study of the impacts of population subdivision on the mountain brushtail possum, *Trichosurus caninus* Ogilby (Phalangeridae: Marsupialia), in southeastern Australia. I. Demographic stability and population persistence. *Biol. Conserv.* 73:119–129.

Lindenmayer, D.W. and H.P. Possingham. 1996. Ranking conservation and timber management options for Leadbeater's possum in southeastern Australia using population viability analysis. *Conserv. Biol.* 10:1–18.

Lindenmayer, D.B., R.C. Lacy, V.C. Thomas, and T.W. Clark. 1993. Predictions of the impacts of changes in population size and environmental variability on Leadbeater's possum, *Gymnobelideus leadbeateri* McCoy (Marsupialia: Petauridae) using population viability analysis: an application of the computer program VORTEX. *Wildl. Res.* 20:67–86.

Lindenmayer, D., M. Burgman, H.R. Akçakaya, R. Lacy, and H. Possingham. 1995. A review of generic computer programs ALEX, RAMAS/space, and VORTEX for modelling the viability of wildlife metapopulations. *Ecol. Model.* 82:161–174.

Linder, E., G.P. Patil, G.W. Suter II, and C. Taillie. 1986. Effects of Toxic Pollutants on Aquatic Resources using Statistical Models and Techniques to Extrapolate Acute and Chronic Effects Benchmarks. Pennsylvania State University, Department of Statistics, Center for Statistical Ecology and Environmental Statistics, University Park, PA.

Lindner, M. 1998. Wirkung von Klimaveränderungen in mitteleuropäischen Wirtschaftswäldern. Dissertation, Mathematisch-Naturwissenschaftliche Fakultät, Universität Potsdam (as cited at http://www.pik-potsdam.de/cp/chief/forska.htm).

Lindner, M. 2000. Developing adaptive forest management strategies to cope with climate change. *Tree Physiol.* 20:299–307.

Lindner, M., R. Sievanen, and H. Pretzsch. 1997. Improving the simulation of stand structure in a forest gap model. *Forest Ecol. Manage.* 95:183–195.

Lindner, M., P. Lasch, and M. Erhard. 2000. Alternative forest management strategies under climatic change — prospects for gap model applications in risk analyses. *Silva Fennica* 34(2):101–111.

Litvaitis, A.J. and R. Villafuerte. 1996. Factors affecting the persistence of New England cottontail metapopulations. *Wildl. Soc. Bull.* 24:686–693.

Liu, J. and P.S. Ashton. 1998. FORMOSAIC: an individual-based spatially explicit model for simulating forest dynamics in landscape mosaics. *Ecol. Model.* 106(2–3):177–200.

LMS. 1999. Landscape Management System Web Page. http://lms.cfr.washington.edu, updated June 30, 1999. University of Washington, College of Forest Resources, Silviculture Laboratory, Seattle.

Loehle, C. and P. Johnson. 1994. A framework for modelling microbial transport and dynamics in the subsurface. *Ecol. Model.* 73:31–49.

Lomolino, M.V. and R. Davis. 1997. Biogeographic scale and biodiversity of mountain forest mammals of western North America. *Global Ecol. Biogeogr. Lett.* 6(1):57–76.

Long, T.F., M.A. Katona, and K.R. Trowbridge. 1997. Four Episodes of Apparent Secondary Poisoning of Raptors by an Avicide, Fenthion, in Illinois. Poster presented at the 18th SETAC annual meeting, San Francisco. November 16–20, 1997.

Lorek, H. and M. Sonnenschein. 1995. Using Parallel Computers to Simulate Individual-Oriented Models in Ecology: A Case Study. In Proc. ESM'95 European Simulation Multiconference, Prague.

Lotka, A.J. 1924. *Elements of Physical Biology.* Williams & Wilkins, Baltimore, MD, reprinted 1956, Dover, NY.

MacArthur, R.H. and E.O. Wilson. 1967. *The Theory of Island Biogeography.* Princeton University Press, Princeton, NJ.

MacCormick, A.J.A., O.L. Loucks, J.F. Koonce, J.F. Kitchell, and P.R. Weiler. 1975. An ecosystem model for the pelagic zone of a lake. *Environ. Sys. Sci.* 2:339–382.

MacIntosh, D.L., G.W. Suter II, and F.O. Hoffman. 1994. Uses of probabilistic exposure models in ecological risk assessments of contaminated sites. *Risk Anal.* 14(4):405-419.

Mace, G.M. and R. Lande. 1991. Assessing extinction threats: toward a reevaluation of IUCN threatened species categories. *Conserv. Biol.* 5:148–157.

Mackay, D., M. Joy, and S. Paterson. 1983. A quantitative water, air, sediment interaction (QWASI) fugacity model for describing the fate of chemicals in rivers. *Chemosphere* 12:1193–1208.

Madenjian, C.P., D.W. Schloesser, and K.A. Krieger. 1998. Population models of burrowing mayfly recolonization in western Lake Erie. *Ecol. Appl.* 8:1206–1212.

Maguire, L.A., G.F. Wilhere, and Q. Dong. 1995. Population viability analysis for red-cockaded woodpeckers in the Georgia Piedmont. *J. Wildl. Manage.* 59:533–542.

Malanson, G.P., W.E. Westman, and Y-L. Yan. 1992. Realized vs. fundamental niche functions in a model of chaparral response to climate change. *Ecol. Model.* 64:261–277.

Malthus, T.B. 1798 (reprinted 1926). First essay on population (*An Essay on the Principle of Population as It Affects the Future Improvement of Society with remarks on the Speculation of Mr. Godwin M. Condorcet and Other Writers*). Macmillan & Company, London.

Marschall, E.A. and L.B. Crowder. 1996. Assessing population responses to multiple anthropogenic effects: a case study with brook trout. *Ecol. Appl.* 6:152–167.

Maschinsky, J., R. Fyre, and S. Rutman. 1997. Demography and population viability of an endangered plant species before and after protection from trampling. *Conserv. Biol.* 11:990–999.

Master, L.L. 1991. Assessing threats and setting priorities for conservation. *Conserv. Biol.* 5:559–563.

MathWorks. 2001. Mathematical Computing and Visualiztion Software. Natick, MA. http://www.mathworks.com/products/matlab/description/overview.shtml

Matsuda, H. 1997. Is tuna critically endangered? Extinction risk of a large and overexploited population. *Ecol. Res.* 12:345–356.

Maurer, B.A. 1990. *Dipodomys* populations as energy-processing systems: regulation, competition, and hierarchical organization. *Ecol. Model.* 50:157–176.

May, R.M. 1974. *Stability and Complexity in Model Ecosystems.* Princeton University Press, Princeton, NJ.

Mayer, F.L., G.F. Krause, D.R. Buckler, M.R. Ellersieck, and G. Lee. 1994. Predicting chronic lethality of chemicals to fishes from acute toxicity test data: concepts and linear regression analysis. *Environ. Toxicol. Chem.* 13(4):671–678.

Mayer, F.L., K.S. Mayer, and M.R. Ellersieck. 1986. Relation of survival to other endpoints in chronic toxicity tests with fish. *Environ. Toxicol. Chem.* 5:737–748.

Maynard Smith, J. 1974. Models in Ecology. Cambridge University Press, Cambridge, U.K..

Maywald, G.F. and R.W. Sutherst. 1991. User's Guide to CLIMEX a Computer Program for Comparing Climates in Ecology. Report No. 48. 2nd ed. CSIRO Australia, Division of Entomology.

McCauley, E., W.W. Murdoch, R.M. Nisbet, and W.S.C. Gurney. 1990. The physiological ecology of *Daphnia*: development of a model of growth and reproduction. *Ecology* 71(2):703–715.

McFadden, J.T., G.R. Alexander, and D.S. Shetter. 1967. Numerical changes and population regulation in brook trout *Salvelinus fontinalis*. *J. Fish. Res. Board Can.* 24:1425–1459.

McIntire, C.D. and J.A. Colby. 1978. A hierarchical model of lotic ecosystems. *Ecol. Monogr.* 48:167–190.

McIntire, D. No date. Stream Ecosystem Model Web Page. http://www.fsl.orst.edu/lter/data/models/strmeco.htm.

McIntosh, R.P. 1985. *The Background of Ecology.* Cambridge University Press, Cambridge, U.K.

McNaughton, S.J. and L.L. Wolf. 1979. *General Ecology.* Holt, Rinehart and Winston, New York.

Meir, E. 1997. EcoBeaker Version 1.0 — Ecological Simulation Software. BeakerWare, Ithaca, NY.

Menzie, C.A., D.E. Burmaster, J.S. Freshman, and C.A. Callahan. 1992. Assessment methods for estimating ecological risk in the terrestrial component: a case study of the Baird & McGuire Superfund site in Holbrook, MA. *Environ. Toxicol. Chem.* 11:245–261.

Metz, J.A.J. and O. Diekman (Eds.). 1986. *The Dynamics of Physiologically Structured Populations.* Lecture Notes in Biomathematics 68. Springer-Verlag, Berlin.

Mineau, P., B.T. Collins, and A. Baril. 1996. On the use of scaling factors to improve interspecies extrapolation of acute toxicity in birds. *Regul. Toxicol. Pharmacol.* 24:24–29.

Minns, C.K., J.R.M. Kelso, and M.G. Johnson. 1986. Large-scale risk assessment of acid rain impacts on fisheries: models and lessons. *Can. J. Fish. Aquat. Sci.* 43:900–921.

Miyamoto, K., S. Masunaga, J. Nakanishi, and S. M. Bartell. 1997. Assessing the Ecological Risk of Chemicals in the Japanese Aquatic Environment. 1997 Annual Meeting of the Society for Risk Analysis, Washington, D.C.

Mladenoff, D.J., G.E. Host, J. Boeder, and T.R. Crow. 1996. LANDIS: a spatial model of forest landscape disturbance, succession, and management. pp. 175–179. In *GIS and Environmental Modeling.* M.F. Goodchild, L.T. Steyaert, B.O. Parks, C. Johnston, D. Maidment, M. Crane, and S. Glendining (Eds.). GIS World Books, Fort Collins, CO.

Mladenoff, D.J. and H.S. He. 1999. Design and behavior of LANDIS, an object-oriented model of forest landscape disturbance and succession. In *Advances in Spatial Modeling of Forest Landscape Change: Approaches and Applications.* D.J. Mladenoff and W.L. Baker (Eds.). Cambridge University Press, Cambridge, U.K.

ModelKinetix. 2001. ModelMaker Software. http://www.modelkinetix.com/.

Montague, C.L., L.W. Lefebvre, D.G. Decker, and N.R. Holler. 1990. Simulation of cotton rat population dynamics and response to rodenticide applications in Florida sugar cane. *Ecol. Model.* 50:177–203.

Morrison, M.L., B.G. Marcot, and R.W. Mannan. 1992. *Wildlife-Habitat Relationships, Concepts and Applications.* The University of Wisconsin Press, Madison.

Mount, D.I. and C.E. Stephan. 1967. A method for establishing acceptable toxicant limits for fish — malathion and butoxyethanol ester of 2,4-D. *Trans. Am. Fish. Soc.* 96:185–193.

Munns, W.R., D.E. Black, T.R. Gleason, K. Salomon, D. Bengston, and R. Gutjahr-Gobell. 1997. Evaluation of the effects of dioxin and PCBs on *Fundulus heteroclitus* populations using a modeling approach. *Environ. Toxicol. Chem.* 16:1074–1081.

Myers, R.A., B.R. MacKenzie, and K. Bowen. 1999. Empirical Models of Carrying Capacity, Maximum Reproductive Rate, and Species Interactions using a Meta-Analytic Approach. ICES CM 1999/Y:18. International Council for the Exploration of the Sea, Copenhagen, Denmark.

Nagarajan, K., R.J. O'Neil, C.R. Edwards, and J. Lowenberg-DeBoer. 1994. Indiana soybean system model (ISSM): II. Mexican bean beetle model development, integration and evaluation. *Agric. Sys.* 45:291–313.

Nagel, J. 1999. Forest Growth Model for Northwest Germany Web Page. http://merlin.uni-forst.gwdg.de/nfvabw02.htm, updated March 1999. Department of Growth and Yield, Forest Research Station of Lower Saxony, Göttingen.

Naito, W., K. Miyamoto, J. Nakanishi, and S.M. Bartell. 1999. Assessing Ecological Risks of Chemicals in Lake Suwa: A Modeling Approach. pp. 107–116. In Proc. 2nd International Workshop on Risk Evaluation and Management of Chemicals, Yokohama National University, Yokohama, January 22–23, 1999.

Nakaoka, M. 1997. Demography of the marine bivalve *Yoldia notabilis* in fluctuating environments: an analysis using a stochastic matrix model. *Oikos* 79:59–68.

Namkoong, G. and J.H. Roberds. 1974. Extinction probabilities and the changing age-structure of redwood forests. *Am. Nat.* 108:355–368.

Nantel, P., D. Gagnon, and A. Nault. 1996. Population viability analysis of American ginseng and wild leek harvested in stochastic environments. *Conserv. Biol.* 10:608–621.

NASA and UNH. 1999a. EOS-IDS (Earth Observing System — Interdisciplinary Science Investigation) Web Page: The Drainage Basin Model. www.eos-ids.sr.unh.edu/ drainage.html, updated May 14, 1998. National Aeronautics and Space Administration and University of New Hampshire.

NASA and UNH. 1999b. EOS-IDS (Earth Observing System — Interdisciplinary Science Investigation) Web Page: The Terrestrial Ecosystem Model (TEM) and Beyond. www.eos-ids.sr.unh.edu/tem.html, updated May 14, 1998. National Aeronautics and Space Administration and University of New Hampshire.

Nestler, J.M. 2001. Personal Communication (E-mail to Scott Ferson, Applied Biomathematics, Setauket, NY, on April 30, 2001, regarding the CEL HYBRID Modeling Approach. U.S. Army Engineer Research and Development Center, Vicksburg, MS.

Nestler, J. M. and R.A. Goodwin. 2000. Simulating Population Dynamics in an Ecosystem Context using Coupled Eulerian-Lagrangian Hybrid Models (CEL HYBRID Models). ERDC/EL TR-00-4. U.S. Army Engineer Research and Development Center, Vicksburg, MS.

Nisbet, R.M. and W.S.C. Gurney. 1982. *Modeling Fluctuating Populations.* John Wiley & Sons, Chichester, U.K.

Nixon, S.W. and J.N. Kremer. 1977. Narragansett Bay—the development of a composite simulation model for a New England estuary. Chapter 25. In *Ecosystem Modeling in Theory and Practice.* C.A.S. Hall and J.W. Day, Jr. (Eds.). John Wiley & Sons, New York.

Noda, T. and S. Nakao. 1996. Dynamics of an entire population of the subtidal snail *Umbonium costatum*: the importance of annual recruitment fluctuation. *J. Anim. Ecol.* 65:196–204.

Norton, T.W. (Ed.). 1995. Special issue: applications of population viability analysis to biodiversity conservation. *Bio. Conserv.* 73:91–176.

Nott, P., J. Comiskey, and L. Gross. 1998. SIMSPAR Version 1.2: A Spatially Explicit Individual-Based Model for the Cape Sable Seaside Sparrow in the Everglades and Big Cypress Landscapes. Report from The Institute for Environmental Modeling, University of Tennessee, Knoxville. http://www.atlss.org/.

ODFW. 1995. Public Review Draft. Marbled Murrelet: Biological Status Assessment. Oregon Department of Fish and Wildlife, Wildlife Division, Portland.

OECD. 1992. Report of the OECD Workshop on the Extrapolation of Laboratory Aquatic Toxicity Data to the Real Environment. OECD Environ. Monogr. No. 59. Organization for Economic Cooperation and Development, Paris.

OFFIS. 1999. Ecotools: Reimplemented Ecological. Crowns Crowding. http://www.offis.uni-oldenburg.de/pro-jekte/ecotools/project_ecotools2.htm.

Okkerman, P.C., E.J. van de Plassche, W. Slooff, C.J. van Leeuwen, and J.H. Canton. 1991. Ecotoxicological effects assessment: a comparison of several extrapolation procedures. *Ecotoxicol. Environ. Saf. J.* 21:182–193.

Okkerman, P.C., E.J. van de Plassche, H.J.B. Emans, and J.H. Canton. 1993. Validation of some extrapolation methods with toxicity data derived from multiple species environments. *Ecotoxicol. Environ. Saf. J.* 25:341–359.

Okubo, A. 1980. *Diffusion and Ecological Problems: Mathematical Models.* Springer, Berlin.

O'Neill, R.V., R.H. Gardner, and L.W. Barnthouse. 1982. Ecosystem risk analysis: A new methodology. *Environ. Toxicol. Chem.* 1:167–177.

ORNL. 1998. Oak Ridge National Laboratory's Web Site for Ecological Risk Analysis Tools and Applications. http://www.hsrd.ornl.gov/ecorisk/ ecorisk.html.

Ostendorf, B. and J.F. Reynolds. 1993. Relationships between a terrain-based hydrologic model and patch-scale vegetation patterns in an arctic tundra landscape. *Landsc. Ecol.* 8(4):229–237.

OSU. 1999. ORGANON (Oregon Growth Analysis and Projection) Model Web Page. www.cof.orst.edu/ cof/fr/research/organon/, updated April 12, 1999. Oregon State University, Department of Forest Resources.

Paasivirta, J. 1994. Environmental Fate Models in Toxic Risk Estimation of a Chemical Spill. Research Centre of the Defense Forces (Finland). Publications A/4. pp. 11–21.

Park, R.A. 1998. AQUATOX for Windows: A Modular Toxic Effects Model for Aquatic Ecosystems. Contract No. 68-C4-0051, 3-13. U.S. Environmental Protection Agency, Washington, D.C.

Park, R.A., R.V. O'Neill, J.A. Bloomfield, H.H. Shugart, R.S. Booth, R.A. Goldstein, J.B. Mankin, J.F. Koonce, D. Scavia, M.S. Adams, L.S. Clescer, E.M. Colon, E.H. Dettman, J.A. Joppes, and D.D. Huff. 1974. A generalized model for simulating lake ecosystems. *Simulation* 23(2):33–50.

Park, R.A., C.I. Connolly, J.R. Albanese, L.S. Clesceri, and G.W. Heitzman. 1982. Modeling the Fate of Toxic Organic Materials in Aquatic Environments. Report EPA-600/3-82-028, Rensselaer Polytechnic Institute, Troy, NY.

Park, R.A., B. Firlie, R. Camacho, K. Sappington, M. Coombs, and D. Mauriello. 1995. AQUATOX, a general fate and effects model for aquatic ecosystems. pp. 3.7–3.17. In Toxic Substances in Water Environments Proceedings. Water Environment Federation, Alexandria, VA.

Park, S.S. 1985. Mathematical Modeling of Mixing Zone Characteristics in Natural Streams. Dissertation. Rutgers University, New Brunswick, NJ.

Park, S.S. and C.G. Uchrim. 1988. A numerical mixing zone model for water quality assessment in natural streams: conceptual development. *Ecol. Model.* 42:233–234.

Pascual, M.A. and M.D. Adkinson. 1994. The decline of the Steller sea lion in the northeast Pacific: demography, harvest or environment? *Ecol. Appl.* 4:393–403.

Patten, B.C. 1968. Mathematical models of plankton production. *Int. Revue Ges. Hydrobiol.* 53:357–408.

Patten, B.C., D.A. Egloff, and T.H. Richardson. 1975. Total ecosystem model for a cove in Lake Texoma. pp. 205–421. In *Systems Analysis and Simulation in Ecology*, Volume III. B.C. Patten (Ed.). Academic Press, New York.

Paxton, J.W. 1995. The allometric approach for interspecies scaling of pharmacokinetics and toxicity of anticancer drugs. *Clin. Exp. Pharmacol. Physiol.* 22:851–854.

Pearl, R. 1925. *The Biology of Population Growth*. Alfred Knopf, New York.

Pearl, R. and L.J. Reed. 1920. On the rate of growth of the population of the United States since 1790, and its mathematical representation. *Proc. Nat. Acad. Sci. (U.S.A.)* 6:275–288.

Pearson, S.M., M.G. Turner, L.L. Wallace, and W.H. Romme. 1995. Winter habitat use by large ungulates following fire in Northern Yellowstone National Park. *Ecol. Appl.* 5(3):744–755.

Peters-Volleberg, G.W.M., E.J. De Waal, and J.W. Van Der Laan. 1994. Interspecies extrapolation in safety evaluation of human medicines in the Netherlands (1990–1992): practical considerations. *Regul. Toxicol. Pharmacol.* 20:248–258.

Petzoldt, T. and F. Recknagel. 1992. Monte Carlo simulation with an ecological lake model. pp. 189–196. In *Assessment of Modelling Uncertainties and Measurement Errors in Hydrology*. D. Graillt et al. (Eds.). Proc. Fourth Junior Scientists' Course, Saint-Etienne, France.

Pielou, E.C. 1977. *Mathematical Ecology*. John Wiley & Sons, New York.

Pimm, S.L. 1991. *The Balance of Nature? Ecological Issues in the Conservation of Species and Communities*. University of Chicago Press, Chicago.

Pimm, S.L. and J.H. Lawton. 1977. Number of trophic levels in ecological communities. *Nature* 268:329–331.

Pokras, M.A., A.M. Pokras, J.K. Kirkwood, and C.J. Sedgwick. 1993. An introduction to allometric scaling and its uses in raptor medicine. Chapter 37. pp. 211–244. In *Raptor Biomedicine*. P.T. Redig, J.E. Cooper, J.D. Remple, and D.B. Hunter (Eds.). University of Minnesota Press, Minneapolis.

Poole, R.W. 1974. *An Introduction to Quantitative Ecology*. McGraw-Hill, New York.

Possingham, H.P. and I. Davies. 1995. ALEX: a population viability analysis model for spatially structured populations. *Biol. Conserv.* 73:143–150.

Priestly, A. 1993. A quasi-conservative version of the semi-Lagrangian advection scheme. *Month. Weather Rev.* 121:621–629.

Pulliam, H.R., J.B. Dunning, Jr. and J. Liu. 1992. Population dynamics in complex landscapes: a case study. *Ecol. Appl.* 2:165–177.

Radeloff, V.C., D.J. Mladenoff, and M.S. Boyce. 2000. Effects of interacting disturbances on landscape patterns: budworm defoliation and salvage logging. *Ecol. Appl.* 10:233–247.

Railsback, S.F. and B.C. Harvey. In prep. Comparison of Salmonid Habitat Selection Objectives in an Individual-Based Model. Lang, Railsback and Associates, Arcata, CA.

Railsback, S.F., R.H. Lamberson, B.C. Harvey, and W.E. Duffy. 1999. Movement rules for individual-based models of stream fish. *Ecol. Model.* 123(2–3):73–89.

Rajar, R. and A. Kryzanowski. 1994. Self-Induced Opening of Spillway Gates on the Mavcic Dam, Slovenia. Q71. In Proc. 18th Congress of the International Commission on Large Dams.

Ramos-Jiliberto, R. and E. Gonzalez-Olivares. 2000. Relating behavior to population dynamics: a predator–prey metaphysiological model emphasizing zooplankton diel vertical migration as an inducible response. *Ecol. Model.* 127(2–3):221–233.

Reed, J.M., C.S. Elphick, and L.W. Oring. 1998. Life-history and viability analysis of the endangered Hawaiian stilt. *Biol. Conserv.* 84:35–45.

Regan, T., K. Bonham, H. Regan, R. Taylor, D. Tuson, and M. Burgman. 1999. Forest Management and Conservation of *Tasmaphena lamproides* in Northwest Tasmania: Use of Population Viability Analysis to Evaluate Management Options. Final report to Forestry Tasmania. University of Melbourne and Forestry Tasmania.

Reichle, D.E., R.V. O'Neill, S.V. Kaye, P. Sollins, and R.S. Booth. 1973. Systems analysis as applied to modeling ecological processes. *Oikos* 24:337–343.

Reinert, K.H., S.M. Bartell, and G.R. Biddinger. 1998. Ecological Risk Assessment Decision-Support System: A Conceptual Design. SETAC Press, Pensacola, FL.

Reyes, E. 1994. Ecosystem models of aquatic primary production and fish migration in Laguna de Terminos, Mexico. pp. 519–536. In *Global Wetlands, Old World and New*. W.J. Mitsh (Ed.). Elsevier, Amsterdam.

Reynolds, C. 2001. Individual-Based Models. http://www.red3d.com/cwr/ibm.html.

Ricker, W.E. 1954. Stock and Recruitment. *J. Fish. Res. Board Can.* 11:559–623.

Ricker, W.E. 1975. Computation and Interpretation of Biological Statistics of Fish Populations. Bulletin 191. Department of Fisheries and Oceans, Fisheries Research Board of Canada, Ottawa.

Riechert, S.E., L. Provencher, and K. Lawrence. 1999. The potential of spiders to exhibit stable equilibrium point control of prey: tests of two criteria. *Ecol. Appl.* 9(2):365–377.

Riley, G.A. 1965. A mathematical model of regional variations in plankton. *Limnol. Oceanogr.* 10 (Suppl.):R202–R215.

Riley, M.J. 1998. User's Manual for the Dynamic Lake Water Quality Simulation Program "MINLAKE." External Memorandum No. 213, St. Anthony Falls Hydraulic Laboratory, University of Minnesota.

Riley, M. and H.G. Stefan. 1987. Dynamic Lake Water Quality Simulation Model "MINLAKE." Project Report No. 263, St. Anthony Falls Hydraulic Laboratory, University of Minnesota.

Riley, G.A., H. Stommel, and D.F. Bumpus. 1949. Quantitative ecology of the plankton of western North Atlantic. *Bull. Binghamton Oceanog. Collection* 12:1–169.

Roex, E.W.M., C.A.M. Van Gestel, A.P. Van Wezel, and N.M. Van Straalen. 2000. Ratios between acute aquatic toxicity and effects on population growth rates in relation to toxicant mode of action. *Environ. Toxicol. Chem.* 19(3):685–693.

Root, K.V. 1998. Evaluating the effects of habitat quality, connectivity and catastrophes on a threatened species. *Ecol. Appl.* 8:854–865.

Root, K.V., C.S. Grogan, and H.R. Akçakaya. 1997. RAMAS Ecological Risk Model for Assessing Temperature Effects on Redhorse Sucker Species in the Muskingum River. Applied Biomathematics, Setauket, NY.

Rose, K.A. 1985. Evaluation of Nutrient-Phytoplankton–Zooplankton Models and Simulation of the Ecological Effects of Toxicants Using Laboratory Microcosm Ecosystem. Dissertation. University of Washington, Seattle.

Rose, K.A., G.L. Swartzman, A.C. Kindig, and F.B. Taub. 1988. Stepwise iterative calibration of a multispecies phytoplankton–zooplankton simulation model using laboratory data. *Ecol. Model.* 42:1–32.

Rose, K.A., J.H. Cowan, E.D. Houde, and C.C. Coutant. 1993. Individual-based modeling of environmental quality effects on early life stages of fish: a case study using striped bass. *Am. Fish. Soc. Symp.* 14:125–145.

Rose, K.A., J. Tyler, D. SinghDermot, and E. Rutherford. 1996. Multispecies modelling of fish populations. pp. 194–222. In *Computers in Fisheries Research*. B. Megrey and E. Moksmess (Eds.). Chapman & Hall, London.

Rose, K.A., J.A. Tyler, R.C. Chambers, G. Klein-MacPhee, and D. Danila. 1996. Simulating winter flounder population dynamics using coupled individual-based young-of-the-year and age-structured adult models. *Can. J. Fish. Aquat. Sci.* 53:1071–1091.

Royama, T. 1992. *Analytical Population Dynamics*. Chapman & Hall, London.

Running, S., J.M. Melillo, J. Borchers, J. Chaney, H. Fisher, S. Fox, A. Haxeltine, A. Janetos, D.W. Kicklighter, T.G.F. Kittel, A.D. McGuire, R. McKeown, R. Neilson, R. Nemani, D.S. Ojima, T. Painter, Y. Pan, W.J. Parton, L. Pierce, L. Pitelka, C. Prentice, B. Rizzo, N. Rosenbloom, D.S. Schimel, S. Sitch, T. Smith, and F.I. Woodward. 1995. Vegetation/ecosystem modeling and analysis project (VEMAP): comparing biogeography and biogeochemistry models in a continental-scale study of terrestrial ecosystem responses to climate change and CO_2 doubling. *Global Biogeochem. Cycles* 9:407–437; http://www.cgd.ucar.edu/vemap/users_guide.html, last updated March 3, 2000.

Russell, G.J., J.M. Diamond, S.L. Pimm, and T.M. ReEd. 1995. A century of turnover: community dynamics at three timescales. *J. Anim. Ecol.* 64(5):628–641.

SAIC. 1996. Use of Predictive Models in Aquatic Habitat Restoration. Prepared for U.S. Army Corps of Engineers, Evaluation of Environmental Investments Research Program, Washington, D.C., and Waterways Experiment Station, Vicksburg, MS. Science Applications International Corporation, Bothell, WA.

Salencon, M.J. and J.M. Thebault. 1994. Simulation model of a mesotrophic reservoir (Lac Pareloup): MELODIA, an ecosystem management model. *Ecol. Model.* 84:163–187.

Salencon, M.J. and J.M. Thebault. 1996. Simulation model of a mesotrophic reservoir (Lac de Pareloup, France): MELODIA, an ecosystem reservoir management model. *Ecol. Model.* 84:163–187.

Salencon, M.J. and J.Y. Simonot. 1989. Proposition de Modèle Bidimensionnel (x-z) Dynamique et Thermique de Retenue. Rapport HE31/89/21. Electricité de France, Paris.

Saltz, D. and D.I. Rubenstein. 1995. Population dynamics of a reintroduced asiatic wild ass (*Equus hemionus*) herd. *Ecol. Appl.* 5:327–335.

Sample, B.E. and C.A. Arenal. 1999. Allometric models for interspecies extrapolation of wildlife toxicity data. *Bull. Environ. Contam. Toxicol.* 62:653–663.

Sample, B.E., H. Regan, S. Ferson, R. Pastorok, M. Butcher, P. Rury, R. Ryti, J. Bascietto, and S. Ells. 1999. Ecological Soil Screening Levels for Wildlife: Development and Comparison of Deterministic and Probabilistic Approaches. Poster presented at the 1999 annual meeting of the Society of Environmental Toxicology and Chemistry, Philadelphia.

Sample, B.E. and G.W. Suter II. 1999. Ecological risk assessment in a large river-reservoir. 4. Piscivorous wildlife. *Environ. Toxicol. Chem.* 18:610–620.

Sample, B.E., K.A. Rose, and G.W. Suter II. 2001. Estimation of population-level effects on wildlife based on individual-level exposures: influence of life history strategies. pp. 225–244. In *Environmental Contaminants and Terrestrial Vertebrates: Effects on Populations, Communities, and Ecosystems*. P.H. Albers, G.H. Heinz, and H.M. Ohlendorf. (Eds.). SETAC Special Publication Series. Society of Environmental Toxicology and Chemistry, Pensacola.

Scheiner, S.M. and J.M. Rey-Benayas. 1997. Placing empirical limits on metapopulation models for terrestrial plants. *Evol. Ecol.* 11(3):275–288.

Schoener, T.W. 1973. Population growth regulated by intraspecific competition for energy or time: some simple representations. *Theor. Popul. Biol.* 4:56–84.

Schumaker, N.H. 1998. A Users Guide to the PATCH Model. EPA/600/R-98/135. U.S. Environmental Protection Agency, Environmental Research Laboratory, Corvallis, OR.

Shoemaker, C.A. 1977. Mathematical construction of ecological models. pp. 76–114. In *Ecosystem Modeling in Theory and Practice: An Introduction with Case Histories*. C.A.S. Hall (Ed.). John Wiley & Sons, New York.

Sciandra, A. 1986. Study and modelling of a simple planktonic system reconstituted in an experimental microcosm. *Ecol. Model.* 34:61–82.

Shugart, H.H., R.A. Goldstein, R.V. O'Neill, and J.B. Mankin. 1974. TEEM: a terrestrial ecosystem energy model for forests. *Ecol. Plant.* 9(3):231–264.

Sibley, R.M. 1996. Effects of pollutants on individual life histories and population growth rates. pp. 197–223. In *Ecotoxicology: A Hierarchical Treatment*. M.C. Newman and C.H. Jagoe (Eds.). Lewis Publishers, Boca Raton.

Silow, E.A. and D.J. Stom. 1992. *Model Ecosystems and Models of Ecosystems in Hydrobiology*. University Press, Irkutsk.

Sjögren-Gulve, P. and C. Ray. 1996. Using logistic regression to model metapopulation dynamics: large-scale forestry extirpates the pool frog. pp. 111–137. In *Metapopulations and Wildlife Conservation*. D.R. McCullough (Ed.). Island Press, Washington, D.C.

Skalski, J.R. and S.G. Smith. 1994. Risk assessment in avian toxicology using experimental and epidemiological approaches. pp. 467–488. In *Wildlife Toxicology and Population Modeling, Integrated Studies of Agroecosystems*. R.J. Kendall and T.E. Lacher, Jr. (Eds.), Proceedings of the Ninth Pellston Workshop, July 22–27, 1990. SETAC Special Publication Series. Society of Environmental Toxicology and Chemistry. Lewis Publishers, Boca Raton.

Slade, N.A. 1994. Models of structural populations: Age and mass transition matrices. pp. 189–199. In *Wildlife Toxicology and Population Modeling, Integrated Studies of Agroecosystems*. R.J. Kendall and T.E. Lacher, Jr. (Eds.), Proceedings of the Ninth Pellston Workshop, July 22–27, 1990. SETAC Special Publication Series. Society of Environmental Toxicology and Chemistry. Lewis Publishers, Boca Raton.

Slobodkin, L.B. 1961. *Growth and Regulation of Animal Populations*. Holt, Rinehart and Winston, New York.

Sloof, W., J.A.M. van Oers, and D. de Zwart. 1986. Margins of uncertainty in ecotoxicological hazard assessment. *Environ. Toxicol. Chem.* 5:841–852.

Smith, F.E. 1963. Population dynamics in *Daphnia magna* and a new model for population growth. *Ecology* 44:651–663.

Smith, E.P. and J. Cairns, Jr. 1993. Extrapolation methods for setting ecological standards for water quality: statistical and ecological concerns. *Ecotoxicology* 2:203–219.

Snell, T.W. and M. Serra. 2000. Using probability of extinction to evaluate the ecological significance of toxicant effects. *Environ. Toxicol. Chem.* 19(9):2357–2363.

Solbreck, C. 1991. Unusual weather and insect population dynamics: *Lygaeus equestris* during an extinction and recovery period. *Oikos* 60:343–350.

Sonnenschein, M., F. Köster, H. Lorek, and U. Vogel. 1999. OFFIS ECOTOOLS (High-Level Tools for Modeling and Simulation of Individual-Oriented Ecological Models). Short description. University of Oldenburg, Germany. http://www.offis.uni-oldenburg.de/ projekte/ecotools/project_ecotools.htm.

Southgate, R. and H.P. Possingham. 1995. Population viability analysis of the greater bilby, *Lagotis macrotis*. *Biol. Conserv.* 73:151–160.

Speed, T. 1993. Modelling and managing a salmon population. pp. 267–292. In *Statistics for the Environment*. V. Barnett and K.F. Turkman (Eds.). John Wiley & Sons, Chichester, U.K..

Spencer, M. and S. Ferson. 1997a. *RAMAS® Ecosystem*. Applied Biomathematics, Setauket, New York.

Spencer, M. and S. Ferson. 1997b. RAMAS Ecotoxicology: Ecological Risk Assessment for Structured Populations. Applied Biomathematics, Setauket, NY. (Windows software and 81-page manual.)

Spencer, M. and S. Ferson. 1997c. RAMAS Ecotoxicology: Ecological Risk Assessment for Food Chains and Webs. Applied Biomathematics, Setauket, NY. (Software and 199-page manual; see http://www.ramas.com/ecotox.htm.)

Spencer, M., N.S. Fisher, and W. Wang. 1999. Exploring the effects of consumer-resource dynamics on contaminant bioaccumulation by aquatic herbivores. *Environ. Toxicol. Chem.* 18:1582–1590.

Spencer, M., N.S. Fisher, W.-X. Wang, and S. Ferson. 2001. Temporal variability and ignorance in Monte Carlo contaminant bioaccumulation models: a case study with selenium in *Mytilus edulis. Risk Anal.* 21:383–394.

Spromberg, J.A., B.M. Johns, and W.G. Landis. 1998. Metapopulation dynamics: indirect effects and multiple discrete outcomes in ecological risk assessment. *Environ. Toxicol. Chem.* 17:1640–1649.

Stage, A.R. 1973. Prognosis Model for Stand Development. Res. Pap. INT-137. U.S. Department of Agriculture, Forest Service, Intermountain Research Station, Ogden, UT.

Stark, J.D. and U. Wennergren. 1995. Can population effects of pesticides be predicted from demographic toxicological studies. *J. Econ. Entomol.* 88:1089–1096.

Steele, J.H. and E.W. Henderson. 1984. Modeling long-term fluctuations in fish stocks. *Science* 224:985–987.

Stephan, C.E., D.I. Mount, D.J. Hansen, J.H. Gentile, G.A. Chapman, and W.A. Brungs. 1985. Guidelines for Deriving Numerical National Water Quality Criteria for the Protection of Aquatic Organisms and their Uses. U.S. Environmental Protection Agency, Office of Research and Development, Environmental Research Laboratories, Duluth, MN.

Straskraba, M. 1999. GIRL: General Simulation of Reservoirs and Lakes Web Page. http://dino.wiz.uni-kassel.de/model_db/mdb/girl.html, last modified 2/22/99. Biomathematical Laboratory, Academy of Sciences, Czech Republic.

Sturtevant, R. and R.T. Heath. 1995. LEEM: The Base of the Food Web. Final Report. Water Resources Research Institute, Kent State University, Kent, OH. http://www.epa.state.oh.us/oleo/lepf/sg16-95.html.

Suárez, L.A. and M.C. Barber. 1995. Modeling bioaccumulation in aquatic organisms. pp. 343–354. In *Agrochemical Environmental Fate: State of the Art*. M.L. Leng, E.M. Leovey, and P.L. Zubkoff (Eds.). Lewis Publishers, Boca Raton.

Sudo, R., K. Kobayashi, and S. Aiba. 1975. Some experiments and analysis of a predator–prey model — interaction between *Coplidium campylum* and *Alcaligenes faecalis* in continuous and mixed culture. *Biotechnol. Bioeng.* 17(2):167–184.

Sugihara, G. 1984. Graph theory, homology and food webs. In *Proc. Symposia Appl. Math.* 30:83–101.

Sullivan, T.J., J.A. Bernert, R.S. Turner, D.F. Charles, and B.F. Cumming. 1991. *Use of Historical Assessment for Evaluation of Process-Based Model Projections of Future Environmental Change: Lake Acidification in the Adirondack Mountains, New York*. Oak Ridge National Laboratory, Oak Ridge, TN.

Suter, G.W. II and A.E. Rosen. 1988. Comparative toxicology for risk assessment of marine fishes and crustaceans. *Environ. Sci. Technol.* 22(5):548–556.

Suter, G.W. II, D.S. Vaughan, and R.H. Gardner. 1983. Risk assessment by analysis of extrapolation error: a demonstration for effects of pollutants on fish. *Environ. Toxicol. Chem.* 2:369–378.

Suter, G.W. II. (Ed.) 1993. *Ecological Risk Assessment*. Lewis Publishers, Boca Raton.

Suter, G.W. II and L.W. Barnthouse. 2001. Modeling toxic effects on populations: Experience from aquatic studies. pp. 177–188. In *Environmental Contaminants and Terrestrial Vertebrates: Effects on Populations, Communities, and Ecosystems*. P.H Albers, G.H. Heinz, and H.M. Ohlendorf (Eds.). SETAC Special Publication Series. Society of Environmental Toxicology and Chemistry. Pensacola.

Swart, J.H. and K.J. Duffy. 1987. The stability of a predator-prey model applied to the destruction of trees by elephants. *S. Afr. J. Sci.* 83(3):156–158.

Swartzman, G. and S. Kaluzny. 1987. *Ecological Simulation Primer*. Macmillan Publishing Company, New York.

Swartzman, G.L., F.B. Taub, J. Meador, C. Huang, and A. Kindig. 1990. Modeling the effect of algal biomass on multispecies aquatic microcosms response to copper toxicity. *Aquatic Toxicol.* 17:93–118.

Tanaka, Y. 1998. Extinction Probability and the Ecological Risk Assessment. In Proc. First International Workshop of Risk Evaluation and Management of Chemicals, Japan Science and Technology Corporation, Yokohama.

Tanaka, Y. 1999. On the Ecological Risk Assessment of Endocrine Disrupting Chemicals. In Proc. Second International Workshop of Risk Evaluation and Management of Chemicals. Japan Science and Technology Corporation, Yokohama.

Thebault, J.M. and M.J. Salencon. 1983. Simulation model of a mestrophic reservoir (Lac de Pareloup): biological model. *Ecol. Model.* 65:1–30.

Thomann, R.V. 1998. The future "golden age" of predictive models for surface water quality and ecosystem management. *J. Environ. Eng.* 124:94–103.

Tikhonova, I.A., O. Arino, G.R. Ivanitskii, H. Malchow, and A.B. Medvinskii. 2000. The dependence of fish school movement and plankton spatial distributions on the phytoplankton growth rate. *Biofizika* 45:352–359.

Traas, T.P., J.A. Stäb, P.R.G. Kramer, W.P. Cofino, and T. Aldenberg. 1996. Modeling and risk assessment of tributyltin accumulation in the food web of a shallow freshwater lake. *Environ. Sci. Technol.* 30(4):1227–1237.

Traas, T.P., J.H. Janse, T. Aldenberg, and T.C.M. Broek. 1998. A food web model for fate and direct and indirect effects of Dursban® 4E (active ingredient chlorpyrifos) in freshwater microcosms. *Aquatic Ecol.* 32:179–190.

Trumper, E.V. and J. Holt. 1998. Modeling pest population resurgence due to recolonization of fields following an insecticide application. *J. Appl. Ecol.* 35(2):273–285.

Tuljapurkar, S. 1990. *Population Dynamics in Variable Environments*. Lecture Notes in Biomathematics 85, Springer-Verlag, New York.

Turchin, P. 1995. Population regulation: old arguments and a new synthesis. pp. 19–40. In *Population Dynamics: New Approaches and Synthesis*. N. Cappuccino and P.W. Price (Eds.). Academic Press, San Diego.

Turchin, P. and S.D. Taylor. 1992. Complex dynamics in ecological time series. *Ecology* 73:289–305.

Turner, M.G., Y. Wu, L.L. Wallace, W.H. Romme, and A. Brenkert. 1994. Simulating winter interactions among ungulates, vegetation, and fire in Northern Yellowstone Park. *Ecol. Appl.* 4(3):472–496.

Tyler, J.A., K.A. Rose, and R.C. Chambers. 1997. Compensating for chronic stress in the first year of life: an examination using an individual-based model of winter flounder. In *Early Life History and Recruitment in Fish Populations*. R.C. Chambers and E.A. Trippel (Eds.). Chapman & Hall, London.

Tyler, J.A., K.A. Rose, and R.C. Chambers. 1997. Compensatory responses to decreased young-of-the-year survival: an individual-based modeling analysis of winter flounder. pp. 391–422. In *Early Life History and Recruitment of Fish Populations*. R.C. Chambers and E.A. Trippel (Eds.). Chapman & Hall, London.

UFZ and OFFIS. 2000. Meta-X. http://www.oesa.ufz.de/meta-x/english/overview.html. Last updated January 24, 2000. Centre for Environmental Research Leipzig Halle Ltd. and Oldenburg Research and Development Institute for Computer Science Tools and Systems, Denmark.

Ulrich, M. 1991. Modeling of Chemicals in Lakes — Development and Application of User-Friendly Simulation Software (MASAS & CHEMSEE) on Personal Computers. Dissertation ETH No. 9632.

Urban, D.L. 2000. Using model analysis to design montoring programs for landscape management and impact assessment. *Ecol. Applic.* 10(6):1820–1832.

Urban, D.L. and T.M. Smith. 1989. Microhabitat pattern and the structure of forest bird communities. *Am. Nat.* 133(6):811–829.

Uryasev, S. 1995. Derivatives of probability functions and some applications. *Ann. Operations Res.* 56:287–311.

USDA. 1999. Forest Vegetation Simulator Web Page. http://www.fs.fed.us/fmsc/fvs.htm. Last updated August 10, 1999. USDA Forest Service, Forest Management Service Center.

U.S. EPA. 1992. Framework for Ecological Risk Assessment. EPA/630/R-92/001. U.S. Environmental Protection Agency, Risk Assessment Forum, Washington, D.C.

U.S. EPA. 1998. Guidelines for Ecological Risk Assessment; Notice. U.S. Environmental Protection Agency, Washington, D.C. Federal Register Vol. 63, No. 93, May 14, 1998, pp. 26846–26924.

U.S. EPA. 2000a. AQUATOX for Windows: A Modular Fate and Effects Model for Aquatic Ecosystems — Volume 1: User's Manual. EPA-823-R-00-006.

U.S. EPA. 2000b. AQUATOX for Windows: A Modular Fate and Effects Model for Aquatic Ecosystems — Volume 2: Technical Documentation. EPA-823-R-00-007.

U.S. EPA. 2000c. AQUATOX for Windows: A Modular Fate and Effects Model for Aquatic Ecosystems — Volume 3: Model Validation Reports. EPA-823-R-00-008.

USFWS. 1996a. Black-Capped Vireo Population and Habitat Viability Assessment Report. U.S. Fish and Wildlife Service, Austin.

USFWS. 1996b. Golden-Cheeked Warbler Population and Habitat Viability Assessment Report. U.S. Fish and Wildlife Service, Austin.

USGS. 1997. ATLSS: Across-Trophic-Level System Simulation, an Approach to Analysis of South Florida Ecosystems. Progress report from the South Florida/Caribbean Ecosystem Research Group. U.S. Geological Survey, Biological Resources Division.

USGS. 1999. Comdyn–Community Dynamics Web Page. http://www.mbr.nbs.gov/ software. U.S. Geological Survey, Patuxent Wildlife Research Center.

van de Plassche, E.J., J.H.M. De Bruijn, R. Stephenson, S.J. Marschall, T.C.J. Feijtel, and S.E. Belanger. 1999. Predicted no-effect concentrations and risk characterization of four surfactants: linear alkyl benzene sulfonate, alcohol ethoxylates, alcohol ethoxylated sulfates, and soap. *Environ. Toxicol. Chem.* 18:2653–2663.

van Horssen, P.W., P.P. Schot, and A. Barendregt. 1999. A GIS-based plant prediction model for wetland ecosystems. *Landsc. Ecol.* 14:253–265.

van Leeuwen, K. 1990. Ecotoxicological effects assessment in the Netherlands: recent developments. *Environ. Manage.* 14(6):779–792.

van Leeuwen. 2000. Personal Communication (E-mail to R. Pastorok, Exponent, Bellevue, WA, on October 20, 2000, regarding CATS-4 Model and PC Lake). Division Chemical Substances Bureau, RIVM/CSR, Netherlands.

van Leeuwen, C.J., M. Rijkeboer, and G. Niebeek. 1986. Population dynamics of *Daphnia magna* as modified by chronic bromide stress. *Hydrobiologia* 133:277–285.

van Straalen, N.M. and C.A.J. Denneman. 1989. Ecotoxicological evaluation of soil quality criteria. *Ecotoxicol. Environ. Safety* 18:241–251.

van Straalen, N.M. and C.J. van Leeuwen. (In progress). The use of species-sensitivity distributions in ecotoxicology — European history.

Vassiliou, P-C.G. 1984. Cyclic behavior and asymptotic stability of non-homogeneous Markov systems. *J. Appl. Prob.* 21:315–325.

Verboom, J., A. Schotman, P. Opdam, and J.A. Metz. 1991. European nuthatch metapopulations in a fragmented agricultural landscape. *Oikos* 61:149–156.

Verhulst, P.F. 1838. Notice sur la loi que la population suit dans son accroisssement. *Corresp. Math. Phys.* 10:113–121. [Translation in Kormandy, E.J. 1965. *Readings in Ecology*. Prentice-Hall, Englewood Cliffs, NJ].

Villa, F., R. Orazio, and F. Sartore. 1992. Understanding the role of chronic environmental disturbance in the context of island biogeographic theory. *Environ. Manage.* 16(5):653–666.

Virtanen, M. J. Koponen, K. Dahlbo, and J. Sarkkula. 1986. Three-dimensional water-quality transport model compared with field observations. *Ecol. Model.* 31:185–199.

Voinov, A., H. Voinov, and R. Costanza. 1999a. Landscape modeling of surface water flow: 2. Patuxent watershed case study. *Ecol. Model.* 119:211–230.

Voinov, A., R. Costanza, L. Wainger, R. Boumans, F. Villa, T. Maxwell and H. Voinov. 1999b. Patuxent landscape model: integrated ecological economic modeling of a watershed. *J. Ecosys. Model. Software* 14:473–491.

Volterra, V. 1926. Variations and fluctuations of the numbers of individuals in animal species living together. In *Animal Ecology*. R.N. Chapman (Ed.). McGraw-Hill, New York.

Vos, C.C., C.F.J. ter Braak, and W. Nieuwenhuzen. 2000. Incidence function modelling and conservation of the tree frog (*Hyla arborea*) in the Netherlands. *Ecol. Bull.* 48:165–180.

Wagner, C. and H. Løkke. 1991. Estimation of ecotoxicological protection levels from NOEC toxicity data. *Water Res.* 25(10):1237–1242.

Waller, W.T., M.L. Dahlberg, R.E. Sparks, and J. Cairns, Jr. 1971. A computer simulation of the effects of superimposed mortality due to pollutants on populations of fathead minnows (*Pimephales promelas*). *J. Fish. Res. Board Canada* 28:1107–1112.

Walters, C. J. 1986. *Adaptive Management of Renewable Resources*. McMillan, New York.

Walthall, W.K. and J.D. Stark. 1997. Comparison of two population-level ecotoxicological endpoints: the intrinsic (r_m) and instantaneous (r_i) rates of increase. *Environ. Toxicol. Chem.* 16:1068–1073.

Wania, F. 1996. Modelling the interaction of eutrophication and hydrophobic organic contaminant behaviour in aquatic systems. In *Interactions of Nutrients and Toxicants in the Foodchain of Aquatic Ecosystems*. P.R.G. Kramer, D.A. Jonkers, and L. Van Liere (Eds.). RIVM Report 703715001. The Netherlands.

Wardle, D.A., O. Zackrisson, G. Hoernberg, and C. Gallet. 1997. The influence of island area on ecosystem properties. *Science* 277(5330):1296–1299.

Warren-Hicks, W.J. and D.R.J. Moore (Eds.). 1998. Uncertainty Analysis in Ecological Risk Assessment. Proceedings from the Pellston Workshop on Uncertainty Analysis in Ecological Risk Assessment. 23–28 August 1995, Pellston, MI. Society of Environmental Toxicology and Chemistry (SETAC), Pensacola.

Watt, K.E.F. 1959. A mathematical model for the effect of densities of attacked and attacking species on the number attacked. *Can. Entomol.* 92:129–144.

Weishample, J.F., R.G. Knox, and E.R. Levine. 1999. Soil saturation effects on forest dynamics: scaling across a southern boreal/northern hardwood landscape. *Landsc. Ecol.* 14(2):121–135.

Wenk, G. 1994. A yield prediction model for pure and mixed stands. *Forest Ecol. Manage.* 69:259–268.

West, D.C., H.H. Shugart, and D.B. Botkin (Eds.). 1981. *Forest Succession: Concepts and Applications*. Springer-Verlag, New York.

White, E.G. 1984. A multispecies simulation model of grassland producers and consumers. I. Validation. *Ecol. Model.* 24:137–157.

Whittaker, R.J. 1995. Disturbed island ecology. *Trends Ecol. Evol.* 10(10):421–425.

Wiegert, R.G., R.R. Christian, J.L. Gallagher, J.R. Hall, R.D.H. Jones, and R.L. Wetzel. 1975. A preliminary ecosystem model of coastal Georgia Spartina marsh. pp. 583–601. In *Estuarine Research. Volume I: Chemistry, Biology, and the Estuarine System.* L.E. Cronin (Ed.). Academic Press, New York.

Williams, M.R. 1995. An extreme-value function model of the species incidence and species-area relations. *Ecology* 76(8):2607–2616.

Wilson, E.O. and W.H. Bossert. 1971. *A Primer of Population Biology.* Sinauer Associates, Sunderland, MA.

With, K.A. 1997. The application of neutral landscape models in conservation biology. *Conserv. Biol.* 11(5):1069–1080.

With, K.A. and A.W. King. 1997. The use and misuse of neutral landscape models in ecology. *Oikos* 79(2):219–229.

With, K.A., R.H. Gardner, and M.G. Turner. 1997. Landscape connectivity and population distributions in heterogeneous environments. *Oikos* 78(1):151–169.

Witteman, G.J. and M. Gilpin. 1995. Review of RAMAS/metapop. *Q. Rev. Biol.* 70:381–382.

Wolfe, J.R., R.D. Zweig, and D.G. Engstrom. 1986. A computer simulation model of the solar-algae pond ecosystem. *Ecol. Model.* 34:1–59.

Wolff, W.F. 1994. An individual-oriented model of a wading bird nesting colony. *Ecol. Model.* 72:75–114.

Woodward, I.O. 1982. Modelling the lineage of a growing population as an age-dependent branching process. *Aust. J. Ecol.* 7:91–96.

World in a Box. 1999. Landlord Version 1.2. hhtp://www.worldinabox.co.uk/Landlord index.html.

Wu, J. and J.L. Vankat. 1991. An area-based model of species richness dynamics of forest islands. *Ecol. Model.* 58:249–271.

Wykoff, W.R., N.L. Crookston, and A.R. Stage. 1982. User's Guide to the Stand Prognosis Model. General Technical Report INT-133. U.S. Department of Agriculture, Forest Service, Intermountain Research Station, Ogden, UT.

Yearsley, J.R. 1989. State Estimation and Hypothesis Testing: A Framework for the Assessment of Model Complexity and Data Worth in Environmental Systems. Report No. 116. University of Washington, Seattle.

York, A.E. 1994. The population dynamics of northern sea lions, 1975–1985. *Mar. Mammal Sci.* 10:38–51.

Zabel, R.W., J.J. Anderson, and P.A. Shaw. 1998. A multiple-reach model describing the migratory behavior of Snake River yearling chinook salmon (*Oncorhynchus tschawytscha*). *Can. J. Fish. Aquatic Sci.* 55:658–667.

Fish Population Modeling: Data Needs and Case Study

Stan J. Pauwels

GENERAL INTRODUCTION

A limitation in ecological risk assessment is the difficulty of predicting the effects of toxic chemicals on the long-term stability of exposed fish populations. As a result, analysts tend to rely on screening-level approaches in which measured or estimated contaminant levels in surface waters are compared with published toxicity data (such as LOEC, MATC, LC50, etc.). Risk is assumed to be unacceptable if toxic chemical concentrations exceed their respective toxicity thresholds. These approaches may help rank chemicals for further evaluation, but they provide no information on the population-level impacts of contaminant exposures. A key challenge is to determine the probability that a target fish population — such as a recreational or commercial species — will drop below a minimum acceptable level because of exposure to a chemical contaminant.

In risk-based population modeling, a target fish species is first defined in terms of its demographics, which include age-specific initial abundances, survival, and fecundity with the measured or estimated annual variations in those values. These and other species- or site specific information (such as density dependence or migration rates) are obtained from the literature or other data sources. The model is first run in the absence of a stressor to ensure that it can predict a plausible population over time. If required, some of the input variables are modified to achieve this goal.

To assess the impacts of a contaminant, a species-specific dose–response function, which reflects life-cycle exposures, is specified for each potentially affected vital rate in the model population. This information typically comes from laboratory toxicity testing. The model is run again to determine what effect, if any, the presence of a contaminant could have on long-term population levels. At each time-step, usually each year, new values for each parameter are randomly selected from the probability distributions to calculate that year's population level. A distribution of outcomes is generated during the simulation. This distribution provides a probabilistic estimate of the likelihood of reaching a given abundance within the time limits imposed on the model. Using such an approach, one can compare the effects of stressors with the background of normal variation in the abundance of the target population. One can also compare and contrast the population responses with different intensities of a particular stressor.

This appendix discusses the types of information typically required to parameterize a fish population model. It also presents a case study to show how the response of a brook trout (*Salvelinus fontinalis*) population exposed continually to a contaminant — in this case the pesticide toxaphene — can be assessed and quantified.

MODEL COMPONENTS

Introduction

Initially, one must determine whether the model should represent a structured or unstructured population (Spencer and Ferson, 1997). In unstructured populations, one assumes that no systematic differences exist in the rates of survival or fecundity between individuals of different ages. This assumption is not tenable for detailed modeling of most fish species; for example, larvae or fry typically have higher mortalities than adults. Fecundity also increases with age because older fish lay more eggs that can hatch and grow into juveniles. The only exception is for species in which most individuals reproduce and die by the end of age one year. Under those circumstances, the numeric and reproductive contribution of surviving older fish is assumed to be negligible and can be ignored. Such a life history is relatively rare in fish and will not be discussed further.

The quality and predictiveness of a model is only as good as the quality of the input data. Great care must be taken to collect, obtain, or develop population-specific input variables that are realistic and defensible. This step is even more important when only limited or incomplete data exist for a fish species of interest. Typically, a population model requires information on the following input variables:*

- Model structure (age classes within the fish population)
- Age-specific survival rates for target fish species
- Age-specific fecundity for target fish species
- Estimates of variability/stochasticity for vital rates
- Initial abundance for each age
- Density dependence

These inputs are discussed below.

Model Structure

In nature, populations show substantial variation in fecundity and survival rates. Differences among the vital characteristics of individuals are often related to age. Hence, dividing individuals into discrete age classes and assigning age-specific demographic characteristics to each class is an easy and accurate way to model populations. This simplification assumes that the within-age differences in fecundity and survival are relatively small compared with the between-age differences. Because fish usually reproduce annually, age classes can be defined on a *per annum* basis. For example, if all the fish in a population are dead by age 5, the population could be divided into as many as five age classes (age 0, 1, 2, 3, and 4 years).

It can be a challenge to determine age-specific survivals and fecundities of older individuals because of the difficulty of separating such individuals based on age. The survival and fecundity of older individuals may not differ much from one year to the next (growth from one year to the next is minimal). In such circumstances, lumping two or more age classes into a composite age class and assigning a common survival and fecundity rate may be appropriate.

* Other variables not discussed here may include rates of immigration or emigration, probability of catastrophe, etc.

Age-Specific Survival Rates

Introduction

The age-specific survival rate $[S_i(t)]$ for a given fish species is the proportion of individuals present in a given year (t) within a given age class (i) that survives into the next age class $(i+1)$ in the following year $(t+1)$ (Spencer and Ferson, 1997).

Estimates of age-specific survival rates can be calculated by the following equation:

$$S_i(t) = N_{i+1}\ (t+1)/N_i(t)$$

where

$$S_i(t) = \text{survival rate at time } t$$
$$N_{i+1}(t+1) = \text{number of individuals in age class } i+1 \text{ at time } t+1$$
$$N_i(t) = \text{number of individuals in age class } i \text{ at time } t$$

Determining age-specific survival rates requires a minimum of two *consecutive* yearly field censuses in which the ages of the captured fish are determined. Results are more reliable if data from three or more consecutive years are available because they will better reflect the natural year-to-year variations. In any case, the censuses should be consecutive to follow the age classes from one year to the next and to estimate age-specific survival rates.

The census data can be obtained from an area or region of particular concern (for site specific population modeling) or from one or more bodies of water that reflect the general conditions to be modeled (for generic population modeling). Census data collected from different locations on the same body of water could be pooled to increase the sampling size if the physical–chemical characteristics of the sampled areas are similar.

The survival rates calculated from field-collected target populations integrate all the biotic and abiotic factors that impinge on their members throughout the census years; these rates account for spawning success, hatching success, and mortalities due to predation, competition, diseases, extremes in temperatures or water flow, and any other factor that affects survival of populations in the wild. When constructing a population model for chemical risk assessment, one should establish survival rates from a reference population similar to the potentially impacted population except for the absence of toxicologically important levels of the contaminant of concern.

For highly variable systems — such as smaller streams that may experience annual swings in populations due to extremes in flow rates, temperatures, or dissolved oxygen — obtaining many years of consecutive census data may be necessary so that the average survival rates used in the model reflect the range of intrinsic variation. If those data are not available, assuming a reasonable coefficient of variation (CV) to account for this higher stochasticity may be necessary (see below for details).

Calculating Annual Survival Rates

Several approaches exist to calculate annual survival rates depending on the quality of the census data. In the best-case situation, the published information provides age-abundance data (for an example, see the brook trout data developed by McFadden et al. 1967 and discussed in the Case Study below). Mark–recapture studies using tagged fish can also be an effective tool for estimating survival. More often than not, the available data are incomplete; and one or more of the steps outlined below may be required to calculate survival rates (for specific applications, see for example Ferson et al. 1991; Oines et al. 1997; Root and Akçakaya 1997; Root et al. 1997; Root 1998).

*Step 1: Obtain Age-Specific Length Data from the Target Populations**

Age-specific length data are usually empirically derived from published field studies on the species of interest: fish from a target population are censused from rivers or lakes and are measured for length (and often weight); scales are collected from a representative subsample; the scales can be used to determine the age of the fish by counting the yearly growth rings; the ages are cross-referenced to the fish from which the scales were collected to assign age classes based on length. One can then determine the range of lengths that correspond to a given age class and assign a specific age to each member of the censused population. The year-to-year sampling effort must be constant for the results to be comparable. If the sampling effort is unequal, then survival data will be skewed and cause bias. Error can be introduced if different sampling gear or fishing effort were used from one year to the next. Additional error can be added if the sampling conditions are unequal (for example, low water one year but high water the next). These factors should be considered before selecting a published study to obtain such data.

Step 2: For Each Census Year, Divide the Fish into Discrete Age Classes

All the fish captured in a given census year are lumped into discrete age classes based on their lengths to follow their survival into the next year.

Step 3: Determine Age-Specific Year-to-Year Survival

The raw age data are converted to age-specific survival data by dividing the number of individuals within a given age class (i) in year one (t) into the number of individuals in the next age class ($i+1$) in year two ($t+1$). For the hypothetical example shown in the table below, the survival rate for age one year fish equals $105/157 = 0.669$.

If data are available for several consecutive years, the age-specific survival rates are calculated for each adjacent pair of census years. The survival rates are then combined within each age class to obtain a mean rate and a standard deviation (SD). The SD is useful for modeling natural variability.

Age-Specific Fecundity Rates

Introduction

Age-specific fecundity, $F_i(t)$, is defined as the number of live offspring per individual in a given age class (i) *that will survive to be counted in the first age class* next year ($t+1$). Like survival, fecundity rates are needed to construct a population projection matrix. If published age- and species-specific fecundity data do not exist, then this information must be estimated or back-calculated. The steps discussed below represent potential strategies developed by Oines et al. (1997), Root and Akçakaya (1997), Root et al. (1997), and Root (1998). One or more of these steps may be required to calculate age-specific fecundities, depending on the available data.**

---

* Length data are not an essential part of estimating survival rates. Such information is often used, however, because relationships based on regressions are available to estimate survival from size.
** Fecundity refers to the number of offspring per individual fish (either male or female) in a given age class. The strategies outlined in this section refer specifically to females; one would therefore need to multiply the final values by the ratio of females over individuals (females + males) in each age class to obtain the age-specific individual fecundity.

Table A.1 Determining Age-Specific Year-to-Year Fish Survival

Age (years)	Standard Length (mm)	Number of Individuals in Year 1 Census	Number of Individuals in Year 2 Census	Survival (S)
0	<95	157	131	0.669
1	95–150	97	105	0.536
2	151–200	35	52	0.257
3	201–245	11	9	0.091
4	>245	3	1	—

Calculating Age-Specific Fecundities

Step 1: Classify Individuals in the Fish Population by Age

This step is necessary when fecundity varies by age. Fish can be lumped into a composite age class in longer-lived species for which the life-history characteristics of the older age classes do not change much from one year to the next.

Step 2: Obtain or Estimate the Average Length for Fish in Each Age Class

Typically, this information is available from the survival rate calculations.

Step 3: Estimate the Weight for an Average Female in Each Age Class

Body weight often determines egg production: heavier fish produce more eggs. The actual relationship is species-specific. For many fish species, length-weight relationships using published regression equations have been developed from field-derived data. These can be used to estimate weight based on a given length. One can estimate an average weight using the age-specific lengths determined in Step 2.

Step 4: Estimate the Potential Number of Eggs That Can Be Produced by an Average Female in Each Age Class

The potential number of eggs per average female is obtained using two variables:

- The weight in kilograms of an average female in each age class (determined in Step 3)
- The number of eggs produced per kilogram of female (empirically derived from field studies or from hatchery data)

The egg potential is determined by multiplying the age-specific weight of the females by the number of eggs produced per kilogram.

For example, assume the following: 3-year-old females weigh an average of 0.31 kg, and the number of eggs produced per kg body weight for females in this species equals 10,000. Therefore, the egg potential for these females = 0.31 kg × 10,000 eggs/kg = 3100 eggs.

Step 5: Estimate the Actual Numbers of Eggs Laid by Age-Specific Females*

Knowing how many eggs an average female in a given age class can potentially produce based on its weight is not sufficient to determine how many eggs will actually be spawned by females within each age class. Three independent pieces of information are needed to determine this value:

* Several other ways to accomplish this goal are available depending on data availability.

1. The egg potential for the average female within each age class — this potential was determined in Step 4 based on weight considerations.
2. The percentage of females that are mature in each age class — depending on the fish species, a cohort of females can either become sexually mature all at the same time in one season or instead over a period of several years. These different strategies can have profound effects on age-specific fecundity rates. The fraction of females within each age class able to reproduce is species-specific and is determined based on empirical field studies. Usually, smaller and shorter-lived species such as minnows tend to become sexually mature at the same time; larger and longer-lived species such as sturgeon or striped bass mature later and over a period of several years.
3. The breeding periodicity of mature females — most fish species reproduce every year after they reach sexual maturity until they die. However, some species such as sturgeons only spawn inter- mittently (several years can go by between spawning events). If spawning is intermittent, then the breeding *periodicity* must be considered. For example, if females of a given species reproduce every year after they reach sexual maturity, then their periodicity is 1.0. If sexually mature females reproduce on average once every three years, then their periodicity is 0.33. This reproductive strategy can affect population dynamics and needs to be accounted for if relevant to the species of interest.

This concept is highlighted in the following example. Assume that the females of a hypothetical fish species reach sexual maturity and reproduce according to the following schedule:

Table A.2. Reproduction Schedule for a Hypothetical Fish Species

Age	Potential Eggs[a]	Sexual Maturity	Breeding Periodicity	Actual Eggs
0	100	0% of females	Not sexually mature	0
1	1,000	0% of females	Not sexually mature	0
2	5,000	50% of females	1.0	2,500
3	10,000	100% of females	1.0	10,000
4	15,000	100% of females	1.0	15,000

[a] These values would be based on body weight and actual egg counts.

The "actual" average number of eggs laid per female in each age class can be determined as follows:

$$\text{Age } 0 = (100 \text{ eggs}) \times (0 \text{ reproductive activity}) \times (0 \text{ periodicity}) = 0 \text{ eggs}$$
$$\text{Age } 1 = (1{,}000 \text{ eggs}) \times (0 \text{ reproductive activity}) \times (0 \text{ periodicity}) = 0 \text{ eggs}$$
$$\text{Age } 2 = (5{,}000 \text{ eggs}) \times (0.5 \text{ reproductive activity}) \times (1.0 \text{ periodicity}) = 2{,}500 \text{ eggs}$$
$$\text{Age } 3 = (10{,}000 \text{ eggs}) \times (1.0 \text{ reproductive activity}) \times (1.0 \text{ periodicity}) = 10{,}000 \text{ eggs}$$
$$\text{Age } 4 = (15{,}000 \text{ eggs}) \times (1.0 \text{ reproductive activity}) \times (1.0 \text{ periodicity}) = 15{,}000 \text{ eggs}$$

Step 6: Estimate the Average Fecundity in Each Age Class

Fecundity (F) is defined as the number of live offspring per individual in a given age class that will survive to be counted in the first age class. Knowing only the actual numbers of eggs laid is not enough to derive F. The reason is that one has to include the probability of hatching and the probability that the newly hatched fry will survive until the next census to recruit into age class 0.

Approach A: Enough Data Are Available to Estimate Survival to Age One Year —

For some fish, enough species-specific empirical data on the first season of development may be available to assess the survival potential of offspring to age one and calculate age-specific fecundities (F_i). If insufficient data exist for the target species, but data are available for a closely related species, then one could extrapolate the information from the related species to the target species.

This approach should only be used after considering the life histories of both species to ensure that they are similar enough to justify transfer of information from one to the other. For some species, even this approach will not provide the required data. Looking at less closely related species and using conservative values in the calculations may then be necessary to obtain a range of hatching or hatchling survival probabilities for use in the equations. The quality of input data required to calculate fecundities varies depending on the target species and available published studies.

Without measured field data, the general equation for calculating F is as follows:

$$F = \text{(actual eggs/female)} \times \text{(probability of hatching)} \times \text{(probability of hatchlings surviving to age 0 year)}$$

where

$$\text{actual eggs/female} = \text{the values developed in Step 5}$$
$$\text{probability of hatching} = \text{empirically derived species-specific value between 0 and 1.0}$$
$$\text{probability of survival to age 0 year} = \text{empirically derived species-specific value between 0 and 1.0}$$

For example, if we assume that the probability of hatching is 0.2 and the probability of survival from hatchling to age 0 = 0.08, then, based on the average number of eggs calculated in Step 5, the age-specific fecundities are as follows:

$$\text{Age 0 year } F = 0 \text{ actual eggs} \times 0.2 \times 0.08 = 0$$
$$\text{Age 1 year } F = 0 \text{ actual eggs} \times 0.2 \times 0.08 = 0$$
$$\text{Age 2 year } F = 2{,}500 \text{ actual eggs} \times 0.2 \times 0.08 = 40$$
$$\text{Age 3 year } F = 10{,}000 \text{ actual eggs} \times 0.2 \times 0.08 = 160$$
$$\text{Age 4 year } F = 15{,}000 \text{ actual eggs} \times 0.2 \times 0.08 = 240$$

Note that fecundity increases with age: older, heavier fish produce more eggs and therefore have a higher fecundity than smaller fish, everything else being equal. Note also that the final F would need to be adjusted to represent individuals — both males and females — in each age class instead of just the females.

Approach B: Insufficient Data to Obtain Reasonable Estimates of Survival from Egg Stage to Age 0 Year

— For some fish species, the data may simply not be available to develop reasonable estimates of egg survival, hatching success, and/or survival to year 0. For river and stream populations, developing a reasonable fecundity estimate indirectly with simplifying assumptions may still be possible. Some fish species in streams or rivers may not live at densities high enough to make space and/or food a limiting factor. In other words, the sizes of their populations are unlikely to be controlled by density-dependent factors alone. The main reason is that stochastic events — such as floods or droughts — prevent such populations from reaching or exceeding their carrying capacities. This assumption may be less reasonable for lakes or reservoirs, which have more stable environments, or for certain fish species in streams and rivers.

Under stochastic conditions, estimating fecundity using the number of eggs produced by each age class requires an empirically derived scaling factor (SF). Such a scaling factor is the ratio between the number of young fish reaching age 0 and the initial number of eggs spawned. This factor is obtained using a trial-and-error approach: a range of potential fecundities is fitted with the age-specific survival rates before running the population model. The goal is to develop estimates in which the median population size is predicted to remain relatively stationary over time in a model that assumes no density dependence and a reasonable coefficient of variation (such as 20 or 30 %) in survival and fecundity. Under these conditions, the population growth rate (a.k.a.

lambda in age-structured models) in the model should equal about 1.0, which reflects a population that is stable over time (Root et al. 1997).

The scaling factor can be "backed out" of the fecundity associated with this median population. The reason is that fecundity equals the number of eggs times the scaling factor; hence, the scaling factor equals fecundity divided by the number of eggs. The scaling factor in fish will typically fall between 10^{-2} and 10^{-5}; this range implies that only 1 to 0.001% of the eggs that are spawned actually produce fish which reach age 0 year. The scaling factor incorporates fertilization rates, hatching success, and survival from sac fry to year 0 — all in a single measure. Unlike the fecundity obtained using approach *A* outlined earlier, however, this fecundity is a lumped average which applies *to all age classes* of the modeled population. It also assumes that the fish population is stable over time (Root et al. 1997).

Estimates of Variability

Introduction

Survival and fecundity rates for fish are typically obtained from field sampling studies and/or hatchery data. Temporal variability is inherent in natural populations; it results from stochastic events such as extreme weather patterns, variable predation pressures, diseases, or changes in food supply, or from man-made factors such as spills, pollution, reservoir draw-downs, or overfishing. These changes can cause large annual fluctuations in reproductive success and/or survival and hence population levels. To mimic such natural variation and develop more realistic predictions requires a measure of randomness in input values for population modeling. Probabilistic population models generate stochasticity by randomly sampling, at each time interval and for each age class, those stochastic variables whose mean, standard deviation and probability distribution have been specified.

Two methods are available to estimate such variability.

Quantitative Estimates of Variability

If census data are available for a sufficient number of consecutive years, one can directly calculate a standard deviation (SD) for each age-specific survival or fecundity rate. This SD then becomes an input variable that is used to generate variability around the mean. This approach was used for the brook trout case study outlined in Section 3. Care must be taken, however, that the number of sampling events is sufficiently robust to generate an SD with a reasonable dispersion. The qualitative estimate of variability outlined in the next section should therefore be used in most modeling situations.

Qualitative Estimates of Variability

If census data are available for an insufficient number of consecutive years, one cannot calculate a meaningful SD for age-specific survival or fecundity rates. In the absence of sufficient data, one could assume a reasonable coefficient of variation (CV) to generate the required randomness. The CV can be higher to reflect the less certain conditions in highly variable systems, such as smaller rivers and streams. The CV of a statistical distribution is its SD divided by its mean ($\bar{x}$) (i.e., CV $= SD/\bar{x}$). Hence, using the formula $SD = CV*\bar{x}$, SD can be calculated by setting the mean as the age-specific survival or fecundity rate and assuming a CV of 20% or some other reasonable estimate based on literature data. These estimated SD values can then be used by the model to randomly sample the vital rates for each age group at each time interval from the statistical distributions that have the mean, SD, and CV discussed above. The SD values are entered when the model is parameterized.

Initial Abundance for Each Age Class

After a fish population has been parameterized in terms of its vital rates (age-specific fecundity and survival) and intrinsic variability, one needs to set a starting population for each age class at time $t = 0$. Most often, initial abundances are based on field sampling data for the target fish species. For example, a published fisheries survey used for calculating the species-specific vital rates may also have age-specific population data (see the case study on brook trout below). If the vital rates were calculated based on other sources of information, then survey data from a closely related species in the same body of water or from the target species in another similar body of water could be used to set reasonable initial age-specific abundances. In the absence of site specific data, this variable requires a great deal of professional judgment. For generic population modeling, obtaining realistic estimates of the *ratios* between the different age classes is probably more important than obtaining the actual starting numbers. These numbers can be deduced from published field studies for the target species or another closely related species in habitats that represent the conditions to be modeled.

Density Dependence

Density dependence represents a phenomenon in which the survival or reproductive success of a population depends on its density. Thus, the size of a density dependent population affects its survival, fecundity, and persistence. For example, higher numbers of fish can attract predators, decrease the amount of available food or nesting sites, increase the chances of diseases, or reduce the amount of available habitat. All of these can lead to lower fecundity and/or survival rates and long-term declines in population levels. Most natural populations in relatively stable habitats are thought to be regulated by some form of density dependence. Otherwise, they would increase without bounds or decline to extinction. Without density dependence, a population could remain stable only when birth rates and death rates balanced exactly.

With density dependence, rates of survival and fecundity decrease as the number of organisms increases. Density dependence is difficult to detect and quantify in natural populations. One needs to know (1) which of the vital rates (fecundity or survival or both) are affected by density, (2) which type of function to use (for many fish populations, the Ricker function is appropriate because it assumes that resources are shared equally among individuals), and (3) the actual parameters of that function. Hence, properly parameterizing density dependence into an age-structured population model can be a challenge (see Ferson et al. 1991 for an example of density dependence in a bluegill sunfish population model).

A simplifying assumption for fish living in small rivers and streams could be that their populations are less likely to be affected by density dependence. The reason is that an intrinsically unstable environment may prevent their numbers from approaching or exceeding their carrying capacity to the point that density dependence would have a measurable effect on population size. Different strategies have been developed to quantify density dependence in fish population modeling (see for example Oines et al. 1997; Root and Akçakaya 1997; Root et al. 1997).

CASE STUDY

Introduction

Using population modeling in regulatory analyses allows one to predict population-level responses in target species exposed to a stressor of concern. This stressor can be physical (hunting, fishing, entrainment in cooling water intake structures, heat stress), chemical (pesticides, metals, petroleum hydrocarbons), or biological (predation, disease). For this example, we will use a population model to examine the effects of a chemical stressor on the long-term dynamics of a brook trout population.

The case study consists of two distinct steps:

- Deriving the vital rates for the target species
- Obtaining and summarizing life-cycle toxicity data for the chemical tested on the target species

These two steps are combined to parameterize RAMAS® Ecotoxicology, a PC-based probabilistic population model developed by Applied Biomathematics. This model is an age-structured model that requires age-specific vital rates. The software also accepts toxicity data from life-cycle exposures to single toxicants; this allows the analyst to forecast the long-term response of a target population when exposed to a chemical stressor of concern.

Deriving Vital Rates for a Target Species

McFadden et al. (1967) provide detailed census data on the yearly egg production and age class abundances for a brook trout (*Salvelinus fontinalis*) population in a Michigan stream sampled over 14 consecutive years between 1949 and 1962. The study section was 1.75 miles long, averaged 18.5 feet in width, and covered 3.91 acres of water surface. The brook trout population was relatively stable during the study period. The data provided by McFadden et al. (1967) are complete enough so that age-specific survival and fecundity rates can be calculated without extraneous assumptions.

The calculations below are based on those presented by McFadden et al. (1967), Spencer and Ferson (1997), and Akçakaya et al. (1999). The brook trout population model derived from these data is representative of relatively small trout streams in the north central U.S.

The raw population data for this brook trout population are provided in Table A.3.

Table A.3 Age-Specific Census Data for Brook Trout in Hunt Creek, Michigan

Census Year	Total Estimated # of Eggs	Age 0 yr Trout	Age 1 yr Trout	Age 2 yr Trout	Age 3 yr Trout	Age 4 yr Trout	Total # of Trout
1949	119,000	4471	2036	287	14	0	6808
1950	120,000	3941	2013	304	13	0	6271
1951	111,000	4287	1851	265	16	1	6820
1952	102,000	5033	1763	261	16	0	7073
1953	80,000	5387	1637	175	13	0	7212
1954	104,000	6325	2035	234	13	0	8607
1955	146,000	4235	2325	383	24	0	6947
1956	134,000	4949	1612	392	51	1	7005
1957	117,000	6703	1796	309	33	1	8842
1958	151,000	5097	2653	355	26	2	8133
1959	212,000	4038	2395	685	68	0	7186
1960	166,000	5057	2217	473	47	1	7795
1961	144,000	2809	2017	409	23	0	5258
1962	141,000	5052	1589	448	52	2	7143
Average	—	4715	1973	337	25	1	—

From McFadden et al. 1967. Numerical changes and population regulation in brook trout *Salvelinus fontinalis*. *J. Fish Res. Board Can.* 24:1425–1459.

Age-Specific Survival Rates

An age-specific survival rate (S_i) can be calculated because age-specific abundance data are available for many consecutive years. This rate allows one to calculate year-by-year survival rates, which can then be averaged to determine a mean survival rate (and a standard deviation) for each age class. Using such an overall mean is important because it accounts for changes in yearly survival rates due to environment variation or other factors. The standard deviation provides the variability necessary for a probabilistic risk assessment.

To calculate the survival rate from one year to the next, the abundance in a given age class is divided by the abundance in the previous age class in the preceding year, as follows:

$$S_i\ (t) = N_{i+1}\ (t + 1)/N_i\ (t)$$

where

$$S_i(t) = \text{survival rate at time } t$$
$$N_{i+1}(t + 1) = \text{number of individuals in age class } i + 1 \text{ at time } t + 1$$
$$N_i(t) = \text{number of individuals in age class } i \text{ at time } t$$

Example calculations for the brook trout censused between 1949 and 1950 are as follows (data are from Table A.3):

Age 0 year to age 1 year = 2,013/4,471 = 0.4502
Age 1 year to age 2 year = 304/2,036 = 0.1493
Age 2 year to age 3 year = 13/287 = 0.0453
Age 3 year to age 4 year = 0/14 = 0

Table A.4 shows the year-to-year survival rates for brook trout from age 0 year to age 3 years.

Table A.4 Year-to-Year Survival Rates for Brook Trout in a Michigan Stream

Census Year	S (age 0 yr)	S (age 1 yr)	S (age 2 yr)	S (age 3 yr)
1949 to 1950	0.4502	0.1493	0.0453	0
1950 to 1951	0.4697	0.1316	0.0526	0.0769
1951 to 1952	0.4112	0.1410	0.0604	0
1952 to 1953	0.3252	0.0993	0.0498	0
1953 to 1954	0.3778	0.1429	0.0743	0
1954 to 1955	0.3676	0.1882	0.1026	0
1955 to 1956	0.3806	0.1686	0.1331	0.0417
1956 to 1957	0.3629	0.1917	0.0842	0.0196
1957 to 1958	0.3958	0.1977	0.0841	0.0606
1958 to 1959	0.4699	0.2582	0.1915	0
1959 to 1960	0.5490	0.1975	0.0686	0.0147
1960 to 1961	0.3988	0.1845	0.0486	0
1961 to 1962	0.5657	0.2221	0.1271	0.0869

Note that there is no survival rate for age 4 year brook trout because no age 4 year trout survive to become age 5 years; hence, survival for this age class is 0. Based on the year-by-year survival rates presented in Table A.4, the average survival rate (and its standard deviation) can be calculated for each age class. These values are presented in Table A.5.

Table A.5 Mean Age-Specific Survival Rates for the Brook Trout Population

Age	Age-Specific Mean Survival Rate (S_i)	Standard Deviation (SD)
0	0.4249	0.0724
1	0.1748	0.0420
2	0.0863	0.0427
3	0.0231	0.0323
4	0.0000	—

Age-Specific Fecundities

Age-specific fecundity refers to the average number of eggs laid at time t by each fish, which are counted as age 0 year trout at time $t+1$. Data on the proportion of age 0 year trout produced by each age class in each year are required to calculate fecundities. The raw data for this calculation step are presented in Table A.6.

The data in Table A.6 are then used to calculate the *proportion* of age 0 juveniles produced by the age 1 year to age 4 year brook trout. For example, of an estimated 119,000 eggs produced in 1949, the age 1 year group was responsible for 64.71% (77,000/ 119,000) of the total reproductive activity. These proportions are presented in Table A.7.

Note that for this approach to work, one must assume that the probability of hatching and survival to age 0 year is identical for the eggs produced by fish of different age classes.

The data in Table A.7 are combined with those in Table A.3 to calculate the age-specific fecundities for each census year. For example, to calculate the fecundity of 1-year-olds in 1950, the proportion of age 0 year juveniles in 1951 that were produced by 1-year-olds in 1950 is divided by the number of 1-year-olds in 1950, as follows:

$$F_{age1}(1950) = (0.6417 \times 4,287)/2,013 = 1.3666$$

This calculation is made for all age classes and census years to produce the fecundity data set presented in Table A.8.

Table A.6 Estimated Number of Eggs Produced by Brook Trout in a Michigan Stream

Census Year	Age 1 yr	Age 2 yr	Age 3 yr	Age 4 yr	Total
1949	77,000	38,000	4,000	0	119,000
1950	77,000	40,000	3,000	0	120,000
1951	71,000	34,000	5,000	1000	111,000
1952	64,000	34,000	4,000	0	102,000
1953	54,000	22,000	4,000	0	80,000
1954	71,000	30,000	3,000	0	104,000
1955	89,000	50,000	7,000	0	146,000
1956	67,000	50,000	17,000	0	134,000
1957	67,000	39,000	11,000	0	117,000
1958	96,000	45,000	9,000	1000	151,000
1959	104,000	88,000	20,000	0	212,000
1960	90,000	81,000	15,000	0	166,000
1961	84,000	53,000	7,000	0	144,000
1962	68,000	57,000	15,000	1000	141,000

Note: Data are from Table 10 in McFadden et al. (1967); these are estimated values based on regressions of female length over number of eggs and the brook trout size intervals that fall within each age class. Raw data are based on eggs extruded from female brook trout caught downstream from the study area 5 or 6 weeks before the peak of spawning. Using these data, the regression of egg content on length of females was calculated using the following equation: $\log_{10}$ egg content = $\log_{10}a + b\log_{10}$ (total length in inches), where $\log_{10}a = 0.19248$ and $b = 2.69242$.

Table A.7 Proportion of Age 0 Year Juveniles Produced by Age 1 Year to Age 4 Year Brook Trout

Census Year	Age 1 yr	Age 2 yr	Age 3 yr	Age 4 yr
1949	0.6471	0.3193	0.0336	0
1950	0.6417	0.3333	0.025	0
1951	0.6396	0.3063	0.0450	0.0090
1952	0.6275	0.3333	0.0392	0
1953	0.6750	0.2750	0.0500	0
1954	0.6827	0.2885	0.0288	0
1955	0.6096	0.3425	0.0479	0
1956	0.5000	0.3731	0.1269	0
1957	0.5726	0.3333	0.0940	0
1958	0.6358	0.2980	0.0596	0.0066
1959	0.4906	0.4151	0.0943	0
1960	0.5422	0.3675	0.0904	0
1961	0.5833	0.3681	0.0486	0
1962	0.4827	0.4042	0.1064	0.0071

Table A.8 Age-Specific Fecundity by Census Year for Brook Trout in a Small Michigan Stream

Census Year	$F_{(age\ 1\ yr)}$	$F_{(age\ 2\ yr)}$	$F_{(age\ 3\ yr)}$	$F_{(age4\ yr)}$
1949	1.2525	4.3849	9.4622	0
1950	1.3665	4.7007	8.2442	0
1951	1.7392	5.8175	14.1695	45.3423
1952	1.9172	6.8799	13.2034	0
1953	2.6080	9.9393	24.3269	0
1954	1.4207	5.2207	9.3972	0
1955	1.2976	4.4252	9.8867	0
1956	2.0791	6.3804	16.6741	0
1957	1.6252	5.4984	14.5214	0
1958	0.9677	3.3898	9.2567	13.3709
1959	1.0358	3.0644	7.0258	0
1960	0.6869	2.1823	5.4005	0
1961	1.4611	4.5463	10.6775	0

Note that no fecundity is provided for age 0 fish because they are sexually immature and are unable to reproduce. Based on the year-by-year fecundity rates presented in Table A.8, one can calculate the average fecundity rate (and its SD) for each age class. These values are presented in Table A.9.

Table A.9 Mean Age-Specific Fecundity Rates for Brook Trout in a Small Michigan Stream

Age	Mean Age-Specific Fecundity Rate (F)	Standard Deviation (SD)
0[a]	0	0
1	1.4967	0.5081
2	5.1100	1.9563
3	11.7112	4.9547
4	29.3566	22.6072

[a] Age 0 year trout do not have a fecundity value because first reproduction does not occur before age 1 year.

Initial Age Class Abundance for the Michigan Brook Trout Population

Before launching the model, the analyst must assign initial fish abundances for each age class. For models that have been parameterized using site specific data (such as the brook trout population described in this case study), one can calculate age-specific average numbers of fish using specific survey data (see the last row in Table A.3).

When a data set is less complete than the one used in this example, more indirect methods may be required to assign initial abundances. Published fisheries studies on the target species or a closely related species from a body of water comparable with the one to be modeled can be used to estimate age-specific abundances. Proper ratios between different age classes are as important as absolute numbers. Again, published studies can provide valuable insight in this matter. In any case, professional judgment may be necessary to determine a reasonable starting population.

Population Dynamics without Pollutant Effects

The brook trout population model can be parameterized in RAMAS Ecotoxicology using the data developed in the previous sections. Table A.10 shows the data that were used to parameterize the population model. The input data include the following:

1. **Age class**: the brook trout population has five age classes: from young of year (age 0 yr) to four-year-olds; such an age class structure is typical for brook trout in small streams.
2. **Initial numbers**: these values represent the age-class-specific means from 14 consecutive yearly surveys (Table A.3). Note that the number of fish drops rapidly with each increasing age class.
3. **Age-specific fecundity**: reproduction starts with age class 1. The numbers in parentheses are the mean, standard deviation, and the lower bound; the lower bound is set at 0 because a negative fecundity makes no sense. The argument "lognormal" tells the program that adult fecundity in fish is approximated by a lognormal statistical distribution. Note that fecundity increases with age because older fish are larger, lay more eggs, and hence produce more offspring that survive into their first year.
4. **Age-specific survival**: the numbers in parentheses are the mean, standard deviation, lower bound, and upper bound for age-specific survival. The lower bound is set at 0 and the upper bound is set at 1 to indicate that survival is a year-to-year ratio that cannot exceed these two extremes. The argument *lognormal* tells the program that age-specific survival in fish is approximated by a lognormal statistical distribution. Note, too, that survival drops rapidly with age, and that no age 4 year fish survive to age 5 years.

Using these input data, the program predicts the brook trout population in the future. For the purposes of this exercise, the model ran 1000 iterations (replicates) for 10 years.

Table A.10 Input Variables to the Brook Trout Population Model

Age Class (yr)	Starting Densities (individuals)	Age-Specific Fecundity Rates	Age-Specific Survival Rates
0	4715	0	lognormal (0.4249, 0.0724, 0, 1)
1	1973	lognormal (1.4967, 0.5081, 0)	lognormal (0.1748, 0.0420, 0, 1)
2	337	lognormal (5.11, 1.9563, 0)	lognormal (0.0863, 0.0427, 0, 1)
3	25	lognormal (11.7112, 4.9547, 0)	lognormal (0.0231, 0.0323, 0, 1)
4	1	lognormal (29.3566, 22.6072, 0)	0

Note: The data shown in this table were summarized from Tables A.3, A.5, and A.9.

The age-specific fecundity and survival rates are assumed to have a lognormal distribution.

Specifying a Dose–Response Function

Life-cycle toxicity data are needed for assessing the effects of a toxicant on the book trout population to determine how such a toxicant would affect long-term population stability. For this case study, data on the long-term effects of the pesticide toxaphene on brook trout were used in the population model. These ecotoxicity data are explained, analyzed, and summarized by Spencer and Ferson (1997). The toxicity study was done on brook trout; it looked at the effects of life-cycle toxaphene exposure on the number of eggs, juvenile mortality, and adult mortality in a laboratory setting.

Three pieces of information are needed from such a data set to parameterize the population model (Spencer and Ferson, 1997):

1. **The EC_{50}**, the contaminant concentration at which 50% of the test organisms show a response: for this case study, the response was mortality in juvenile brook trout exposed to different concentrations of toxaphene since fertilization.
2. **The slope at the EC_{50}**, the tangent of the dose–response function at the central EC_{50} value: for this case study, the slope was for mortality in juvenile brook trout exposed to toxaphene since fertilization.
3. **The shape of the dose–response function**, the equation to be used to calculate the proportion of test organisms affected depending on the contaminant dose: several statistical approaches are available to calculate this function, including the probit model, logit model, and Weibull cumulative distribution model. The probit model is widely used in ecotoxicology and was used by Spencer and Ferson (1997) for the toxaphene data set.

The calculations done by Spencer and Ferson (1997) resulted in an EC_{50} for age 0 year trout of 0.065 :μg/L, which indicates extreme toxicity; they also reported the slope of the dose–response curve for age 0 year mortality at the EC_{50} as 5.461 L/μg. The dose–response model selected for this analysis was the probit model.

The last step consists of determining to which vital rate (fecundity or survival) and age class (age 0, 1, 2, 3, or 4 years) these toxicity characteristics should be applied. Because the toxaphene study looked at mortality in juvenile (age 0 year) brook trout, the population model was amended by adding this information to the age 0 year survival cell. Finally, the analyst entered a concentration for the contaminant of concern and ran the model to see how the population changed over time when continuously exposed to the chemical.

One can now predict how different concentrations of toxaphene would affect the model trout population based on its impact on juvenile survival. The software can be used to ask a number of questions, such as:

1. What will the size of the brook trout population be after chronic exposure to each concentration of toxaphene for 10 year?
2. How long would it take for the exposed trout population to decline by a certain value (say 25 or 50%)?
3. What is the probability of extinction in the brook trout population for different toxaphene concentrations?
4. What is the probability of the trout population dipping below a given threshold (say 25 or 50% from the original population) at some point in the 10-year simulation?

RAMAS Ecotoxicology provides the following risk statistics (Spencer and Ferson, 1997):

1. **Interval risk decline** is the probability of a population declining by as much as a given percentage of its initial value *at any time during the simulation*.
2. **Interval extinction risk** is the probability of a population falling as low as a given abundance *at any time during the simulation*.
3. **Interval explosion risk** is the probability of a population equaling or exceeding a given abundance *at any time during a simulation*.
4. **Terminal decline risk** is the probability of a population being as much as a given percentage lower than its initial value *at the end of a simulation*.

5. **Terminal extinction risk** is the probability of a population being as low as a given abundance *at the end of a simulation*.
6. **Terminal explosion risk** is the probability of a population being as high as or higher than a given abundance *at the end of a simulation*.
7. **Time to extinction** is the time required by a population to fall below a given threshold abundance.
8. **Time to explosion** is the time required by a population to exceed a given threshold abundance.

Results of the Brook Trout Population Simulations

Graphical Analysis

The first set of figures (Figure A.1) shows the *abundance* of adult brook trout without toxaphene exposure and in the presence of 0.02 and 0.04 µg/L toxaphene, respectively. (NOTE: *any* toxaphene concentration could have been used because the dose–response function of this chemical is specified in the model; for this case study, the contaminant concentrations were kept constant over time to facilitate data comparison and discussion.)

Figure A.1, top, shows the abundance of adult (age 1+ years) brook trout over 10 years without toxaphene exposure. The population at t = 0 equals 2336 fish but reaches an average of 3360 fish at the end of the simulation. This represents a small yearly increase and is not surprising because density dependence was not included in the model. If it had been included, the trout population would fluctuate around a central tendency. The relatively small increase over time is not expected to skew the overall result of the analysis. Its effects could be further minimized by reducing the simulation period, say from 10 to 5 years. Without density–dependence, however, the original trout population would double after about 20 years. Our simplifying assumption may not hold for longer simulation periods.

Figure A.1, middle, shows the average trout population slowly declining over time when exposed to 0.02 µg/L toxaphene. A rapid and dramatic decline in the average trout population occurs in the presence of 0.04 µg/L toxaphene. In fact, after 10 years, the trout numbers are about 75% below their starting point, suggesting the potential for extirpation.

Figure A.2 shows the *interval decline risk* for the average adult brook trout population. Without toxaphene, the risk of, for example, a 30% decline in the population at any time during the simulation is about 30% (Figure A.2, no toxaphene). This surprisingly high risk is due to the noisiness implied by the SDs used to parameterize the model. When exposed to 0.02 µg/L toxaphene, the risk of a 30% decline increases to about 85% (Figure A.2, 0.02 µg/L toxaphane). When exposed to 0.04 µg/L toxaphene, this risk increases to 100% (Figure A.2, 0.04 µg/L toxaphene).

Figure A.3 shows the *interval extinction risk* for the average adult brook trout population. Without toxaphene, the risk of the population falling lower than, for example, 1500 individuals at any time during the simulation is about 20% (Figure A.3, no toxaphene). When exposed to 0.02 µg/L toxaphene, the risk of the population falling below 1500 individuals increases to about 75% (Figure A.3, 0.02 µg/L toxaphene). When exposed to 0.04 µg/L, this risk increases to 100% (Figure A.3, 0.04 µg/L toxaphene).

Figure A.4 shows the *terminal decline risk* for the average adult brook trout population. Without toxaphene, the risk of the population after 10 years falling, for example, 30% below the starting population is 10% (Figure A.4, no toxaphene). When exposed to 0.02 µg/L toxaphene, the risk of the population after 10 year falling 30% below the starting population increases to 55% (Figure A.4, 0.02 µg/L toxaphene). When exposed to 0.04 µg/L, this risk increases to 100% (Figure A.4, 0.04 µg/L toxaphene).

Figure A.5 shows the *terminal extinction risk* for the average adult brook trout population. Without toxaphene, the risk of the population falling to, for example, 1,500 fish at the end of the 10-year simulation is 5 percent (Figure A.5, no toxaphene). When exposed to 0.02 µg/L toxaphene, this risk increases to 45% (Figure A.5, 0.02 µg/L toxaphene). Finally, when exposed to 0.04 µg/L, this risk increases to 100% (Figure A.5, 0.04 µg/L toxaphene).

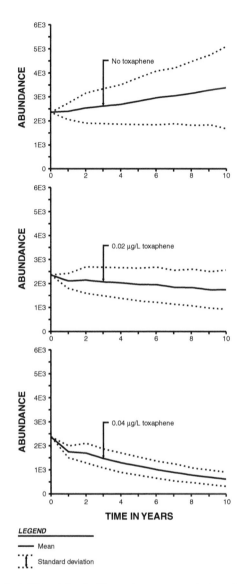

Figure A.1   Abundance of adult brook trout with varying exposure to toxaphene.

Discussion

Figures A.1 through A.5 provide an example of the kind of analysis that can be done using probability-based population simulations. Data interpretation depends on the questions asked. Clearly, the presence of 0.04 μg/L toxaphene in stream water over a 10-year period would result in unacceptable responses in the adult brook trout population. But even 0.02 μg/L toxaphene over 10 years produces some surprising results. For example, the probability that the original trout population would decrease by 30% at the end of 10 generations is about 55% (Figure A.4, 0.02 μg/L toxaphene). The risk of the number of adult trout decreasing to less than 1500 individuals at any time during the simulation is about 75% (Figure A.3, 0.02 μg/L toxaphene). The trout population is responding negatively even at that low concentration.

A clear set of population targets must be identified before a safe toxaphene threshold level can be calculated using this approach. For example, the target might be that the risk of the population

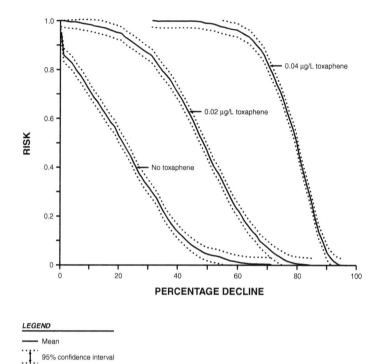

Figure A.2 Interval decline risk for the average adult brook trout population.

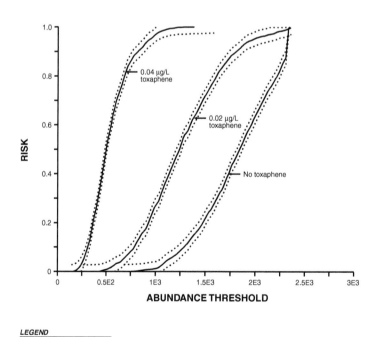

Figure A.3 Interval extinction risk for the average adult brook trout population.

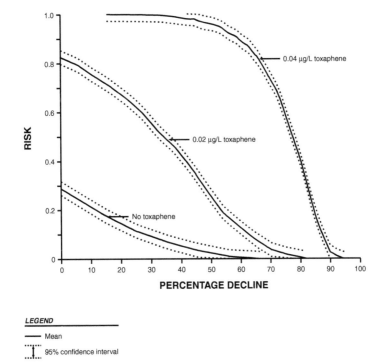

Figure A.4 Terminal decline risk for the average adult brook trout population.

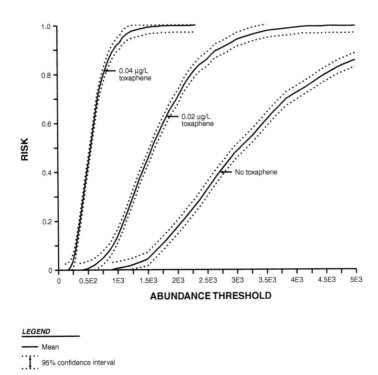

Figure A.5 Terminal extinction risk for the average adult brook trout population.

declining by one third during the simulation cannot exceed 25%. Or the target might be that the population at the end of the simulation cannot be less than 75% of the initial population. With these restrictions, one can enter various toxaphene concentrations and determine at which level these conditions are met.

Some Limitations of Population Modeling

Computer models in general are a powerful approach to forecasting complex events into the future. For example, modeling tools are used to predict the fate and transport of contaminants in the environment, including the movement and concentration of hazardous air pollutants over time, or the effect of various remedial actions on the movement and concentration of contaminants dissolved in groundwater. The output of modeling, however, is only as good as the quality of the input data. As is clear from this study, substantial uncertainty can be associated with population modeling.

Data probably do not get any better than those used in the brook trout case study: the fisheries survey used to parameterize the model was extensive and unusually thorough; the toxaphene ecotoxicity data specifically dealt with the impact of the pesticide on juvenile brook trout mortality. More often than not, the available population data is much more sketchy and incomplete; numerous assumptions must be made to calculate age-specific survival and fecundity.

Several limitations associated with developing fish population models are noted below:

1. It takes a considerable amount of time and effort to identify, gather, organize, and analyze fisheries data to determine vital rates and develop a defensible species-specific population model; this represents a major hurdle in applying fish population modeling in ecological risk assessments.
2. A species-specific population model may also need to be habitat specific. For instance, a large mouth bass model developed for a large river cannot be used for a reservoir. The reason is that the vital rates in the river population are likely to differ meaningfully from those of a reservoir population. Hence, to perform an assessment at the national level requires developing several distinct models for each species of concern that are representative of broad habitats. This, once again, is time consuming and labor intensive.
3. Density dependence can be difficult to quantify using field data. Data for several consecutive years and at different densities are required to draw a meaningful regression to estimate the density-dependence parameters required for population modeling. If density dependence is present, then there comes a point at which increasing numbers of young result in fewer offspring surviving (due to crowding, more competition or predation, lack of food, etc.). Such a relationship was not found in the brook trout case study. This finding suggests that the population might actually have been strongly regulated by density dependence and remained at its carrying capacity during the 14-year study period.
4. Population modeling in general can only be done on a chemical-by-chemical basis. This is a fundamental limitation because fish in their natural environment are usually exposed to *mixtures* of contaminants. To be conservative, an assessment would have to focus on the most toxic contaminant in the mixture (assuming that toxicity data were available for that compound) modeled at a relatively high concentration. Even then, the projected population responses could not be applied directly to field conditions; they would only be indicative of the type of response that might be expected depending on which regulatory scenario is assessed.

Toxicity Assessment

To assess the effects of toxicants on vital rates in fish populations requires data on life-cycle toxicity testing. A review of the literature by Jarvinen and Ankley (1999) suggests that such data are relatively rare. This is not surprising because most fish develop rather slowly (i.e., the test organisms must be maintained in the laboratory and properly exposed to the toxicant for months to years). As a result, such tests are quite expensive and are rarely done. Life-cycle toxicity data for fish have been published for (1) some metals (see, for example, McKim and Benoit 1971; Benoit 1975;

Benoit et al. 1976; Holcombe et al. 1976, 1979; McKim et al. 1976; Pickering 1980; Snarski and Svensson 1982; Brown et al. 1994); (2) a few pesticides (see, for example, Mayer et al. 1975, 1977; Hansen et al. 1977; Hermanutz 1978; Goodman et al. 1979, 1982; Buckler et al. 1981; Jarvinen et al. 1983; Hermanutz et al. 1985; Call et al. 1987); and (3) several other chemicals (see, for example, Nebeker et al. 1974; DeFoe et al. 1978; Bengtsson 1980; Ward et al. 1981; Virtanen and Hattula 1982).

The paucity of published life-cycle ecotoxicity data limits the application of fish population modeling. One exception may be pesticides: these chemicals undergo extensive testing as part of the product registration process. Depending on the compound, life-cycle toxicity testing with fish may be required. Even though such data are usually not published in the peer-reviewed scientific literature, this information may be available and useful to support population modeling and decision-making.

Even with these kinds of limitations, the case study shows that when the data exist, modeling represents a highly flexible approach to forecasting population-level responses. This approach is useful not only with fish but also with other types of animals such as birds and mammals because they, too, can be modeled in terms of age-structured vital rates. Use of such models can enhance most risk assessments and provide valuable information for environmental decision-making.

REFERENCES

Akçakaya, H.R., M.A. Burgman, and L.R. Ginzburg. 1999. *Applied Population Ecology. Principles and Computer Exercises Using RAMAS® Ecolab 2.0*. Sinauer Associates, Inc. Sunderland, MA.

Bengtsson, B.E. 1980. Long-term effects of PCB (Clophen A50) on growth, reproduction and swimming performance in the minnow, *Phoxinus phoxinus. Water Res.* 14:681–687.

Benoit, D.A. 1975. Chronic effects of copper on survival, growth and reproduction of the bluegill (*Lepomis macrochirus*). *Trans. Am. Fish. Soc.* 104:353–358.

Benoit, D.A., E.N. Leonard, G.M. Christensen, and J.T. Fiandt. 1976. Toxic effects of cadmium on three generations of brook trout (*Salvelinus fontinalis*). *Trans. Am. Fish.* Soc. 105:550–560.

Brown, V., D. Shurben, W. Miller, and M. Crane. 1994. Cadmium toxicity to rainbow trout *Oncorhynchus mykiss* Walbaum and brown trout *Salmo trutta* L. over extended exposure periods. *Ecotoxicol. Environ. Saf.* 29:38–46.

Buckler, D.R., A. Witt, F.L. Mayer, and J.N. Huckin. 1981. Acute and chronic effects of kepone and mirex on the fathead minnow. *Trans. Am. Fish. Soc.* 110:270–280.

Call, D.J., L.T. Brooke, R.J. Kent, M.L. Knuth, S.H. Poire, J.M. Huot, and A.R. Lima. 1987. Bromacil and diuron herbicides: toxicity, uptake and elimination in freshwater fish. *Arch. Environ. Contam. Toxicol.* 16:607–613.

DeFoe, D.L., G.D. Veith, and R.W. Carlson. 1978. Effects of Aroclor 1248 and 1260 on the fathead minnow (*Pimephales promelas*). *J. Fish. Res. Board Can.* 35:997–1002.

Ferson, S., R. Akçakaya, L. Ginsburg, and M. Krause. 1991. Use of RAMAS to Estimate Ecological Risk: Two Fish Species Case Studies. Technical Report EN-71-76. Electric Power Research Institute, Palo Alto, CA.

Goodman, L.R., D.J. Hansen, D.L. Coppage, J.C. Moore, and E. Matthews. 1979. Diazinon: chronic toxicity to, and brain acetylcholinesterase inhibition in the sheepshead minnow, *Cyprinodon variegatus. Trans. Am. Fish. Soc.* 108:479–488.

Goodman, L.R., D.J. Hansen, C.S. Manning, and L.F. Faas. 1982. Effects of kepone on the sheepshead minnow in an entire life-cycle toxicity test. *Arch. Environ. Contam. Toxicol.* 11:335–342.

Hansen, D.J., S.C. Schimmel, and J. Forester. 1977. Endrin: effects on the entire lifecycle of a salt water fish, *Cyprinodon variegatus. J. Toxicol. Environ. Health* 3:721–733.

Hermanutz, R.O. 1978. Endrin and malathion toxicity to flagfish (*Jordanella floridae*). *Arch. Environ. Contam. Toxicol.* 7:159–168.

Hermanutz, R.O., J.G. Eaton, and L.H. Mueller. 1985. Toxicity of endrin and malathion mixtures to flagfish (*Jordanella floridae*). *Arch. Environ. Contam. Toxicol.* 14:307–314.

Holcombe, G.W., D.A. Benoit, E.N. Leonard, and J.M. McKim. 1976. Long-term effects of lead exposures on three generations of brook trout (*Salvelinus fontinalis*). *J. Fish. Res. Board Can.* 33:1731–1741.

Holcombe, G.W., D.A. Benoit, and E.N. Leonard. 1979. Long-term effects of zinc exposures on brook trout (*Salvelinus fontinalis*). *Trans. Am. Fish. Soc.* 108:76–87.

Jarvinen, A.W., B.R. Nordling, and M.E. Henry. 1983. Chronic toxicity of Dursban (chlorpyrifos) to the fathead minnow (*Pimephales promelas*) and the resultant acetylcholinesterase inhibition. *Ecotoxicol. Environ. Saf.* 7:424–434.

Jarvinen, A.W. and G.T. Ankley. 1999. Linkage of Effects to Tissue Residues: Development of a Comprehensive Database for Aquatic Organisms Exposed to Inorganic and Organic Chemicals. Society of Environmental Toxicology and Chemistry (SETAC), Pensacola, FL.

Mayer, F.L., P.M. Mehrle, and W.P. Dwyer. 1975. Toxaphene Effects on Reproduction, Growth, and Mortality of Brook Trout. EPA-600/3-75/013. U.S. Environmental Protection Agency, Duluth, MN.

Mayer, F.L., P.M. Mehrle, and W.P Dwyer. 1977. Toxaphene: Chronic Toxicity to Fathead Minnows and Channel Catfish. EPA-600/3-77/069. U.S. Environmental Protection Agency, Duluth, MN.

McFadden, J.T., G.R. Alexander, and D.S. Shetter. 1967. Numerical changes and population regulation in brook trout *Salvelinus fontinalis*. *J. Fish. Res. Bd. Can.* 24:1425–1459.

McKim, J.M. and D.A. Benoit. 1971. Effects of long-term exposures to copper on survival, growth, and reproduction of brook trout (*Salvelinus fontinalis*). *J. Fish. Res. Board Can.* 28:655–662.

McKim, J.M., G.F. Olson, G.W. Holcombe, and E.P. Hunt. 1976. Long-term effects of methylmercuric chloride on three generations of brook trout (*Salvelinus fontinalis*): toxicity, accumulation, distribution, and elimination. *J. Fish. Res. Board Can.* 33:2726–2739.

Nebeker, A.V., F.A. Puglisi, and D.L. DeFoe. 1974. Effect of polychlorinated biphenyl compounds on survival and reproduction of the fathead minnow and flagfish. *Trans. Am. Fish. Soc.* 103:562–568.

Oines, G., K.V. Root, L. Ginzburg, and S. Ferson. 1997. Assessment of Population-Level Threat from Entrainment at Russell Dam on Thurmaond Reservoir Fishes. Final Report prepared for U.S. Army Corps of Engineers Waterways Experiment Station. Applied Biomathematics, Setauket, NY.

Pickering, Q.H. 1980. Chronic Toxicity of Trivalent Chromium to the Fathead Minnow (*Pimephales promelas*) in Hard Water. Manuscript, ES EPA, Cincinnati, OH.

Root, K.V. and R.H. Akçakaya. 1997. Ecological Risk Analysis for the Shortnose Sturgeon Populations in the Connecticut River. Final Report prepared for Northeast Utilities and Electric Power Research Institute. Applied Biomathematics, Setauket, NY.

Root, K.V., C.S. Grogan, and R.H. Akçakaya. 1997. RAMAS Ecological Risk Model for Assessing Temperature Effects on Redhorse Sucker Species in the Muskingum River. Final Report prepared for American Electric Power and Electric Power Research Institute. Applied Biomathematics, Setauket, NY.

Root, K.V. 1998. Ecological Risk Analysis as a Management Tool for Native Fish Species in the Sierra Nevada Range of California. Final Report prepared for Southern California Edison and Electric Power Research Institute. Applied Biomathematics, Setauket, NY.

Snarski, V.M. and B.J. Svensson. 1982. Chronic toxicity and bioaccumulation of mercuric chloride in the fathead minnow (*Pimephales promelas*). *Aquatic Toxicol.* 2:143–156.

Spencer, M. and S. Ferson. 1997. RAMAS® Ecotoxicology Version 1.0a. User's Manual Volume II: Ecological Risk Assessment for Structured Populations. Applied Biomathematics, Setauket, NY.

Virtanen. M.T. and M.L. Hattula. 1982. The fate of 2,4,6-trichlorophenol in an aquatic continuous-flow system [guppy reproduction]. *Chemosphere* 11:641–649.

Ward, G.S., G.C. Cramm, P.R. Parrish, H. Trachman, and A. Shlesinger. 1981. Bioaccumulation and Chronic Toxicity of Bis(tributyltin)oxide (TBTO): Tests with a Saltwater Fish [Sheepshead Minnow Reproduction]. In Branson, D.R. and Dickson, K.L. (Eds.). Aquatic Toxicology and Hazard Assessment, Fourth Conference. STP 737. American Society for Testing and Materials (ASTM), Philadelphia, pp.183–200.

APPENDIX B

Classification Systems

Karen V. Root

The classification systems reviewed here categorize or rank species according to some direct estimate of the risk of extinction or to some indirect estimate obtained by using factors associated with the risk of extinction. These systems (or models) are particularly useful for prioritizing resource allocation for protection and management of species and are used by a variety of organizations and government agencies. Classification systems vary considerably depending of the objectives of the system, the data available, and the units and scale under consideration. These factors need to be considered explicitly when evaluating or developing a classification system.

Five classification systems will be reviewed: Mace and Lande (1991), IUCN (The World Conservation Union) (1994) and a modification (RAMAS Red List; Akçakaya and Ferson 1999), Millsap et al. (1990), Master (1991), and the U.S. Fish and Wildlife Service (USFWS) (1983). All of these classification systems except Millsap et al. (1990) categorize species or populations into discrete classes of risk. Millsap et al. (1990) assigns a discrete integer rank value. Several important features of a classification system should be considered in an evaluation. The objective of classification varies. For example, classification may be used to estimate risk or to develop management guidelines. The unit of consideration may be species or populations. The scale of consideration ranges from local to global depending on the model. Some of the classifications have soft boundaries between categories, namely ranges, and some use firm thresholds. Ranking of species or populations may be as discrete classes, as an absolute ranking, or as continuous classes. These models also deal with uncertainty in many different ways: ignore it, assume security or imperilment until proven otherwise, use expert opinions, and so forth. These features are reviewed individually below for the five classification systems; a summary of similarities and differences and a recommendation follow.

MACE AND LANDE (1991)

Mace and Lande (1991) proposed three categories — critical, endangered, and vulnerable — for classifying species' risk of extinction. Their stated intent was to make the system "scientific and completely objective." Although the scale of consideration was not mentioned, it is presumably global, or the entire range of the species of consideration. Mace and Lande assigned species, the primary unit of consideration, to these categories on the basis of simple quantitative criteria that utilize population biology theory. Six variables are considered: effective population size; degree of fragmentation; population growth trends; probability of catastrophes; estimated habitat alteration; and estimated exploitation or ecological interactions. For each of the six variables considered, there are threshold values for the three classes. As an example, critical species might have an effective population size of less than 50 individuals; endangered species, one of less than 500 individuals; and vulnerable species, less than 2000 individuals. Species do not have to satisfy all of the criteria to be placed in a class — only a minimum number of them (e.g., two out of four). Consideration is given to variation among taxonomic groups; generation time rather than absolute time may be used for estimations. The potential effects of chemicals on a given species are not considered directly in these criteria. This system may be applied to most species, although the authors acknowledge that it might be most appropriate for vertebrate species and at a global scale (the entire range of a species). Some flexibility is inherent in the system, such as the use of ranges for some threshold criteria. Although uncertainty is discussed, no specific methods for incorporating it into the classification system are provided.

IUCN (1994)

The IUCN (1994) classification system, which builds on the Mace and Lande (1991) system, classifies species on the basis of their risk of extinction in a scientific and objective way. The scale is global, and the original goal was to classify species for the regulation of international trade. IUCN classification is made on the basis of four categories with a specified risk of extinction: critically endangered, endangered, vulnerable, and lower risk. The criteria are quantitative and derived from principles of population biology. Species are classified by population size and number, population growth trends, area of occupancy, degree of fragmentation, and estimate of risk. Threshold values are provided for each criterion, and these values are usually specified as a range. Species do not have to satisfy all of the criteria to be placed in a class — a minimum number of them (e.g., two out of four). Consideration is given to variation among taxonomic groups; generation time rather than absolute time may be used for estimations. The potential effects of chemicals on a given species are not considered directly, although the user may designate specific causes, such as pollution, for the observed values. Although uncertainty is discussed, no specific methods for incorporating it into the classification system are provided, although separate categories (data deficient and not evaluated) are given for dealing with species for which little or no data for classification exist.

RAMAS RED LIST

RAMAS Red List (Akçakaya and Ferson 1999) implements the IUCN (1994) classification system with two major modifications in a software program for Windows-based personal computers. It is designed to classify species on a variety of scales (specified by user) on the basis of their risk of extinction. The four categories of risk are critically endangered, endangered, vulnerable, and lower risk. The criteria are quantitative and derived from principles of biology. Species are classified on the basis of population size and number; population growth trends; areas of occupancy; degree of fragmentation, and an estimate of risk. Species do not have to satisfy all of the criteria to be placed in a class — a minimum number of them (e.g., two out of four). Time is specified as the generation time of the species. The potential effects of chemicals on a given species are not considered directly, although the user may designate specific causes, such as pollution, for the observed values.

The two major modifications to the IUCN system are the direct incorporation of uncertainty and the consideration of nonbiological factors. RAMAS Red List allows users to specify input parameters, such as the number of populations, as a single number, a range of numbers, or a range plus a best estimate. This uncertainty is propagated through the classification system and results in the designation of species in multiple classes if appropriate. For example, a species may be classified as "vulnerable, with plausible categories of endangered and vulnerable." User attitudes may affect their parameter estimates; therefore, the user specifies a "dispute tolerance" and a "risk tolerance," which reveal the underlying assumptions they may have made. A person who prefers to rely only on consensus data and who requires proof before classifying a species as at risk would have a high dispute tolerance and a high risk tolerance.

MASTER (1991)

Master (1991) described a classification system adopted by The Nature Conservancy (TNC), a private organization devoted to conservation. This classification system is designed to classify species on the basis of their risk of extinction to prioritize conservation and management actions. The scale of application is considered global, but most applications of the model have occurred within the U.S. On the basis of the view that species are not all equally comparable and that adequate data are often lacking, TNC adopted a less rigorous, more qualitative system that uses guidelines rather than strict numeric criteria or thresholds. Five classes exist: critically imperiled globally (G1); imperiled globally (G2); rare or uncommon (G3); apparently secure (G4); and secure (G5). Species are classified on the basis of information on their number, the quality and conditions of their occurrences (populations); the number of individuals; narrowness of range and habitat; trends in populations and habitats; threats to the species; species fragility; and other considerations. Species do not have to satisfy all of the criteria to be placed in a class — one of them. According to Master (1991), this system uses the same criteria as Mace and Lande (1991) "without numerical cutoffs." Consideration is given to variation among taxonomic groups by allowing the user considerable latitude in interpreting the criteria. The potential effects of chemicals on a given species are not considered directly, although the user may designate specific causes, such as pollution, for the observed values. Although uncertainty is discussed, no specific methods for incorporating it into the classification system are provided. This classification system is integrated into a national computer database that is reviewed regularly. Some of the additional unique features of this classification system are the explicit recommendations on protection, inventory, research, and stewardship needs for each species evaluated. The TNC classification system is designed to be flexible across many scales.

USFWS (1983)

The USFWS (USFWS 1983) developed guidelines for classifying species as either endangered or threatened as defined by the Endangered Species Act of 1973 (ESA). This classification system was designed primarily to consider species within the U.S. to determine listing status under the ESA and developing recovery plans. In this classification system, two or three choices are offered for each criterion: magnitude of threat (high or moderate to low); immediacy of threat (imminent or nonimminent); and taxonomy (montotypic genus, species, or subspecies). These categories are nested, which leads to a 1 through 12 priority rank. No numeric thresholds or criteria are used; the classification system is primarily subjective. The user is instructed to utilize all appropriate, available data for the species evaluated to determine its rank. This system is intended to be quick and straightforward to expedite determination of species status. It requires little or no data. The potential effects of chemicals on a given species may be considered directly in the ranking. Uncertainty is discussed in the text that accompanies a rank. A separate similar ranking system is used to prioritize conservation action for species and incorporates recovery potential as a function of the species' limiting factors, threats, and management needs.

MILLSAP ET AL. (1990)

The Millsap et al. (1990) classification system was designed to categorize vertebrate species in Florida on the basis of their risk of extinction. The scale, therefore, is regional rather than global or national. It uses a different approach than the previous four main systems discussed. Species are ranked according to two different sets of criteria and assigned an integer value; no discrete risk classes are used as they are in the other classification systems. The ranks are the sum of the points assigned for each criterion. Each criterion, such as population size, has a number of subclasses that specify a value or range of values. For example, population size may be characterized as (a) 0–500 individuals; (b) 501–1000 individuals, or unknown but suspected to be small; (c) 1001–3000 individuals; (d) 3001–10,000 individuals; (e) 10,001–50,000 individuals, or unknown but suspected to be large; and (f) >50,000 individuals. In this case, species are assigned 10 points for (a), 8 points for (b), 6 points for (c), 4 points for (d), 2 points for (e), and 0 points for (f). The *biological score* a species receives reflects its scores for population size, population trend, range size, distribution trend, population concentration, reproductive potential, and ecological specialization. A higher biological score indicates a greater vulnerability to extinction. The number of points given for a separate set of criteria that include rankings developed on the basis of the current state of knowledge of distribution, population trend, limiting factors, and the current extent of conservation efforts is designated an *action score*. A higher action score indicates a poorly known, unmanaged species. This system, which is integrated into a computer database, separates information on risk classification from that which is important for the conservation and management of endangered species. The potential effects of chemicals on a given species are considered directly as part of the action score but not the biological score; trends in the biological score are not specifically attributed to a particular cause. Although uncertainty is discussed, no specific methods for incorporating it into the classification system are provided.

DISCUSSION

All of these classification systems address the degree of risk a species faces. Each of them was designed to prioritize species on the basis of their vulnerability to extinction. The scale of application among the five models ranges from global (IUCN, RAMAS Red List, Mace and Lande) to national (Master and USFWS) to regional (Millsap et al. 1990). Although IUCN, RAMAS Red List, and Master discuss environmental factors, none of these models explicitly addresses the effects of chemicals on the overall risks. Also, none of these models is designed specifically for application to populations — the most likely unit of management under consideration. Despite these limitations, these classification models could be tailored to address specific categorization needs. For example, additional factors, such as exposure to chemical agents, could be added in the assessment of risk. In this way, the classification would incorporate specific risks of interest. These models require less data, in general, than a population-level stochastic simulation model. Therefore, in cases in which the amount of data is insufficient for a complete risk analysis, at least the general level of risk for a species or population can be assessed by using *proxy* variables (e.g., population size, growth trends, habitat condition, etc.).

What are the overall similarities? All of these models:

- Assess the risk of extinction for a species
- Use quantitative *and* qualitative criteria
- Have modest data requirements (USFWS has the smallest data requirements)
- Consider threats facing a species
- Consider factors such as degree of fragmentation, habitat alteration, exploitation, and population growth trends

- Can be readily applied to vertebrate species
- Would require modification to make them suitable for assessing the effects of toxic chemicals

What are the major differences?

- Each has different data requirements.
- Flexibility and simplicity vary (RAMAS Red List, IUCN, and Master are the most flexible, and USFWS and Master are the simplest).
- Only USFWS and Millsap suggest management options.
- Only RAMAS Red List and Master identify data sources as part of the classification process.
- Only RAMAS Red List explicitly addresses and incorporates uncertainty.
- Millsap does not use discrete classes as the others do.

Table B.1 compares the five models on the basis of their features, such as factors considered in the classification and rankings of utility, applicability, flexibility, simplicity, and cost on a scale of 1 to 5 (5 being the most desirable). On the basis of this review, RAMAS Red List and IUCN stand out as the most flexible, applicable, readily adaptable to specific goals, and widely accepted of the classification models. RAMAS Red List has the added advantage of dealing directly with uncertainty and can be implemented by using Windows-based software.

REFERENCES

Akçakaya, H.R. and S. Ferson. 1999. *RAMAS Red List: Threatened Species Classifications Under Uncertainty.* Applied Biomathematics, Setauket, NY.
USFWS. 1983. Endangered and Threatened Species Listing and Recovery Priority Guidelines. Federal Register, 48 (184): 43098-43105. U.S. Fish and Wildlife Service.
IUCN. 1994. *IUCN Red List Categories.* IUCN, Gland, Switzerland.
Mace, G.M. and R. Lande. 1991. Assessing extinction threats: toward a reevaluation of IUCN threatened species categories. *Conserv. Biol.* 5:148–157.
Master, L.L. 1991. Assessing threats and setting priorities for conservation. *Conserv. Biol.* 5:559–563.
Millsap, B.A., J.A. Gore, D.E. Runde, and S.I. Cerulean. 1990. Setting priorities for the conservation of fish and wildlife species in Florida. *Wildl. Monogr.* 111:1–57.

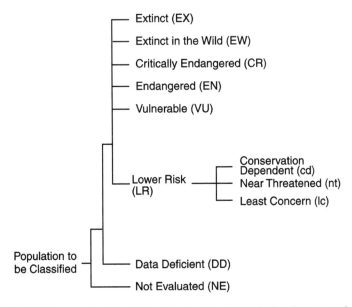

Figure B.1   IUCN (World Conservation Union) classification tree for assigning the status of a population. With permission of Applied Biomathematics.

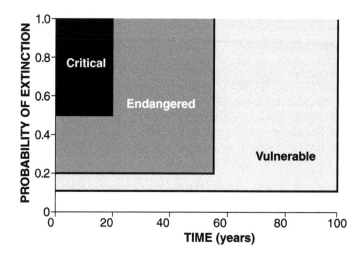

Figure B.2 Graphic approach to classifying population status. With permission of Applied Biomathematics.

Table B.1 Factors Considered in the Five Classification Systems and Factor Rankings (on a Scale of Plus or Minus or a Scale from 1 through 5, with 5 being the Most Desirable) for Several Key Qualities

System	Adoption	Discrete Classes	Effective Population Size	Genetic Considerations	Uniqueness	Threats	Degree of Fragmentation	Population Growth Trends	Probability of Catastrophes	Habitat Alteration	Exploitation	Ecological Interactions	Environmental Factors	Quantitative	Qualitative Criteria	Quantitative Criteria	Uncertainty	Data Sources Identified	Management Options	Data Required	Taxonomic Applicability	Utility (general)	Utility for Chemical Industry	Flexibility	Simplicity	Cost
Mace and Lande (1991)	Modified by IUCN	3	+	−	−	+	+	+	+	+	−	+	−	+	+	+	−	−	−	3	3	4	2	3	3	3
USFWS (1983)	U.S. ESA Law	2	−	+	+	+	+	+	−	+	+	+	−	−	+	−	−	−	+	4	4	2	1	4	4	2
IUCN (1994)	IUCN, CITES, Canada	4	+	−	−	+	+	+	+	+	−	+	+	+	+	+	−	−	−	3	5	4	3	5	3	3
RAMAS Red List Akçakaya and Ferson (1999)	IUCN (provisional)	4	+	−	−	+	+	+	+	+	−	+	+	+	+	+	+	+	−	3	5	4	3	5	3	3
Master (1991)	TNC (states)	5	−	+	+	+	+	+	−	+	−	−	+	+	+	+	−	+	+	3	5	4	3	5	4	3
Millsap et al. (1990)	Florida	n/a	−	+	+	+	+	+	−	+	−	+	−	+	+	+	−	−	−	3	3	4	2	4	3	3

Note: CITES - Convention on International Trade in Endangered Species
ESA   - Endangered Species Act
IUCN   - The World Conservation Union
n/a   - not applicable
TNC   - The Nature Conservancy
USFWS - U.S. Fish and Wildlife Service

APPENDIX C

Results of the Initial Screening of Ecological Models — by Model Analysis Team

1-56670-574-4-6/01/$0.00+$1.50
© 2002 by CRC Press LLC

Table C.1 Population Models Selected for Detailed Review

Category	Model Name	Endpoint	Reference
Simple scalar abundance	Malthusian growth	Abundance	Alstad et al. (1994a,b); Alstad (2001); Poole (1974); May (1974)
Simple scalar abundance	Logistic growth	Abundance	Poole (1974); May (1974); Alstad et al. (1994a,b); Alstad (2001)
Simple scalar abundance	Stock-recruitment	Abundance	Ricker (1954; 1975); Beverton and Holt (1959); Alstad et al. (1994a,b); Alstad (2001)
Simple scalar abundance	Stochastic differential equation	Risk, abundance	Poole (1974); Goel and Richter-Dyn (1974); May (1974); Ginzburg et al. (1982); Iwasa (1998); Tanaka (1998); Hakoyama and Iwasa (1998)
Simple scalar abundance	Stochastic discrete-time	Risk, abundance	Poole (1974); Capocelli and Ricciardi (1974); Goel and Richter-Dyn (1974); May (1974); Ginzburg et al. (1982); Ferson (1999)
Simple scalar abundance	Equilibrium exposure	Abundance	Hallam et al. (1983a,b)
Simple scalar abundance	Bioaccumulation and population growth	Abundance	Bledsoe and Megrey (1989); Bledsoe and Yamamoto (1996); Spencer and Ferson (1997b);
Life history	Deterministic age- or stage-based matrix	Abundance by age or stage	Caswell (2001)
Life history	Stochastic age- or stage-based matrix	Risk, abundance by age or stage	Caswell (2001)
Life history	RAMAS age, stage, Metapop, or Ecotoxicology	Risk, abundance by age or stage	Ferson (1993); Spencer and Ferson (1997c); Applied Biomathematics (2000)
Life history	Unified Life Model (ULM)	Risk, abundance	Legendre and Clobert (1995)
Individual-based	SIMPDEL	Abundance by location	Abbot et al. (1995)
Individual-based	SIMSPAR	Abundance by age, sex, and location	Nott et al. (1998)
Individual-based	CompMech	Abundance by age, sex, and location	EPRI (1982, 1996); Jager et al. (1999)
Individual-based	EcoBeaker	Abundance	Meir (1997)
Individual-based	Daphnia	Abundance	Gurney et al. (1990); McCauley et al. (1990)
Individual-based	CIFSS	Abundance by age, sex, and location	Langton et al. (1999); Railsback et al. (1999); Railsback and Harvey (in prep. 1999); Humboldt State University (1999)
Individual-based	WESP & ECOTOOLS	Risk, abundance	Lorek and Sonnenschein (1995); Sonnenschein et al. (1999)
Individual-based	GAPPS	Abundance by age and sex	Harris et al. (1986)
Individual-based	PATCH	Abundance by age or stage	Schumaker (1998)
Individual-based	NOYELP	Abundance and foraging by location	Turner et al. (1994); Pearson et al. (1995)
Category	Model Name	Endpoint	Reference
Individual-based	Wading bird nesting colony	Abundance and foraging by location	Hallam et al. (1996)

Table C.1 (cont.)

Metapopulation	Occupancy — incidence function	Abundance by location	Hanski (1994); Hanski and Gilpin (1991)
Metapopulation	Occupancy — state transition	Abundance by location	Verboom et al (1991); Sjögren-Gulve and Ray (1996)
Metapopulation	RAMAS Metapop and RAMAS GIS	Risk, abundance	Akçakaya (1998); Akçakaya et al. (1995); Kingston (1995)
Metapopulation	VORTEX	Risk, abundance	Lacy (1993); Lindenmayer et al. (1995)
Metapopulation	ALFISH	Abundance by age, sex, and location	DeAngelis et al. (1998a)
Metapopulation	ALEX	Risk, abundance	Possingham and Davies (1995)
Metapopulation	Meta-X	Species viability	UFZ and OFFIS (2000)

Table C.2 Population Models Excluded from Detailed Review[a]

Category	Model Name	Endpoints	Reference	Reason for Exclusion
Scalar abundance	Cotton rat population dynamics	Abundance	Montague et al. (1990)	Special case represented by other models reviewed
Scalar abundance	Tephritid fly populations	Abundance	Jeltsch et al. (1992)	Special case represented by other models reviewed
Scalar abundance	Stochastic model of seagrass	Abundance	Bearlin et al. (1999)	Special case represented by other models reviewed
Scalar abundance	Zebra mussel	Abundance	Akçakaya and Baker (1998)	Special case represented by other models reviewed
Scalar abundance	Porcupine caribou harvest	Abundance	http://www.taiga.net	Special case represented by other models reviewed
Scalar abundance	Branching processes	Abundance	Lebreton (1982); Gosselin and Lebreton (2000)	Chaos models not realistic
Scalar abundance	Chaos due to density dependence	Abundance	May (1974)	Chaos models not realistic
Scalar abundance	Time-delay	Abundance	Maynard Smith (1974); Royama (1992)	Model needs further development
Life history model	Physiological based coupled with chemical effect	Population structure	Hallam and Funasaki (1999)	Not an assessment tool
Life-history model	Waller's	Abundance by age	Waller et al. (1971)	Too specific
Life-history model	Population effects	Abundance	Emlen (1989); Barnthouse (1993)	Review papers
Life-history model	Population effects with toxicity, predation, and harvesting	Abundance	Landis and Yu (1995)	Textbook, no new models
Life-history model	Population effects with toxicity, predation, and harvesting	Abundance	Stark and Wennergren (1995)	No new models
Life-history model	DEB[b] structured populations	Abundance	Kooijman and Metz (1984)	Too specific
Life-history model	DEB structured populations	Abundance	Kooijman (1993)	Not an assessment tool
Life-history model	DEB structured populations	Abundance	Kooijman et al. (1998)	Not an original model
Individual-based model	BACHMAP	Population structure	Pulliam et al. (1992)	Too specific
Branching process	Lattice structured population models	Abundance	Lebreton (1982); Woodward (1982); Iwasa (1998)	Too general
Time series	Complex dynamics	Abundance	Turchin and Taylor (1992)	Not a realistic assessment tool
Time series	Analytical population dynamics	Abundance	Royama (1992)	Not a realistic assessment tool
Time series	PAS-P1a	Abundance	Berryman and Millstein (1992)	Not a realistic assessment tool
Metapopulation	Balanced dispersal	Distribution in a landscape	Diffendorfer (1998)	Model of dispersal, endpoint irrelevant
Metapopulation	Source-sink dynamics	Distribution in a landscape	Diffendorfer (1998)	Model of dispersal, endpoint irrelevant
Metapopulation	Comparative island biogeography and static Metapopulation	Species abundance	Scheiner and Rey-Benayas (1997)	Represented by other models reviewed

[a] This table includes models that were excluded based on screening criteria discussed in text (see Chapter 2, Methods). This table does not contain all documents published that address model applications.
[b] DEB - Dynamic Energy Budget.

Table C.3   Ecosystem Models Selected for Detailed Review

Category	Subcategory	Model Name	Endpoint	Reference
Food web	N/A	Predator–Prey	Abundances of predator and prey	Lotka (1924); Volterra (1926); DeAngelis et al. (1975); Ivlev (1961); Gallopin (1971); Watt (1959); Holling (1959, 1966); Arditi and Ginzburg (1989); Hassell and Varley (1969)
Food web	N/A	Population-Dynamic food-chain	Species abundances	Spencer et al. (1999)
Food web	N/A	RAMAS ecosystem	Species abundances	Spencer and Ferson (1997a,c); Spencer et al. (1999)
Food web	N/A	Populus	Species abundances	Alstad et al. (1994a,b); Alstad (2001)
Food web	N/A	Ecotox	Species abundances	Bledsoe and Megrey (1989)
Aquatic	Estuaries	Transfer of impacts between trophic levels	Multitrophic abundance	Horwitz (1981)
Aquatic	Lake	AQUATOX	Multitrophic biomass	Park et al. (1974); Park (1998); U.S. EPA (2000a,b,c)
Aquatic	Lake	ASTER/EOLF	Microbial and plankton distribution	Salencon and Thebault (1994); Thebault and Salencon (1983)
Aquatic	Lake	DYNAMO pond model	Multitrophic biomass	Wolfe et al. (1986)
Aquatic	Lake	EcoWIN	Biomass, productivity	Ferreira (1995); Duarte and Ferreira (1997)
Aquatic	Lake	LEEM	Multitrophic biomass and abundance	Koonce and Locci (1995)
Aquatic	Lake	LERAM	Plankton, zoo, and fish community structure	Hanratty and Stay (1994); Hanratty and Liber (1996)
Aquatic	Lake	Modified SWACOM/CASM	Multitrophic biomass	DeAngelis et al. (1989); Bartell et al. (1992, 1999)
Aquatic	Lake	PCLAKE	Biomass productivity	Janse and Van Liere (1995)
Aquatic	Lake	PH-ALA	Phytoplankton productivity	Jorgensen (1976); Jørgensen et al. (1981)
Aquatic	Lake	SALMO	Phytoplankton and zooplankton biomass	Benndorf and Recknagel (1982); Benndorf et al. (1985)
Aquatic	Lake	SIMPLE	Interspecies population dynamics	Jones et al. (1993)
Aquatic	River	FLEX/MIMIC	Multitrophic biomass	McIntire and Colby (1978)
Aquatic	River	IFEM	Biomass, body burden, risk of decline	Bartell et al. (1988)
Aquatic	General	INTASS	Species abundances, fitness	Emlen et al. (1989, 1992)
Terrestrial	Desert	Hierarchical model of *Dipodomys*	Mammalian productivity	Maurer (1990)
Terrestrial	Forest	FVS	Community structure, forest productivity	USDA (1999 and references therein)
Terrestrial	Forest	FORCLIM	Community structure and population dynamics	Bugmann (1997); Bugmann and Cramer (1998)
Terrestrial	Forest	FORSKA	Forest community structure	Lindner et al. (1997, 2000); Lindner (2000)

Table C.3 (cont.)

Category	Subcategory	Model Name	Endpoint	Reference
Ecosystem Type	Habitat	Model Name	Endpoint	Reference
Terrestrial	Forest	HYBRID	Forest production	Friend et al. (1993, 1997)
Terrestrial	Forest	ORGANON	Community structure	OSU (1999)
Terrestrial	Forest	SIMA	Community structure	Kellomaki et al. (1992)
Terrestrial	Forest	TEEM	Productivity	Shugart et al. (1974)
Terrestrial	Grassland	Energy flow in short grass prairie	Biomass	Jeffries (1989)
Terrestrial	Grassland	SAGE	Community structure	Heasley et al. (1981)
Terrestrial	Grassland	SWARD	Multi-trophic biomass	White (1984)
Terrestrial	Rangeland	SPUR	Plant-consumer production	Hanson et al. (1988)
Terrestrial	Island	Multi-timescale community dynamics	Community turnover	Russell et al. (1995)
Terrestrial	Island	Nested species sub-set analysis	Species presence	Cook and Quinn (1998)

Note: N/A - not applicable.

Table C.4 Ecosystem Models Excluded from Detailed Review[a]

Category	Subcategory	Model Name	Endpoints	Reference	Reason for Exclusion
Food-web	Population-dynamic food-web model	Dynamic food-web models	Abundances of multiple species	Pimm and Lawton (1977); Swartzman and Kaluzny (1987); Pimm (1991)	Not applicable to ecotoxicological studies
Food-web	Predator–prey model	Hassell	Abundance of predators and prey	Beddington et al. (1976)	Not an assessment tool
Food-web	Predator–prey model	Diffusion models	Abundance of predators and prey	Okubo (1980)	Not applicable to ecotoxicological studies
Food-web	Cascade model	Cascade food-web models	Food web interconnections	Cohen (1978); Sugihara (1984); Cohen and Newman (1985)	Not applicable to ecotoxicological studies
Food-web	Unnamed	Assembly rules	Probability of a species occurrence	Diamond (1975)	Not applicable to ecotoxicological studies
Abiotic/biotic ecosystem	Aquatic	Lappter and Rocter	Primary production/fish distribution	Reyes (1994)	
Abiotic/biotic ecosystem	Aquatic	MICMOD	Algal biomass	Swartzman et al. (1990)	Microcosm based; endpoint too limited
Abiotic/biotic ecosystem	Aquatic	Multicompartment kinetic transfer	Algal/zoo production	Rose (1985)	Microcosm based; endpoint too limited
Abiotic/biotic ecosystem	Aquatic	Multicompartment kinetic transfer	Community structure	Rose et al. (1988)	Microcosm based; endpoint too limited
Abiotic/biotic ecosystem	Aquatic	Multiple compartment percolation	Multitrophic biomass	Chen and Orlob (1975)	Obsolete
Abiotic/biotic ecosystem	Aquatic	Time-series dynamic	Algal/zoo production	Sciandra (1986)	Partial ecosystem model
Abiotic/biotic ecosystem	Aquatic	A simple eutrophication lake	Microbial productivity	Jørgensen (1994)	Microcosm based; endpoint too limited
Abiotic/biotic ecosystem	Aquatic	CHEMSEE	Geochemistry	Ulrich (1991)	Chemical endpoints only
Abiotic/biotic ecosystem	Aquatic	GIRL: general simulation of reservoirs and lakes	Productivity	Straskraba (1999)	Represented by other models reviewed
Abiotic/biotic ecosystem	Aquatic	MINLAKE	Succession change	Riley and Stefan (1987); Riley (1998)	Represented by other models reviewed
Abiotic/biotic ecosystem	Aquatic	Model of lake baikal ecosystem disturbance	Interspecies population dynamics	Silow and Stom (1992)	Lack of information
Abiotic/biotic ecosystem	Aquatic	The glumso-model	Microbial productivity	Jørgensen (1976); Jørgensen et al. (1981)	Microcosm based; endpoint too limited
Abiotic/biotic ecosystem	Aquatic	Microbial benthos method	Biomass and turnover	Cazelles et al. (1991)	Endpoint too limited

Table C.4 (cont.)

Category	Subcategory	Model Name	Endpoints	Reference	Reason for Exclusion
Abiotic/biotic ecosystem	Aquatic	Stream ecosystem (M & C stream model)	Primary production	McIntire (no date)	Represented by other models reviewed
Abiotic/biotic ecosystem	Aquatic	Old Woma Creek Phosphorus Model: nutrient retention of a freshwater coastal wetland	Phosphorus distribution	Silow and Stom (1992)	Chemical endpoints only
Abiotic/biotic ecosystem	Aquatic	Modeling seasonal changes of biotic and abiotic variables	Community structure	Vassiliou (1984)	Too specific
Abiotic/biotic ecosystem	Aquatic	GIS-based plant prediction	Spatial pattern of plants	van Horssen et al. (1999)	Inputs limited to nutrition; specific to Netherlands database
Abiotic/biotic ecosystem	Aquatic	Narragansett Bay	Multitrophic biomass	Nixon and Kremer (1977)	Obsolete
Abiotic/biotic ecosystem	Aquatic	Nutrient/biomass model for Liberty Lake, Washington	Primary production	Funk et al. (1982)	Lack of information; likely represented by other models reviewed
Abiotic/biotic ecosystem	Aquatic	Marine multi-compartment transfer	Impact dispersal model	Harris et al. (1983)	Nutrient model
Abiotic/biotic ecosystem	Aquatic	RBM10, a dynamic river basin model	Aquatic community compositon/impact	Yearsley (1991); Bowler et al. (1992)	Nutrient model
Abiotic/biotic ecosystem	Aquatic	Lake ecosystem	Corporate production accounting	Yearsley (1989)	Lack of information
Abiotic/biotic ecosystem	Aquatic	Georgia spartina marsh	Algae/macro productivity	Wiegert et al. (1975)	Model output unclear
Abiotic/biotic ecosystem	Aquatic	3DWFGAS: Three-dimensional water flow and quality model with air and soil modules for optional forcing functions (FINNWIND - FINNSHED - FINNFLOW - FIINQAU - FIINWAVE - FINNBIO - FINNFISH - FINNESOC)	Water cycle	Virtanen et al. (1986); Koponen and Alasaarela (1993)	Physical and chemical endpoints only
Abiotic/biotic ecosystem	Aquatic	Mixing zone (MIXZON)	Water quality	Park (1985); Park and Uchrim (1988)	Chemical endpoints only
Abiotic/biotic ecosystem	Aquatic	Pigeon River allocation model (PRAM)	Resource distribution	Brown and Barnwell (1987)	Endpoint irrelevant
Abiotic/biotic ecosystem	Aquatic	Segment travel river Ecosystem autograph model (STREAM)	Water quality	Jørgensen et al. (1996)	Chemical endpoints only

Table C.4 (cont.)

Abiotic/biotic ecosystem	Aquatic	Segment travel river Ecosystem stochastic simulation (STRESS)	Water quality	Jørgensen et al. (1996)	Chemical endpoints only
Abiotic/biotic ecosystem	Aquatic	Lake Erie	Trophic conductance, food chain	Heath et al. (1999)	Represented by other models reviewed
Abiotic/biotic ecosystem	Aquatic	OPHELIE: a two-dimensional (x-z) laterally-averaged reservoir model	Thermal and hydrodynamic	Salencon and Simonot (1989)	Physical endpoints only
Abiotic/biotic ecosystem	Aquatic	reward: regional water quality and resources decision	Surface water profile and nutrient distribution	Koncsos and Somlyody (1991)	Chemical endpoints only
Abiotic/biotic ecosystem	Aquatic	Environmental periodic time series	Productivity	Dejak et al. (1989)	Modeling platform only
Abiotic/biotic ecosystem	Terrestrial	None	Forest productivity	Li et al. (1993)	Specific to cutting practices
Abiotic/biotic ecosystem	Terrestrial	BIO-DAP	Species diversity	Etter and van Deelen (1999)	Software for calculating diversity indices
Abiotic/biotic ecosystem	Terrestrial	COMDYN	Species richness	USGS (1999)	Not quantitative
Abiotic/biotic ecosystem	Terrestrial	Population analysis system	Population density, species interaction	ESA (1994)	Lack of information
Abiotic/biotic ecosystem	Terrestrial	Species-area curves	Species diversity	Hill et al. (1994)	Speculative and specific to East Africa
Abiotic/biotic ecosystem	Terrestrial	Harvard forest	Forest productivity, decomposition	Amthor and Davidson (1997)	Endpoint represented in other models reviewed
Abiotic/biotic ecosystem	Terrestrial	BIOME-BGC (biogeochemistry cycles)	Biogeochemistry	OSU (1999)	Chemical endpoints only
Abiotic/biotic ecosystem	Terrestrial	Boreal Ecosystem	Productivity, respiration	Frolking (1999)	Endpoint represented in other models reviewed
Abiotic/biotic ecosystem	Terrestrial	BWIN	Forest productivity	Nagel (1999)	Lack of information
Abiotic/biotic ecosystem	Terrestrial	None	Community recovery	Li and Apps (1996)	Specific to contagious stressors
Abiotic/biotic ecosystem	Terrestrial	None	Forest productivity	Wenk (1994)	Specific to thinning
Abiotic/biotic ecosystem	Terrestrial	SIMILE	Community structure	IERM (1999)	Unavailable

Table C.4 (cont.)

Category	Subcategory	Model Name	Endpoints	Reference	Reason for Exclusion
Abiotic/biotic ecosystem	Terrestrial	GEM	Plant production	Hunt (1992)	Models climatic changes
Abiotic/biotic ecosystem	Terrestrial	ISSM crop-insect	Biomass productivity	Nagarajan et al. (1994)	Too specific
Abiotic/biotic ecosystem	Terrestrial	Impacted island biogeography	Various productivity measures	Wardle et al. (1997)	Fire-based model
Abiotic/biotic ecosystem	Terrestrial	True path equilibrium	Community makeup	Whittaker (1995)	Not quantitative
Abiotic/biotic ecosystem	Terrestrial	UK butterfly	Butterfly productivity	Dennis and Shreeve (1997)	High model specificity and no new techniques
Abiotic/biotic ecosystem	Terrestrial	ELM	Productivity, energy, structure	Innis (1978)	Lack of information
Abiotic/biotic ecosystem	Terrestrial	Multiple impact distribution	Isopod populations	Goodman (1982)	Lack of information
Abiotic/biotic ecosystem	Terrestrial	ATLAS: arctic transitions in the land-atmosphere system	Physical parameters of Arctic	LAII (1998)	Physical endpoints only
Abiotic/biotic ecosystem	Terrestrial	LEGOMODEL	Plant community structure	Kleyer (1998)	Effectors primarily climatic
Abiotic/biotic ecosystem	Terrestrial	Growing and grazing season in UK	Plant productivity	Hough (1993)	Effectors primarily climatic
Abiotic/biotic ecosystem	Terrestrial	BIOME2	Landscape pattern	Running et al. (1995)	Endpoint irrelevant
Abiotic/biotic ecosystem	Terrestrial	CENTURY	Biogeochemistry	Running et al. (1995)	Chemical endpoints only
Abiotic/biotic ecosystem	Terrestrial	Dynamic global phytogeography (DOLY)	Landscape pattern	Running et al. (1995)	Spatial scale too large
Abiotic/biotic ecosystem	Terrestrial	Mapped atmosphere-plant soil system (MAPSS)	Landscape pattern	Running et al. (1995)	Spatial scale too large
Abiotic/biotic ecosystem	Terrestrial	Terrestrial ecosystem model (TEM)	C + N fluxes	NASA and UNH (1999b)	Chemical endpoints only
Abiotic/biotic ecosystem	Terrestrial	Multiple succession regressions	Community structure	Huston and Smith (1987)	Limited test of succession
Abiotic/biotic ecosystem	Terrestrial	A spatial model for tropical forests	Forest productivity	Boersma et al. (1991)	Nonpredictive
Abiotic/biotic ecosystem	Terrestrial	FINICH	Community productivity	Malanson et al. (1992)	Specific to fire-based forest disturbances

Table C.4 (cont.)

Abiotic/biotic ecosystem	General	CELSS	Undefined	http://www.kabir.umd.edu/s mp/mrd/o.htm [address no longer in use]	Modeling platform only
Abiotic/biotic ecosystem	General	Information models of natural hierarchical systems	Various	Kirsta (1992)	Evolutionary model
Abiotic/biotic ecosystem	General	EcoAudit: Ecosystem	Corporate production accounting	Gil Friend and Associates (1995)	Endpoint irrelevant
Abiotic/biotic ecosystem	General	Extreme value functions	Species richness	Williams (1995)	Endpoint represented in other models
Abiotic/biotic ecosystem	General	NLM	Multitrophic biomass and diversity	With et al. (1997); With (1997); With and King (1997)	Endpoint represented in other models

[a] This table includes models that were excluded based on screening criteria discussed in text (see Chapter 2, Methods). This table does not contain all documents published that address model applications.

Table C.5 Landscape Models Selected for Detailed Review

Category	Subcategory	Model Name	Endpoint	Reference
Aquatic	Marine	ERSEM	Productivity, respiration, multitrophic biomass	Ebenhoh et al. (1995); Baretta et al. (1995)
Aquatic	Estuarine	Barataria Bay Ecological	Multitrophic biomass	Hopkinson and Day (1977)
Aquatic	River, Lake	CEL HYBRID	Community structure and dynamics	Nestler and Goodwin (2000)
Aquatic	River	Delaware River Basin	Multitrophic biomass	Kelly and Spofford (1977)
Aquatic	River	Patuxent Watershed Model	Landscape pattern	Voinov et al. (1999a,b); Institute for Ecological Economics (2000)
Aquatic	Wetland	ATLSS	Trophic structure, species interaction, Everglades	DeAngelis (1996); DeAngelis et al. (1998b)
Aquatic	Wetland	Disturbances to Wetland Vascular Plants	Plant community structure	Ellison and Bedford (1995)
Terrestrial	Forest	LANDIS, FORMIX, FORMOSAIC, JABOWA, ZELIG	Forest productivity and community structure	Botkin et al. (1972); Burton and Urban (1990); Bossel and Krieger (1991); Botkin (1993a,b); Bugmann et al. (1996); Mladenoff et al. (1996); Liu and Ashton (1998); Mladenoff and He (1999)
Terrestrial	Forest	Regional Landscape	Landscape measures	Graham et al. (1991)
Terrestrial	Forest	Spatial Dynamics of Species Richness	Species richness	Wu and Vankat (1991)
Terrestrial	Grassland	STEPPE	Community structure	Coffin and Lauenroth (1989); Humphries et al. (1996)
Terrestrial	Grassland	Wildlife-Urban Interface	Species richness	Boren et al. (1997)
Terrestrial	Island	SLOSS	Species richness and measure of nestedness	Boecklen (1997)
Terrestrial	Island	Island Disturbance Biogeographic	Community structure	Villa et al. (1992)
Terrestrial	Multi-scale	Multi-scale Landscape	Landcover distribution and dominant vegetation	Johnson et al. (1999)

Table C.6 Landscape Models Excluded from Detailed Review[a]

Category	Subcategory	Model Name	Endpoints	Reference	Reason for Exclusion
Landscape	Aquatic	Crayfish and rice dynamics	Biomass production	Anastacio et al. (1993)	Too specific
Landscape	Aquatic	Multiple compartment percolation	Various productivity measures	Patten et al. (1975)	Obsolete
Landscape	Aquatic	BIOTEST	C + N fluxes	Dejak et al. (1989)	Lack of information
Landscape	Aquatic	EXWAT and RIVER	Contaminant distribution	Paasivirta (1994)	Chemical endpoints only
Landscape	Aquatic	LAXBREAK	Hydrological endpoints	Rajar and Kryzanowski (1994)	Physical endpoints only
Landscape	Aquatic	MAGIC	Species richness/ distribution	Sullivan et al. (1991)	Index-based model
Landscape	Aquatic	Multiple percolation	Water quality	Thomann (1998)	Ecological component is not described; relies on other undefined model systems
Landscape	Terrestrial	Landlord	Landscape pattern	World in a Box (1999)	Index-based model
Landscape	Terrestrial	Index of habitat patch	Proximity	Gustafson and Parker (1994)	Index-based ecosystem model
Landscape	Terrestrial	SmartForest	Community structure, landscape pattern	Imaging Systems Laboratory (2000)	Graphical interface
Landscape	Terrestrial	Multiple models	Species distribution	Fries et al. (1998)	Review paper
Landscape	Terrestrial	Visual management system	Landscape pattern	Daniel and Vining (1983)	Modeling platform only
Landscape	Terrestrial	Conifer	Forest productivity	CFBMG (1977)	Only carbon and water movement predicted
Landscape	Terrestrial	Landscape management system	Spatial distribution	Burnham and Overton (1979); LMS (1999)	Too specific
Landscape	Terrestrial	TREEDYN3	Forest production	Bossel and Schafer (1989)	Specific to monocultures
Landscape	Terrestrial	CLIMEX	Community structure	Maywald and Sutherst (1991)	Index-based model
Landscape	Terrestrial	NDVI	Community structure	Ostendorf and Reynolds (1993)	Index-based model
Landscape	Terrestrial	Chaparral-specific Island	Rodent distribution	Bolger et al. (1997)	Too specific
Landscape	Terrestrial	Island biogeography	Community structure	MacArthur and Wilson (1967)	Obsolete
Landscape	Terrestrial	Multiscale island biogeography	Insect community structure	Lomolino and Davis (1997)	Too specific
Landscape	Terrestrial	Carbon percolation	Carbon distribution	Reichle et al. (1973)	Obsolete
Landscape	Terrestrial	FORET, FORECE, FORMAN, FORMIND, FORENA, MOSEL	Forest productivity	Dale and Gardner (1987); Kellomaki and Vaisanen (1991); Kohler and Huth (1998)	Precursors of or similar to other reviewed forest productivity models
Landscape	General	Drainage basin	Water cycle	NASA and UNH (1999a)	Physical endpoints only
Landscape	General	Undefined	Community distribution	Congleton et al. (1997)	Modeling platform only
Abiotic	Terrestrial	ECOLECON	Forest productivity	ECOLECON (1993)	Model specific to commercial productivity
Abiotic	Terrestrial	Modified Markov	Forest community structure	Hobbs (1994)	Effectors primarily climatic
Geochemistry	Terrestrial	FIRESUM	Forest production	Keane et al. (1989)	Specific to fire-based forest disturbances
Geochemistry	Terrestrial	Microbial subsurface transport	Microbial community structure	Loehle and Johnson (1994)	Endpoint too limited

[a] This table includes models that were excluded based on screening criteria discussed in text (see Chapter 2, Methods). This table does not contain all documents published that address model applications.

Table C.7 Extrapolation Models Selected for Detailed Review

Category	Subcategory	Model Name	Endpoint	Reference
Species sensitivity distribution	Log-triangular extrapolation	Final chronic value	Similar to van Straalen and Denneman (1989) but assumes a log-triangular distribution for a population of the four lowest genus mean chronic values	Stephan et al. (1985)
Species sensitivity distribution	Logistic extrapolation	HCS	Developed a hazardous concentration for the most sensitive species; assumes that species sensitivity follows log-logistic distribution	Kooijman (1987)
Species sensitivity distribution	Log-normal extrapolation	HCp	AKA Kooijman method. Similar to van Straalen and Denneman (1989) but assumes a log-normal distribution for the population of NOAELs; bases acceptable impact at level less than 95% of all NOAELs for all potential receptors	Aldenberg and Slob (1993)
Species sensitivity distribution	Logistic extrapolation	Modified HCp1	Assumes a log-logistic distribution of NOAELs and defines ecological impact based on the proportion of NOAELs exceeded	van Straalen and Denneman (1989)
Species sensitivity distribution	Log-normal extrapolation	Modified HCp2	Uses NOEC values in calculation of HCp and applies tolerance limits	Wagner and Locke (1991)
Interendpoint extrapolation	Quantitative UF	ACR and AF	Acute-to-chronic ratio (ACR) equal to the LC50 derived by the MATC; the application factor (AF) is the inverse of the ACR	Kenaga (1979; 1982)
Interendpoint extrapolation	Quantitative UF	Acute-to-chronic UF	UFs for acute-to-chronic extrapolation derived from regression of LC50 or EC50 data.	Sloof et al. (1986)
Interendpoint extrapolation	Scaling	NOEC for survival to other endpoints	Regression method for estimating NOAELs across all species within a class based on fish size and weight	Mayer et al. (1986)
Interendpoint extrapolation	Quantitative UF	Acute Lethality to NOEC	Similar to Kooijman (1987) but uses direct ratios and assumes no distribution	Mayer et al. (1994)
Intertaxon extrapolation	Scaling	Scaling between bird species	Derived an exponential scaling coefficient of 1.0 to 1.55 for birds based on the regression of LD50s for a series of pesticides	Mineau et al. (1996)
Intertaxon extrapolation	Transformed regression	Interspecies toxicity	Examined the relationship in LD50s between various aquatic species for different classes of toxicants	LeBlanc (1984)
Intertaxon extrapolation	Scaling	Allometric scaling	Provided scaling relations for toxicity based on body mass for mammalian receptors	Davidson et al. (1986)

Table C.7 (cont.)

Category	Subcategory	Model Name	Endpoint	Reference
Intertaxon extrapolation	Species Sensitivity Ratio	Species sensitivity ratios	Quantitative species sensitivity relations; a model to estimate relative sensitivity based on the taxonomic position of the particular receptor	Hoekstra et al. (1994)
Intertaxon extrapolation	Intertaxon regression	AEE	Derives an acceptable concentration by regressing intraorder toxicity (for fish) and then applying the variance about the regression to the geometric mean to derive a distribution estimate about the GMATC	Suter et al. (1983); Linder et al. (1986)
Intertaxon extrapolation	Intertaxon regression	Errors-in-variables regression	Errors-in-variables regression model for extrapolation between freshwater and marine taxa and between LC50 and MATC	Suter and Rosen (1988)

Table C.8 Extrapolation Models Excluded from Detailed Review

Interendpoint extrapolation	Quantitative UF (ADI)	Derives an acceptable daily intake rate based on LD50s by applying a regression-derived uncertainty factor	Layton et al. (1987)	Human health risk assessment approach, not ecological
Intertaxon extrapolation	Transformed regression	Comparison of LC50 data for three fish species and Daphnia using regression	Doherty (1983)	Too specific
Intertaxon extrapolation	Scaling	A body mass-based scaling model specific to raptors for the scaling of drug dosage based on basal metabolic activity	Pokras et al. (1993)	Pharmacokinetic, not ecological, model
Species sensitivity distribution	Bootstrap	A partial application that uses bootstrap analysis of available NOEL data to derive the distribution, thus bypassing the need to assume proportional distributions	Jagoe and Newman (1997)	Could not be obtained

Note: ADI - Acceptable daily intake
LC50 - Median lethal concentration
NOEL - No-observed-effects level
UF   - Uncertainty factor

Glossary

λ (lambda) — see finite rate of increase.

abiotic — nonliving, usually referring to physical and chemical components of an ecosystem. See also biotic.

abundance — the total number or density (number per unit area or unit volume) of organisms in a given location.

acute toxicity — the ability of a chemical to cause a toxic response in organisms immediately or shortly after exposure to the chemical.

adaptive management — an approach to natural resource management that includes adjusting management strategies and actions in response to information gained by monitoring previous results. It often involves ecological modeling and experimentation to develop and evaluate management approaches.

age class — a category comprising individuals of a given age within a population.

age (or stage) structure — the relative proportions of different age- (or stage-) classes in the population.

age-specific fecundity — the number of eggs or offspring produced per unit time by an individual of a specified age.

age-specific mortality — the death rate for a given cohort or age class of a population. Calculated as the number of individuals in a cohort who die in the interval t to $t+1$.

age-specific survival — the proportion of individuals of age x alive at time t who will be alive at time $t+1$.

allee effect — a positive-feedback effect that occurs when population abundance or density becomes small enough to negatively affect mate finding or mating. This effect increases in intensity as population abundance decreases and is destabilizing in the sense that populations experiencing this effect tend to go extinct.

allometric growth — differential growth of body parts (x and y), expressed by the equation:

$$y = bx^a$$

where a and b are fitted constants. Change of shape or proportion with increase in body size.

assessment endpoint — an explicit expression of the environmental value that is to be protected. An assessment endpoint includes both an ecological entity and specific attributes of that entity. For example, salmon are a valued ecological entity; reproduction and population maintenance of salmon form an assessment endpoint (U.S. EPA 1998).

background level — the natural level (or rate) of some system component (or process) that exists in the absence of anthropogenic effects. Often the background level is considered to be the level or rate of some process before an additional factor is added to the system.

bell curve — see normal distribution.

Beverton–Holt function — a mathematical model for density-dependent effects. The Beverton–Holt model relates the parental investment (number of eggs produced) to recruitment (the number of zero-year olds eventually entering the population). Density dependence typically arises from such behavior as cannibalism or uneven resource sharing, sometimes called *scramble competition*. This model has two parameters: ρ and k. If we represent new recruits as Z and parental investment as E, then the density-dependent relationship between Z and E is

$$Z = 1/[\rho + k/E]$$

Both ρ and k are parameters determined by fitting the Beverton–Holt model to data. They are both non-negative.

bioaccumulation — net retention of a chemical in the tissues of an organism as a result of ingestion, respiration, or direct contact with a medium, such as water or soil.

biomass — the total mass (or mass per unit area or volume) of living organisms in the population or community.

biota — living groups of organisms or species.

biotic — living organisms, usually referring to the biological components of an ecosystem. See also abiotic.

calibration — the adjusting of parameters and coefficients in a model so the output more accurately matches observations from the system being modeled.

canopy — the portion of a forest consisting of the tree crowns. It typically refers to the uppermost layer of foliage, but it may also refer to the lower layers in a multi-tiered forest.

carrying capacity, K — the maximum number (or density) of organisms that can be supported in a given unit of habitat. Often computed as the long-term average abundance. See also logistic equation.

cellular automata — a spatially explicit, typically grid-based model in which the state of any given cell depends on the state of other cells.

chronic toxicity — the ability of a chemical to produce a toxic response when an organism is exposed over a long period of time.

chronic value — see maximum acceptable toxic concentration (MATC).

cohort — a group of individuals within a population who were all born within the same time period.

community — an assemblage of populations of different species within a specified location in space and time (U.S. EPA 1998).

compensatory mechanism — a biological process that offsets or counteracts an adverse effect (for example, increased survival of young fish related to reduced competition because egg hatching success was reduced).

cumulative distribution function (CDF) — particularly useful for describing the likelihood that a variable will fall within different ranges of x. $F(x)$ (i.e., the value of y at x in a CDF plot) is the probability that a variable will have a value less than or equal to x (U.S. EPA 1998).

demographic modeling — the mathematical description and simulation of processes that occur within populations and between members of a population and that determine the population age structure and population growth.

demographic stochasticity — the stochastic features of discrete population models in which birth and death of individuals have specified probabilities. Demographic stochasticity becomes particularly important as population size decreases.

demography — the study of populations, especially their age structure and growth rates.

density dependence — a change in the influence of any factor (a density-dependent factor) that affects population growth as population density changes. Density-dependent factors tend to retard population growth by increasing mortality or emigration or decreasing fecundity as population density increases. They enhance population growth by decreasing mortality or increasing fecundity as population density decreases.

density-independent factor — any factor affecting population density or demographic rates that is not correlated with population density. Density-independent factors include such things as weather.

deterministic model — a mathematical model which has a specified value for each variable and does not include a stochastic component or random variable.

detritus — dead organic carbon, as distinguished from living (organic) cells and inorganic carbon.

dose — the amount of chemical taken into an organism per unit of time.

dose–response — see exposure–response assessment.

ecological model — any mathematical expression used to describe or predict processes and endpoints in populations, ecosystems, and landscapes.

ecological risk assessment — the process that evaluates the likelihood that adverse ecological effects may occur or are occurring as a result of exposure to one or more stressors (U.S. EPA 1998).

ecological-effects models — ecological models (models used to predict population, ecosystem, or landscape endpoints) and toxicity-extrapolation models that describe ecologically relevant responses of organisms (survival, growth, and reproduction).

ecosystem — the biotic community and abiotic environment within a specified location in space and time (U.S. EPA 1998).

ecotoxicology — the study of the effects of toxic chemicals on organisms, populations, communities, ecosystems, and landscapes.

emigration — the movement of an individual or group out of an area or population.

empirical — measured or measurable.

endpoint — the biological or ecological unit or variable being measured or assessed (see also measurement endpoint and assessment endpoint).

environmental stochasticity — the effect of random effects in environmental parameters on population growth rate. Environmental stochasticity affects population growth rate so that population fluctuations have a relative magnitude related to the degree of environmental variance and independent of absolute population size.

epilimnion — the upper mixed layer of a stratified lake.

equilibrium age distribution — an age distribution in which the relative frequencies of the age classes remain the same. The analogue of the stable age distribution for stochastic models.

equilibrium population — a population having an equilibrium age distribution.

Eulerian model — a model that calculates the flux of material or energy at a fixed location (see also LaGrangian model)

exponential growth — growth of a population N that follows the relationship

$$N = N_0 \exp(rt)$$

where N_0 is the initial population abundance, t is time, and r is a constant growth rate.

exponential rate of increase, r — or instantaneous rate of increase, the rate at which a population is growing at a particular instant, expressed as a proportional increase per unit of time.

exposure pathway — the path a chemical takes or could take from a source to exposed organisms. Exposure pathways include the source, the mechanism of release and transport, a point of contact, and the means of contact (for example, ingestion or inhalation).

exposure–response assessment — a description of the relationship between the concentration (or dose) of the chemical that causes adverse effects and the magnitude of the response of the receptor.

exposure — the contact or co-occurrence of a stressor with a receptor (U.S. EPA 1998).

extinction risk — the probability that population abundance will reach and fall below some level of abundance.

extrinsic factors — environmental variables that are outside the biological system (e.g., organism or population) but may influence the behavior, physiology, or structure of that system.

fate and transport model — a description of how a chemical is carried through the environment. This may include transport through biological as well as physical parts of the environment.

fecundity — the potential reproductive capacity of an organism (or a population). Fecundity is measured by the number of gametes produced. With fish, fecundity is the total number of eggs deposited or released. See also age-specific fecundity.

finite rate of increase, λ — the proportion by which the population increases with each time step. It is the dominant eigenvalue of the Leslie matrix.

food chain — a sequence of species at different trophic (feeding) levels that represent a single path of energy within a food web. For example, grasses and seeds are eaten by a mouse, which is then eaten by an owl. The owl is higher up the food chain (at a higher *trophic level*) than the mouse.

food web — interconnected food chains that describe the pathways of energy and matter flow in nature.

fragmentation — isolation of habitat patches due to physical disturbance or intervening human development.

GIS (geographic information system) — software that combines a database and mapping capability; often used in spatially explicit modeling.

growth rate — the rate of change of population abundance. Depending on the context, growth rate could also refer to the rate of change in mass or size of an organism.

habitat — the place where animals and plants normally live, often characterized by a dominant plant form or physical characteristic.

hazard quotient — the ratio of an estimated exposure concentration (or dose) to a toxicity threshold expressed in the same units.

hazard — the ability of a chemical, physical, or biological agent to harm plants, animals, or humans under a particular set of circumstances.

hypolimnion — the lower, cold-water layer of a stratified lake.

immigration — the movement of an individual or group into a new population or geographical region.

indirect effect — an effect by which the stressor acts on supporting components of the ecosystem, which in turn have an effect on the ecological component of interest (U.S. EPA 1998).

individual-based model — a model incorporating variation among individuals by representing each individual separately and explicitly specifying its age, size, spatial location, gender, energy reserves, and so forth.

instantaneous rate of increase — see exponential of increase, *r*.

intrinsic factors — characteristics of the biological system (e.g., organism or population) that may determine the behavior, physiology, or structure of that system.

intrinsic rate of increase, *r* — the maximum instantaneous rate of increase for a population.

keystone species — a predatory species that has a large effect on community structure and usually increases species diversity by selective predation on competitively dominant prey species.

LaGrangian model — a model that calculates the trajectories of individuals, particles, or chemicals through space and time (see also Eulerian model).

landscape — a spatially heterogeneous area containing an ecosystem or group of different ecosystems (generally on the order of hundreds to thousands of acres, although smaller areas may be considered landscapes relative to the smaller scale of the organism or process of interest).

Leslie matrix — a special type of matrix used to model populations in which individuals can be divided, either naturally or arbitrarily, into discrete age classes. See also the text on life-history models for an example of a Leslie matrix.

life history — the temporal pattern and habitat association of life stages (e.g., egg, larva, pupa, and adult in an insect or egg, fry, smolt, juvenile, and adult in a salmon) and the schedule of births and deaths for a species.

life stage — a developmental stage of an organism (for example, juvenile, adult, egg, pupa, larva).

logistic function — a mathematical model for population growth that incorporates density-dependent effects. In the logistic function the percentage rate of increase decreases in linear fashion as population size increases. The logistic function, written in discrete form, is

$$N(t + 1) = N(t) + rN(t)*[1 - [[N(t)]/K]$$

where $N(t)$ is the abundance of individuals at time t, K is the carrying capacity, and r is the intrinsic rate of population growth. This function yields a sigmoid curve for the plot of population abundance vs. time.

log-normal distribution — a frequency distribution of a variable, the logarithm of which is normally distributed.

lowest-observed-adverse-effect level (LOAEL) — the lowest level of a stressor evaluated in a test that causes statistically significant differences from the controls (U.S. EPA 1998).

Malthusian growth — see exponential growth.

Markov model — a matrix in which the columns are the variables at time t and the rows are the states of the variables at time $t + 1$. The entry in each cell of the matrix is the probability that the state will change from A to B. The diagonals of the matrix are the probabilities that the states will remain the same during a single time step.

matrix — a rectangular array of m rows each containing n numbers or elements and arranged in columns. See also Leslie matrix.

maximum acceptable toxic concentration (MATC) — for a particular ecological effects test, this term is used to mean either the range between the NOAEL and the LOAEL or the geometric mean of the NOAEL and the LOAEL for a particular test. The geometric mean is also known as the chronic value (U.S. EPA 1998).

measurement endpoint — a measurable ecological characteristic that is related to the valued characteristic chosen as the assessment endpoint (U.S. EPA 1998).

median lethal concentration (LC$_{50}$) — a statistically or graphically estimated concentration that is expected to be lethal to 50% of a group of organisms under specified conditions (ASTM 1990).

migration — the movement of an individual or group into or out of a new population or geographic region.

model parameterization — the estimation of the values for variables in a model based on data or assumptions guided by professional judgment.

Monte Carlo method — a method for the solution of mathematical or statistical problems by random sampling. For example, for a single run of an ecological model, the value of several or all variables would be randomly selected from their respective probability distributions. Multiple runs of the model in this mode yield a probability distribution for each output variable.

natural mortality — the mortality rate of individuals in the absence of human intervention.

no-observed-adverse-effect level (NOAEL) — the highest level of a stressor evaluated in a test that does not cause statistically significant differences from the controls (U.S. EPA 1998).

nestedness — a measure of the degree to which one group of species is contained within a larger sample of species.

normal distribution — a probability density function that can be represented as

$$z = \left(\frac{1}{\sqrt{2\pi}}\right)\left(\frac{1}{\delta}\right)\exp\left(-(x-\mu)^2 / \delta^2\right)$$

where z indicates the height of the ordinate of the curve, which represents the density of the items. Typically, the ordinate is transformed into frequency (and in this case the area under the curve sums to one). The two parameters μ and _ represent the parametric mean and parametric standard deviation, respectively, and determine the location and shape of the distribution. The normal distribution is indicated when representing a process in which many independent additive effects are acting. The normal distribution is often called the *bell curve* because of its shape.

parameterization — see model parameterization.

population growth rate — the rate at which numbers of individuals are added to the population over time.

population — an aggregate of individuals of a species within a specified location in space and time (U.S. EPA 1998).

probabilistic assessment — a risk assessment that quantifies the likelihood of adverse effects.

probability — the likelihood of an event occurring, expressed as a numerical ratio, frequency, or percentage.

productivity — the rate of production of living biomass in a population or community.

QSARs (quantitative structure-activity relationships) — mathematical or statistical models that estimate the toxicity of a chemical from the known toxicity of a structurally related chemical.

receptor — the organism, population, or community that might be affected by exposure to a stressor.

recovery — the rate and extent of return of a population or community to a condition that existed before the introduction of a stressor. Owing to the dynamic nature of ecological systems, the attributes of a "recovered" system must be carefully defined (U.S. EPA 1998).

recruitment — the influx of new members into a population by reproduction or immigration.

regression — a kind of statistical technique by which the value of a so-called dependent variable is estimated from values of one or more independent variables.

relative risk assessment — a process similar to comparative risk assessment. It involves estimating the risks associated with different stressors or management actions. To some, relative risk connotes the use of quantitative risk techniques, whereas comparative risk approaches more often rely on expert judgment. Others do not make this distinction (U.S. EPA 1998).

remedial action goals — a subset of remedial action objectives consisting of medium-specific chemical concentrations that are protective of human health and the environment.

remediation — action taken to control the sources of contamination and/or to clean up contaminated areas at a hazardous waste site.

reproductive value, v_x — the expected reproductive output of an individual at a particular age (x) relative to that of a newborn individual at the same time.

Ricker function — a mathematical model of density-dependent effects. The Ricker model relates the parental investment (e.g., number of eggs produced) to recruitment (e.g., the number of age-zero individuals eventually entering the population). Density dependence typically arises from such behavior as cannibalism or uneven resource sharing, sometimes called scramble competition. This model has two parameters: α and β. If we represent new recruits as Z and parental investment as E, then the density-dependent relationship between Z and E is:

$$Z = \alpha \, E \, \exp(-\beta E).$$

Both α and β are parameters that are determined by fitting the Ricker model to data. They are both non-negative.

riparian — the habitat along the bank and flood plain of a stream, river, or lake.

risk — the probability of an adverse effect or outcome.

scramble competition — overlapping use of resources by organisms of the same or different species that potentially results in decreased reproduction or increased mortality of the competing individuals. See also Ricker function.

screening-level assessment — a relatively simple analysis that separates sites (or chemicals) that pose no apparent risk from those for which further analysis is necessary. Screening level may also refer to a guideline or criterion used in such an assessment.

sensitivity analysis — the variation of initial conditions or parameter values in a model to determine which variables most influence the model output.

simulation — implementing a model to describe the behavior of a real system.

spatially explicit model — a model that tracks spatial information (e.g., the locations of organisms or the pattern of a landscape).

species richness — the total number of species in a location or the number per unit area or volume.

species — a population or group of populations of interbreeding individuals with reproductively viable offspring. Interbreeding between species is typically limited by reproductive isolating mechanisms (behavioral, morphological, or physiological features that prevent or limit interbreeding and gene exchange).

species-sensitivity distribution — a probability distribution of toxicity values (e.g., median lethal concentrations [LC50s] or no-observed-adverse-effect levels [NOAELs]).

stable age distribution — the proportion of individuals in various age classes whose abundances do not change from time period to time period.

stable population — a population whose abundance remains constant because birth and immigration processes balance death and emigration processes. A population whose abundance is exactly the carrying capacity.

state variable — a variable that describes the condition of a system component. The values of state variables change with time in dynamic models as system components interact with each other and with the environment.

stochastic simulation model — a mathematical model founded on the properties of probability so that a given input produces a range of possible outcomes due to random effects.

stochasticity — the attribute indicating that chance or probability is involved in determining an outcome of some situation. See demographic stochasticity and environmental stochasticity.

stressor — any physical, chemical, or biological entity that can induce an adverse response in an organism (U.S. EPA 1998).

Superfund — a regulatory program of the U.S. Environmental Protection Agency to assess and clean up chemical contamination at high-priority sites throughout the U.S.

threshold — the chemical concentration (or dose) at which physical or biological effects begin to be produced.

toxicity extrapolation model — any mathematical expression for extrapolating toxicity data between species, endpoints, exposure durations, and so forth. Also includes uncertainty factors.

toxicity test — a test in which organisms are exposed to chemicals in a test medium (for example, waste, sediment, soil) to determine the effects of exposure.

trophic levels — a functional classification of taxa within a community that is based on feeding relationships (e.g., aquatic and terrestrial green plants comprise the first trophic level, and herbivores comprise the second) (U.S. EPA 1998). See also food chain.

trophic structure — the relative proportions of different feeding types in the community (e.g., primary producers, herbivores, primary carnivores, secondary carnivores, detritivores, etc.).

uncertainty analysis — evaluation of the information gaps and variability in a model.

uncertainty — lack of knowledge. Uncertainty can be reduced by further observation and measurement.

uncertainty factor — a number used to modify a toxicity value (e.g., a median lethal concentration, or LC50), usually for purposes of extrapolation on the basis of professional judgment.

upland — land usually above the floodplain of a river or stream (contrast with riparian).

validation — the comparison of output from a model with independent data for the modeled system to assess how accurate and precise the model results are.

variability — random variation in nature that cannot be reduced by additional data.

vector — a row or column of numbers. A vector is a special case of a matrix.

von Bertalanffy growth equation — an equation that is typically used to model organism growth processes of various sorts, such as length or volume change over time; for length this model is:

$$L_t = L_\infty \left(1 - \exp[-K(t - t_0)]\right)$$

where t represents age and t_0 represents the hypothetical age at which the organism has length (or volume) zero assuming they had always grown in the manner prescribed by the von Bertalanffy equation. Thus, t_0 can be positive or negative. L represents length, and L_∞ represents the asymptotic length of the organism. K is the growth rate, which is sometimes referred to as a growth coefficient.

Weibull distribution — a random variable has the standard Weibull density with parameter $a > 0$ and $x \geq 0$ when it has density

$$f(x) = a\, x^{[a-1]}\exp(-x^a)$$

The Weibull distribution is often used to predict the occurrence of hazards. It is a generalization of the exponential distribution.

wetlands — the transition zone between terrestrial and aquatic environments. Examples are seasonally inundated floodplains, riparian systems, and permanently flooded swamps and marshes.

worst-case analysis — a screening level assessment in which each variable in a risk model is assigned a value (its maximum or minimum depending on the direction of its effect on the risk estimate) that maximizes the estimated risk.

Index